Thermal Spray Coatings for High-Temperature Conditions

Thermal Spray Coatings for High-Temperature Conditions provides an in-depth analysis of thermal spray coatings covering a wide range of types and applications in aerospace, automotive, and heavy-duty equipment maintenance. It considers the various thermal spray processes available, including high-velocity oxy-fuel spraying, plasma spraying, and flame spraying.

Focusing on the importance of surface preparation for thermal spray coatings, this book demonstrates the significance of establishing a strong bond between the substrate and the coating. It explores a range of surface preparation techniques like grit blasting and laser texturing. This book showcases the wide range of uses for thermal spray coatings, such as protecting against corrosion, enhancing wear resistance, preventing erosion, and prolonging the lifespan of industrial equipment.

This book is intended for researchers and graduate students studying surface engineering, thermodynamics, high-temperature materials, and wear resistance.

Advances in Surface Engineering and Tribology

Series Editors: Hitesh Vasudev, Jashanpreet Singh, and Jeyaprakash Natarajan

Advances in Surface Engineering and Tribology explores the interdisciplinary aspects of surface science and engineering, as well as tribology discussions pertaining to the examination of friction, wear, and lubrication. It covers topics ranging from surface modification techniques, nanomaterials and their applications in nanotechnology, catalytic and photocatalytic processes, synthesis of materials, colloids and surfaces, thin film deposition processes, methods for characterizing surfaces, and surface engineering for diverse industrial purposes. The volumes comprehensively explore the complex mechanisms that regulate surface phenomena, presenting both established principles and emerging developments in the field of surface science and engineering. The series offers a forum for professionals, engineers, and researchers in the energy, automotive, aerospace, and biomedical sectors to delve into innovative developments in these interrelated disciplines.

Should you have an idea for this series, please get in contact with Kyra Lindholm.

Thermal Spray Coatings for High-Temperature Conditions
Edited by Gaurav Prashar, Hitesh Vasudev, and Lalit Thakur

Thermal Spray Coatings for High-Temperature Conditions

Edited by

Gaurav Prashar, Hitesh Vasudev,
and Lalit Thakur

CRC Press
Taylor & Francis Group
Boca Raton London New York

CRC Press is an imprint of the
Taylor & Francis Group, an **informa** business

Designed cover image: Shutterstock

First edition published 2025
by CRC Press
2385 NW Executive Center Drive, Suite 320, Boca Raton FL 33431

and by CRC Press
4 Park Square, Milton Park, Abingdon, Oxon, OX14 4RN

CRC Press is an imprint of Taylor & Francis Group, LLC

ISBN: 978-1-032-89461-4 (hbk)
ISBN: 978-1-032-89462-1 (pbk)
ISBN: 978-1-003-54295-7 (ebk)

DOI: 10.1201/9781003542957

Typeset in Times
by codeMantra

Contents

Preface

This edited book titled *Thermal Spray Coatings for High-Temperature Conditions* provides an in-depth analysis of thermal spray coatings, covering a wide range of applications and types. The central themes focus on the importance of surface preparation for thermal spray coatings, the significance of establishing a strong bond between the substrate and the coating, and the various thermal spray processes available, including high-velocity oxy-fuel spraying, plasma spraying, and flame spraying.

This book showcases the wide range of uses for thermal spray coatings, such as protecting against corrosion, enhancing wear resistance, preventing erosion, and prolonging the lifespan of industrial equipment. Furthermore, it highlights the significance of utilizing thermal spray coatings to prolong the durability of components and improve their overall functionality. This book offers valuable insights into the various types of thermal spray coatings, their applications, and their vital role in safeguarding surfaces against damage such as corrosion, wear, and high-temperature oxidation.

Written by industry professionals, this book is essential for professionals and researchers in aerospace, automotive, heavy-duty equipment maintenance, and industrial part preservation. It provides valuable insights to enhance the understanding and application of thermal spray coatings. It explores a range of surface preparation techniques like grit blasting and laser texturing, delve into the intricate bonding mechanisms that drive thermal spray coatings, and discover the vast selection of coating materials and processes at your disposal, including high-velocity oxy-fuel spraying, high-velocity air-fuel spraying, and plasma spraying.

This book is a valuable resource for gaining a deep understanding of the important role that thermal spray coatings play in improving the durability, efficiency, and lifespan of industrial equipment and components in challenging environments. Through an exploration of the science, technology, and practical applications of these versatile coatings, this book aims to provide readers with the knowledge and tools necessary to make informed decisions and effectively utilize thermal spray coatings to meet their specific needs and overcome challenges.

About the Editors

Dr. Gaurav Prashar is currently working as a Professor and HOD of Mechanical Engineering at Rayat Bahra Institute of Engineering and Nano Technology, Hoshiarpur. He has published 18 SCI-indexed papers, two Scopus-indexed papers, 13 book chapters, and seven international conference papers. He is currently working on the development of Nano-structured || Bimodal alloys coatings for high-temperature oxidation and erosion resistance. He has obtained his Ph.D. degree from Lovely Professional University (LPU) in Phagwara (NACC A++, NIRF: 50), India. He has also worked on the post-processing of thermal spray coatings. He has contributed extensively to thermal spray coatings in reputed journals, which include publications in *Surface and Coatings Technology*, *Journal of Thermal Spray Technology*, *Engineering Failure Analysis*, *Surface Topography: Metrology and Properties*, *Journal of Cleaner Production*, and *Applied Surface Science Advances.* He is serving as a reviewer for internationally reputed journals like *Advances in Materials and Processing Technologies*, *International Journal on Interactive Design and Manufacturing*, *Results in Surfaces and Interfaces*, *Surface Review & Letters*, *Materials Today: Proceedings*, and *Robotics and Autonomous Systems.* He has authored more than 40 research publications in various international journals, books, and conferences of repute. He has a teaching experience of more than 16 years. He has also published one self-authored book named *Artificial Intelligence and Machine Learning in Thermal Spray Industry: Practices, Implementation, and Challenges* (2023; CRC Press – Taylor & Francis).

Prof. (Dr.) Hitesh Vasudev, Ph.D., is working as a Professor in the School of Mechanical Engineering, Lovely Professional University, Phagwara, Punjab, India. He has received his Ph.D. degree from Guru Nanak Dev Engineering College, Ludhiana, India, in 2018 under the guidance of Prof. (Dr.) Harmeet Singh, GNDEC, Ludhiana and Dr. Lalit Thakur, NIT, Kurukshetra. He was listed in the World's Top 2% Scientists published by Stanford University and Elsevier in 2023 and 2024. His research areas include surface engineering/thermal spraying (high-velocity oxy-fuel spraying, flame spray, cold spray, plasma spray, and microwave claddings). Currently, he is working on the development of Nano-structured || Multi-modal || High-entropy alloys coatings for high-temperature oxidation and corrosion resistance-Thermal Barrier Coatings (TBCs) and Microwave processing of materials. He has guided four Ph.D. scholars and is currently supervising four more in the area of Surface Engineering. He has contributed extensively to thermal spray coatings in reputed journals, which include *Surface Coatings and Technology*, *Materials Today Communications*, *Engineering*

Failure Analysis, *Journal of Cleaner Production*, *Surface Topography: Metrology and Properties*, *Journal of Failure Prevention and Control*, *International Journal of Surface Engineering*, and *Interdisciplinary Materials Science* under the flagship of various publication groups such as Elsevier, Taylor & Francis, Springer Nature, IGI Global, and In-Tech Open. Moreover, he is a dedicated reviewer for reputed journals such as *Surface Coatings and Technology*, *Journal of Thermal Spray and Technology*, *Ceramics International*, *Journal of Material Engineering Performance*, *Engineering Failure Analysis*, *Surface Topography: Metrology and Properties*, *Materials Research Express*, *Engineering Research Express*, and *IGI Global Journals*. He received the "Top Cited Paper Awards India 2022" from IOP Publishing, United Kingdom, in the Review Category. He received the Lovely Professional University's *Research Excellence Award* in 2019, 2020, 2021, and 2023. His research contributions include more than 30 SCI-indexed papers, more than 10 Scopus-indexed papers, 15 conference papers, four books, and 18 book chapters. He has filed one patent titled "High-Temperature Oxidation and Erosion Resistant Alloy-718/Al_2O_3 Composite Coatings". He has organized both national and international conferences. He has received seed money for a project titled "Fabrication of Coatings for Bio-medical Applications" (Project Number: LPU/DRD/SEED/SAC/140222/24804) from Lovely Professional University, India. He has also received a grant from the Scientific Engineering and Research Board (SERB), New Delhi, Government of India, to conduct a conference in 2023.

Dr. Lalit Thakur is working as Assistant Professor, Grade-1 in the Department of Mechanical Engineering at the National Institute of Technology Kurukshetra, India. He has obtained his M.Tech. and Ph.D. degrees from the Indian Institute of Technology (IIT) Roorkee, India. For the past 13 years, he has been continuously exploring new possibilities in the fields of Welding Engineering and Thermal Spray Technology. His current research areas include thermal spray coatings and tungsten inert-gas (TIG) weld claddings for wear, corrosion, biomedical applications, ferromagnetic lubricants, advanced composite development by powder metallurgy, stir casting, and friction stir processing. He has authored more than 60 research articles in various international journals (SCI/Scopus), conferences, and edited books published by reputed publishers such as Applied Surface Science, Surface and Coatings Technology, Surface Engineering, Surface Topography, Tribology Transactions, Wear, JMEP, JMST, Materials Today Communications, Engineering Failure Analysis, Materials Research Express, Surface Review & Letters, CRC Press, Springer Nature, and IntechOpen. Moreover, he is a reviewer of most of the reputed journals related to Materials Science, Surface Engineering, and Coatings Technology. He has recently published a book titled *Thermal Spray Coatings* under CRC Press, Taylor & Francis Group. He is also a life member of the Institution of Engineers India (IEI) and the Indian Structural Integrity Society (INSIS). He has published two patents in the fields of thermal spray coatings and friction stir processing. He has also guided many M.Tech. and Ph.D. scholars in emerging areas of Thermal Spraying, Advanced Manufacturing, and Welding Technology.

Contributors

Ganesh Babar
General Science and Humanities Department
Dnyanshree Institute of Engineering & Technology
Satara, India

N. Ramesh Babu
Department of Mechanical Engineering
Nitte Meenakshi Institute of Technology
Bengaluru, India

Guruprasad Balachandran
Department of Mechanical Engineering
Alagappa Chettiar Government College of Engineering and Technology
Karaikudi, India

S. Balaji
School of Mechanical Engineering
REVA University
Bengaluru, India

Saravana Bavan
Department of Mechanical Engineering
Dayananda Sagar University
Bengaluru, India

Yogesh Bhalerao
School of Engineering, Faculty of Science
University of East Anglia
Norwich, United Kingdom

Digvijay Bhosale
Mechanical Engineering Department
Dr. D. Y. Patil Institute of Technology
Pune, Maharashtra, India

Ranjeet Kumar Choudhary
Department of Civil Engineering
Gaya College of Engineering
Gaya, India

Thirumalaikumarasamy Duraisamy
Government College of Engineering
Krishnagiri, India

Umesh Ghorpade
Mechanical Engineering Department
Dnyanshree Institute of Engineering & Technology Satara
Satara, India

Srinivasa Govindaiah
Department of Mathematics
New Horizon College of Engineering
Bengaluru, India

Suresh Gudala
Department of Mechanics and Mechanical Engineering
Siberian State Industrial University
Novokuznetsk, Russia

G. R. Gurunagendra
Department of Mechanical Engineering
Global Academy of Technology
Bengaluru, India

Mohd Hamid Hussain
Department of Mechanical Engineering
University of Technology and Applied Sciences
Muscat, Oman

Natarajan Jeyaprakash
Mechanical Engineering Department
Global College of Engineering and Technology
Muscat, Oman

Anand Ashok Kakhandaki
Department of Engineering
PTC Ltd.
New Delhi, India

Murugan Kaliyamoorty
Department of Mechanical Engineering
Government Polytechnic College
Thiruvarur, India

Praveen Kandasamy
Institute of Advanced Composite Materials
Korea Institute of Science and Technology
Seoul, Republic of Korea

Vanita Kshirsagar
Artificial Intelligence and Data Science Department
Dr. D. Y. Patil Institute of Technology
Pimpri-Chinchwad, India

Hitesh Kumar
Department of Mechanical Engineering
Rayat Bahra Institute of Engineering and Nano Technology
Nangal Shahidan, India

Pardeep Kumar
Yadavindra College of Engineering
Punjabi University Guru Kashi Campus
Talwandi Sabo, India

Mathanbabu Mariappan
Department of Mechanical Engineering
Government College of Engineering
Krishnagiri, India

Puneet Mathur
Department of Mechanical Engineering
Marwadi University
Rajkot, India

Amrinder Mehta
Research and Development Cell
Lovely Professional University
Phagwara, India

Jayatirtha Madhavarao Patil
Department of Mechanical Engineering
Jyothy Institute of Technology
Bengaluru, India

Channabasavaiah Durga Prasad
Department of Mechanical Engineering
RV Institute of Technology and Management
Bengaluru, India

Varun Kuppur Rajendra
Department of Mechanical Engineering
The Oxford College of Engineering
Bengaluru, India

Keshavamurthy Ramaiah
Department of Automobile Engineering
Dayananda Sagar College of Engineering
Bengaluru, India

Aprameya Chokkadi Ramesh
Department of Mechanical Engineering
Government Polytechnic College Ballari
Bellary, India

Kazi Sabiruddin
Department of Mechanical Engineering
Indian Institute of Technology Indore
Indore, India

Dandinashivara Rangadhamaiah Shankara
Department of Mechanical Engineering
SJB Institute of Technology
Bengaluru, India

Hulimane Bhaskara Shivaprasad
Department of Physics
Nitte Meenakshi Institute of Technology
Bengaluru, India

Bhupinder Singh
Department of Chemical and Environmental Engineering
University of Nottingham
Nottingham, United Kingdom

Manpreet Singh
Department of Defense Biotechnology High Performance Computing Software Applications Institute
Telemedicine and Advanced Technology Research Center
Fort Detrick, Maryland

Parvinkal Singh
Department of Mechanical Engineering
Baba Farid College of Engineering and Technology
Deon, India

Rajwinder Singh
Department of Materials Science and Engineering
Uppsala University
Uppsala, Sweden

Divya Gurkar Somashekar
Department of Automobile Engineering
Dayananda Sagar College of Engineering
Bengaluru, India

Setu Suman
Department of Mechanical Engineering
Indian Institute of Technology Indore
Indore, India

Shubhangi Suryawanshi
Artificial Intelligence and Data Science Department
Dr. D. Y. Patil Institute of Technology
Pimpri-Chinchwad, India

Lalit Thakur
Department of Mechanical Engineering
National Institute of Technology
Kurukshetra, India

Hitesh Vasudev
Lovely Professional University
Phagwara, India

G. Veeresha
Department of Mechanical Engineering
New Horizon College of Engineering
Bengaluru, India

Praveen Kumar Verma
Research and Development Cell
Lovely Professional University
Phagwara, India

1 Fundamentals of High-Temperature Corrosion and Wear-Resistant Coatings

Gaurav Prashar, Hitesh Vasudev, and Lalit Thakur

1.1 INTRODUCTION

Coal is a major contributor to electricity generation [1]. Despite the current emphasis on renewable energy sources for generation of electricity, coal-fired power plants continue to play a significant role in electricity production. Therefore, it is crucial to tackle the issues related to material failures in current coal-fired power plants. Coal-based power plants are often upgraded to advanced, supercritical, and ultra-supercritical plants in order to boost thermal efficiency and reduce pollutant emissions caused by burning fossil fuels [2]. These higher operating temperatures can lead to oxidation and hot corrosion of boiler materials, which is a common problem. Corrosion primarily occurs through oxidation in industrial environments. When oxidation occurs, a protective oxide layer is formed to prevent any additional corrosion. Unfortunately, the presence of sulfur, chlorine, and other harmful salts in the fuel renders this protective layer ineffective. As a result, corrosion occurs through the cracked and spalled layer, which provides no protection [3,4].

Erosion and hot corrosion play crucial roles in determining the lifespan of components in gas turbine engines, particularly under conditions of high temperature. Under oxidation conditions, the rate of degradation is slow, but it becomes significantly fast under hot corrosion conditions [5]. If proper materials or suitable coatings are not employed, catastrophic failures are imminent. This chapter reviews high-temperature wear mechanisms for crucial structural components of boilers and gas turbine engines. Corrective measures are also suggested to slow down the rates of high-temperature wear.

1.2 OXIDATION MECHANISM IN COAL-FIRED POWER PLANTS

Figure 1.1 depicts the mechanism of high-temperature oxidation in boiler tube steels. At higher operating temperatures in modern power plants, the diffusion of metal ions in the outward direction (M^{n+}) and the adsorption of oxide ions in the inward direction on the surface take place. After this, the oxide ions undergo a reaction with the metal ions to form metal oxides on the surface of the boiler material. As the oxide

DOI: 10.1201/9781003542957-1

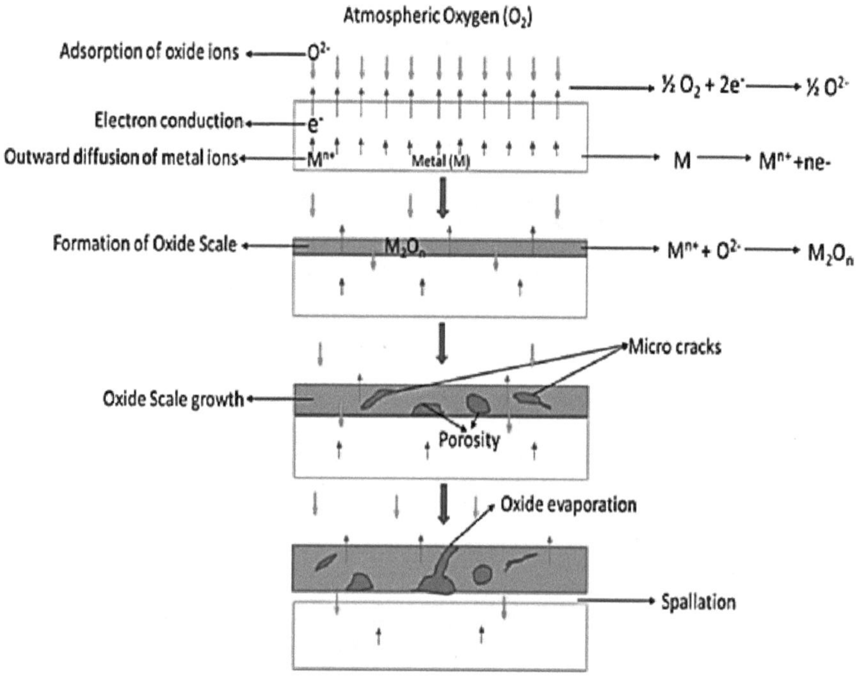

FIGURE 1.1 Mechanism behind high-temperature oxidation [4]. (*Under the terms of the CC-4.0 International License.*)

thickness increases, it can lead to the development of porosity and micro-cracks over time. This, in turn, can result in spallation or delamination of the oxide scales. High-temperature oxidation significantly reduces the lifespan of metallic components, leading to the need for early repairs or replacements of boiler tube materials. As a result, the maintenance costs of the plant are increased [6].

1.3 HOT CORROSION MECHANISM IN COAL-FIRED POWER PLANTS

Hot corrosion develops on structural parts when salt contaminants like Na_2SO_4, NaCl, and V_2O_5 combine to form molten deposits. These deposits can cause damage to the protective surface oxides [3]. Hot corrosion is a type of oxidation that happens when metals are exposed to temperatures between 700°C and 900°C while being surrounded by sulfate deposits. The hot corrosion attack leads to a much more significant deterioration of the alloy compared to when there is no deposit present in the same gas environment. The initial increase in attack is usually minimal, and the alloy shows little signs of deposit effects. Over time, the degradation mechanism undergoes changes due to the deposit, resulting in a significant increase in the rate of attack. Typically, the deposit is in a liquid state. The hot corrosion process is influenced by various factors, including the composition of the alloy, the composition of the deposit, gas composition, fluxing rate, and temperature [7]. According to the type

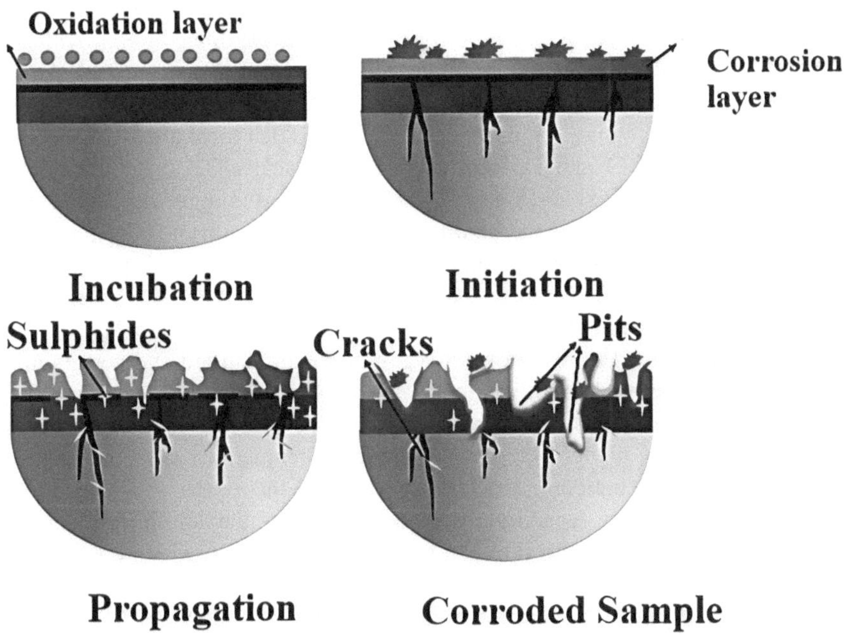

FIGURE 1.2 Hot corrosion mechanism [8]. (Under the terms of CC-4.0 International License.)

of attack, hot corrosion can be divided into high-temperature hot corrosion (Type I) and low-temperature hot corrosion (Type II). For the former, the temperature falls within the range of 825°C–900°C, and for later, the temperature varies from 550°C to 650°C. Fire side corrosion is the example of low-temperature hot corrosion with formation of Na_2SO_4 or K_2SO_4 when NaCl or KCl reacts with SO_2.

The mechanism consists of three stages: the incubation stage, initiation stage, and propagation stage (Figure 1.2). At the beginning of hot corrosion, there is an incubation period that triggers regular oxidation soon after the salt is applied. During this period, stable oxides are formed. After the initial step, the salt is applied and then exposed to the necessary temperature. With the increase in temperature, a harmful sulfide layer is created that damages the protective oxide layer [9]. Over time, the formation of internal sulfide can cause the oxide layer to deplete and dissolve, resulting in a decrease in corrosion resistance and direct contact with the coated surface [10]. During the propagation stage, there is an interaction between the surface oxide scale and the molten salt on the substrate.

1.4 HOT CORROSION MECHANISM IN GAS TURBINE BLADES

Oxidation and hot gas corrosion, commonly referred to as sulfidation, are significant phenomena that decrease the lifespan of components located in high-temperature areas of the gas flow, like combustion chambers and turbine blades. There are two distinct types of damage processes associated with sulfidation. Type I happens at temperatures above 800°C, while Type II occurs within the range of approximately 600°C–800°C. Type I exhibits intergranular attack and the presence of sulfide particles along with a denuded zone of the base metal [11]. Through the utilization of an electron beam microanalyzer, it has been discovered that there are distinct sulfide particles composed of Cr and Ti, located beneath the scale. The surface scales are composed of chromium oxide (Figure 1.3a) [12]. The beginning of the Type I reaction is more likely to occur on the high-pressure side of the turbine blade, as this is where sulfur-rich gas containing contaminants initially strikes [13]. In a separate study, it was verified that Type I takes place at a distance from the edges, specifically in the middle of the airfoil [13].

Type II corrosion is the result of a mixture of sodium sulfate and certain alloy constituents, such as nickel, which forms low melting eutectic compounds [11]. It appears that the level of attack is connected to the amount of mixed salt present, which is stabilized by higher SO_3 pressure. At lower temperatures, the equilibrium Pso_3/Pso_2 ratio tends to increase, which is believed to be the reason behind the higher rates of corrosion that are typically observed. In addition, the equilibrium between SO_3 and SO_2 is highly influenced by the presence of a catalyst, which could potentially be found in operating turbines [14]. There have been observations of high rates of attack with a pitting morphology, especially for gas turbine components that operate in a marine environment [14]. Low-temperature corrosion typically exhibits a lack of denuded zone, minimal intergranular attack, and a corrosion scale that forms in layers [11]. Additionally, there are no subscale sulfide particles [12]. Through the utilization of an electron beam microanalyzer, it has been determined that a continuous layer is formed by the Cr and Ti sulfides. In this case, the surface scale consists of the less protective oxides of Ni and Co (Figure 1.3c) [12]. In the platform area, the temperature is typically lower compared to the airfoil, making these areas susceptible to Type II hot corrosion attack [13]. Another study found that Type II was primarily concentrated in the trailing and leading edge locations. As the height from the platform increased, the extent of this type of corrosion reduced [12].

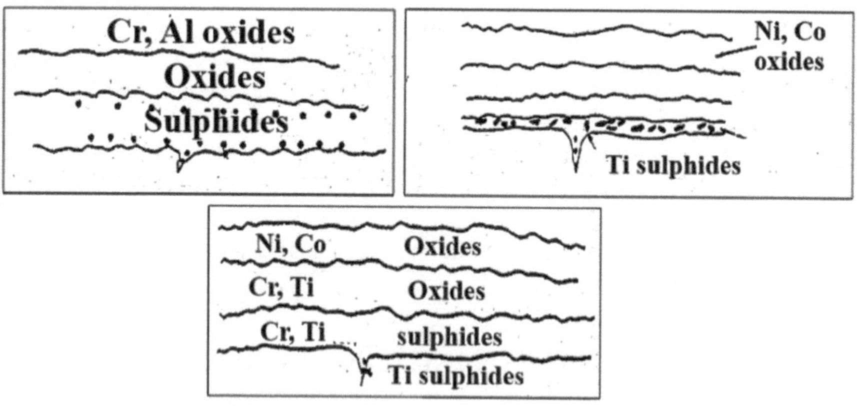

FIGURE 1.3 Hot corrosion mechanism for different temperatures [12]. (a) High-temperature hot corrosion, (b) transition hot corrosion, and (c) low-temperature hot corrosion.

In practice, certain features exhibit characteristics that resemble both types of corrosion, leading to the term transition hot corrosion [14]. With the help of an electron beam microanalyzer, it has been discovered that the sulfides of Cr and Ti form larger interconnected sulfide networks. Additionally, the surface scale mainly consists of the oxides of Ni and Co (Figure 1.3b) [12]. Figure 1.3 depicts the hot corrosion mechanism for gas turbine blades at low and high temperatures.

The temperatures are influenced by the composition of the aggressive deposits, rather than being solely determined by the affected base material. The main cause of damage is the breakdown of the protective oxide conversion coating, allowing harmful salt melts to penetrate. The harmful salt melts contain sodium, vanadium, magnesium, calcium, and sulfur. Due to the use of clean fuel in engines, corrosion-causing deposits typically originate from impurities present in the intake air. Engines in a marine atmosphere and/or exposed to sulfur-containing dusts, like those containing gypsum, are particularly vulnerable due to the unique effects of sodium and sulfur. The sulfidation damage mechanism can be explained in the following way:

Typically, a protective Cr_2O_3 coating is applied to the base material (e.g., Ni-based) to prevent sulfur from penetrating. However, various mechanisms, such as the chemical reaction of the oxide coating with salt melts, can enable sulfur to infiltrate the coating. Through this process, the metal ions in the salt undergo a transformation, causing the Cr_2O_3 coating to become loose and enabling the inward diffusion of sulfur. Once sulfur has infiltrated the base material, it hinders the regeneration of the protective chromium oxide coating. As a result, due to accelerated oxidation, the base material is ultimately destroyed. It seems that even tiny traces of sulfur can cause this damaging effect, as it specifically targets a small area where the material transitions and encourages the formation of significant quantities of Ni oxide. Even tiny traces of nickel sulfide are enough to facilitate the transport of oxygen.

1.5 PREVENTIVE APPROACHES

Diverse techniques have been employed to mitigate the harmful effects of elevated-temperature corrosion and oxidation of boiler components and gas turbine blades in power plants and aero engines [15]. A common practice is to apply coatings to the surface of materials [16]. There are other methods available for depositing coatings onto material surfaces, but thermal spraying is particularly advantageous due to its ability to produce coatings with excellent mechanical and microstructural qualities [17,18]. These techniques possess both advantages and disadvantages and are employed based on their suitability and specific needs [19].

1.5.1 Coating Materials to Resist High-Temperature Oxidation and Corrosion

Experts from around the globe have examined several combinations of coatings [20–24]. Based on the literature [25–27], incorporating elements such as Al, Y, Si, and B into Ni-Cr-based composite coatings enhances their ability to withstand oxidation

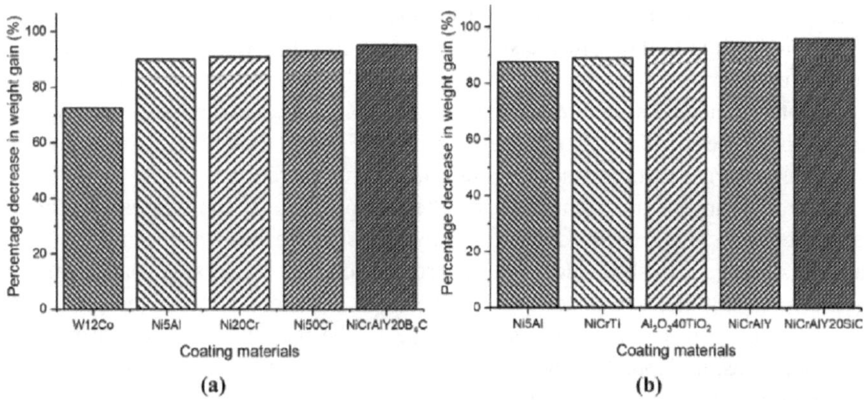

FIGURE 1.4 The study examines the oxidation characteristics of various coatings applied on (a) T22 and (b) T91 boiler steel [4]. (*Under the CC- 4.0 International License.*)

at high temperatures. The Cr_2O_3 and Al_2O_3 oxide layer is principally responsible for safeguarding the substrate surface against the attack of high-temperature oxidation, according to reports [28–34]. Figure 1.4 displays the efficacy of different coating combinations in preventing high-temperature oxidation. Several researchers have performed cyclic oxidation experiments on coating materials applied to T22 and T91 boiler tube steels at a temperature of 900°C for a total of 50 cycles [4]. The authors conducted experimental analysis on several coating combinations to compare the increase in weight of the coated material with that of the uncoated material. The process of high-temperature oxidation results in an increase in the weight of the material. Therefore, the coating combination that exhibits the least amount of weight gain is considered the most effective.

Researchers have studied several thermally sprayed coatings, including Ni20Cr, Ni-50Cr, NiCrAlY, WC-Co, Inconel 625-Al_2O_3, Inconel 718-Al_2O_3, NiCrTi, NiCrAlY-B4C, Stellite-21, Stellite-6, and Cr_3C_2-NiCr, to determine their ability to withstand high-temperature oxidation in boiler tube materials. The researchers employed diverse material characterization techniques to investigate the oxidation behavior of the specimens, both with and without coatings, at high temperatures. Microstructural analyses performed by multiple researchers revealed the existence of non-protective Fe_2O_3 oxide on the surface of uncoated materials (inset Figure 1.5). This oxide is accountable for the material's degradation due to delamination and spallation of the oxide layer. Alternatively, the coated specimens exhibited the presence of different protective oxide phases, including Cr_2O_3, NiO, Al_2O_3, SiO_2, TiO_2, Cr_3C_2, $NiCr_2O_4$, and $NiAl_2O_4$. These oxide phases played a crucial role in safeguarding the boiler tube materials from oxidation at elevated temperatures. Prashar and Vasudev [32] recently tailored a novel concept of bimodal Inconel 625-Al_2O_3 composite coatings and compared their performance with conventional and nano Inconel 625-Al_2O_3 coatings.

After oxidation tests, it was suggested that bimodal Inconel 625-Al_2O_3 heat-treated composite coatings reduced the weight gain of the substrate by 96.7%. Coherent

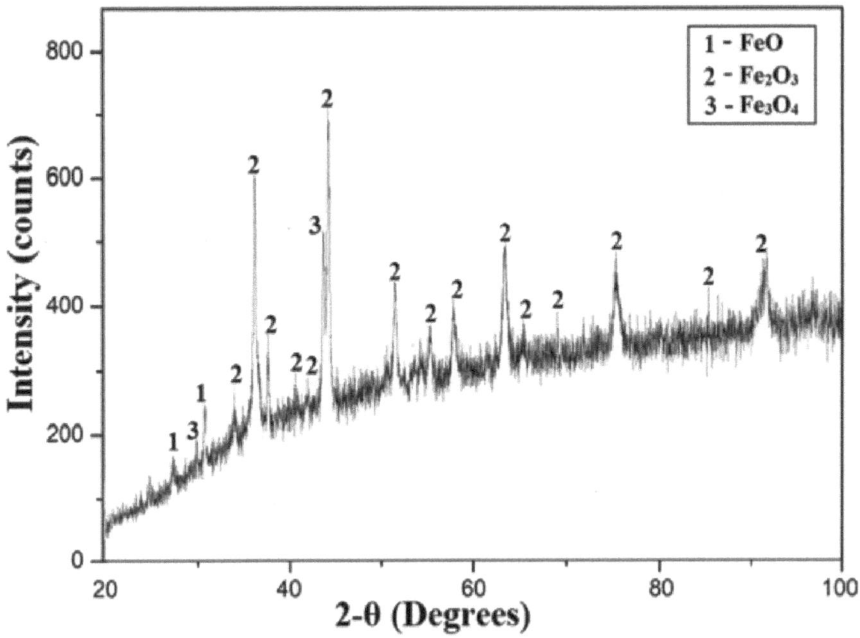

FIGURE 1.5 XRD pattern of uncoated substrate after the cyclic oxidation studies [32]. XRD, X-ray diffraction.

protective phases such as Al_2O_3, Cr_2O_3, TiO_2, and spinels like $NiCr_2O_4$ and $CrNbO_4$ are present within the scale of the coating. Figure 1.6 depicts the weight gain plots for oxidized uncoated and coated samples after oxidation tests at 900°C for 50 cycles. The uncoated sample has the highest weight gain (Figure 1.6a), and bimodal Inconel 625-Al_2O_3 heat-treated composite coating has the least weight gain (Figure 1.6c).

In order for the coatings to perform optimally and withstand hot corrosion in challenging environments, it is crucial that their microstructure is compact, free of flaws, and contains a significant amount of protective elements such as Cr or/and Al [32]. Currently, composite coatings are being used to handle hot corrosion issues [35]. When chromium is combined with nickel, it undergoes oxidation to form Cr_2O_3, which has the potential to be utilized at temperatures as high as 1,200°C [28]. The study conducted by Binal [36] focused on analyzing the hot corrosion behaviors of HVOF-sprayed 80Ni-20Cr coatings on a 316L stainless steel substrate.

1.5.2 Strategies to Resist High-Temperature Corrosion in Gas Turbines

Protective coatings are considered to be the preferred strategy to mitigate hot corrosion in gas turbines. Thermal barrier coatings (TBCs) are particularly noteworthy due to their exceptional capability to preserve blade integrity even in the presence of intense heat. Ceramic TBCs are a popular choice for thermal insulation in hot parts of turbines. They help raise the turbines' operating temperature, resulting in improved engine performance and efficiency [5]. TBCs have porous microstructures

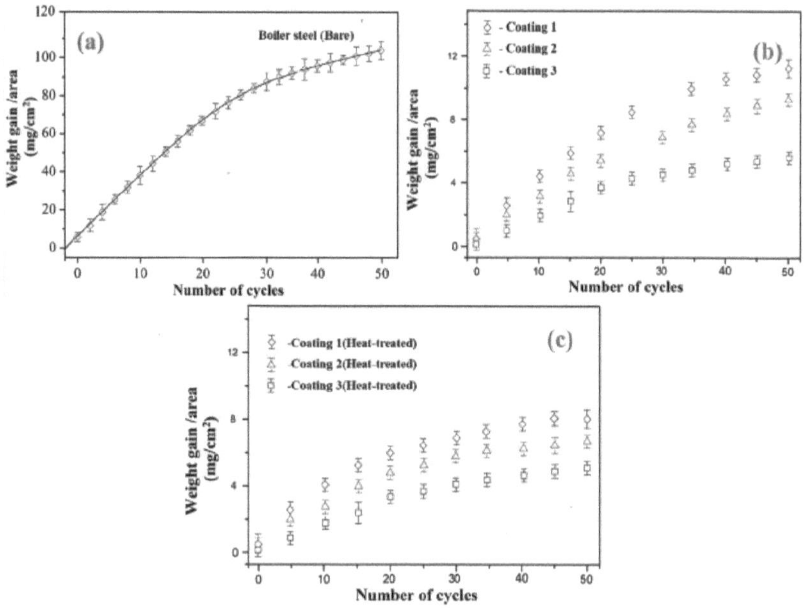

FIGURE 1.6 (a) Weight gain for uncoated boiler tube sample, (b) weight gain for coated tube samples, and (c) weight gain for coated heat-treated samples [32].

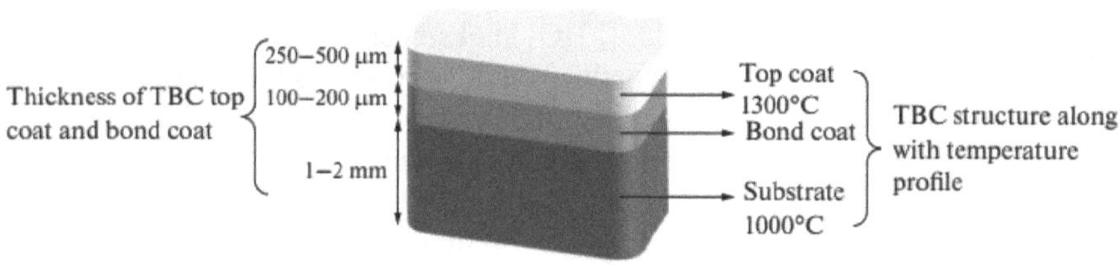

FIGURE 1.7 Typical TBC structure [5]. TBC, Thermal barrier coating.

that offer low thermal conductivity and high strain tolerance [37]. The thermal spray process is commonly used for coating gas turbine components, allowing for the application of coatings on complex shapes as well. Continuous efforts to improve gas turbine efficiency have led to operating temperatures exceeding 1,300°C. This, in turn, necessitates the use of thicker TBCs and the development of new cooling systems [5]. Figure 1.7 displays the typical TBC structure.

However, the exit temperature of the modern-day turbine combustion chamber has significantly increased to enhance performance and turbine work output [38]. As a result, the downstream section of the turbine operates in an extremely high-temperature environment that exceeds the maximum limits of superalloy substrates. To ensure the safety of hot components, especially high-pressure turbine blades, it is crucial to have a range of active cooling and passive thermal protection solutions in place. Figure 1.8 illustrates a TBC system deposited on the turbine blade [39].

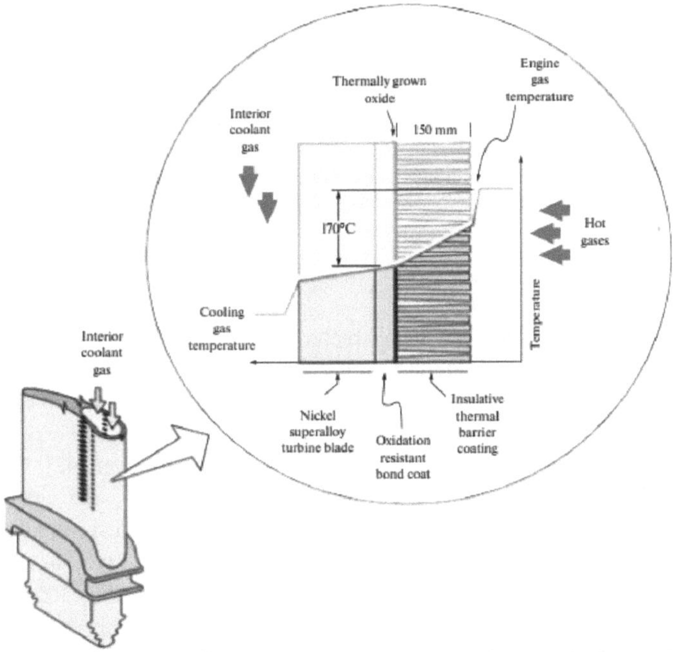

FIGURE 1.8 TBC system deposited on the turbine blade and the temperature drop across the thickness of the TBC [5]. TBC, Thermal barrier coating.

1.5.3 TBC Challenges

Achieving high-performance TBCs can be challenging due to the complexities of optimizing the columnar microstructure. Factors such as overall porosity and distribution of pore size play a significant role in this process. Controlling the porosity of air plasma spraying (APS) microstructures during coating formulation has been found to be crucial. Process parameters play a significant role in tailoring these microstructures. These studies have contributed to the development of reliable APS TBCs for high-performance applications, emphasizing the importance of enhancing suspension plasma spray (SPS) TBC knowledge to encourage industry adoption.

SPS and EB-PVD produced distinct columnar microstructures. The SPS coating has a limited strain tolerance because of the moderate inter-columnar gaps, while EB-PVD has a high thermal conductivity that can be detrimental to TBCs. A perfect columnar microstructure, as depicted in Figure 1.9c, would have a desirable combination of dense columns, moderate porosity between columns, and minimal gaps between them. This combination would result in low thermal conductivity and a long lifespan, thanks to its ability to withstand strain and resist fractures [40]. Undoubtedly, attaining this desired microstructure will pose challenges and may not be technically feasible. Research has been conducted on the production of TBCs using SPS [41]. However, further testing is necessary to fully determine their potential.

The formation of thermally grown oxide (TGOx) layers after prolonged exposure to high temperatures remains a significant concern and obstacle for the TBC research

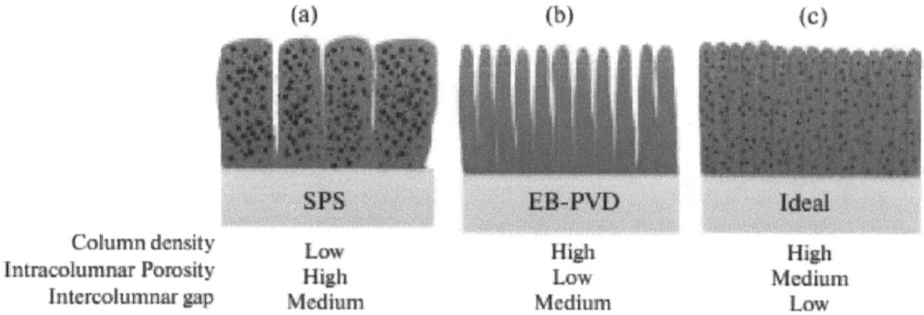

FIGURE 1.9 Comparing the ideal microstructure with the microstructure produced by SPS and EB-PVD [5]. SPS, Suspension plasma spray; EB-PVD, electron beam physical vapor deposition.

community. This issue leads to the delamination of the top layer [42]. This poses a challenge for scientists to increase the turbine inlet temperature, leading to decreased engine efficiency. TGOx layers primarily form due to the diffusion of reactive elements, such as Al, Cr, and Ni, through the substrate matrix, as well as the diffusion of oxygen through the TBC top coat [43,44]. One important barrier for the scientific research community is to lower TGOx layer development rates.

1.5.4 Future Work Suggestions for TBCs

The high-temperature corrosion mechanisms in TBCs can be quite complex. Further research is necessary to gain a better understanding of the degradation mechanisms of TBCs. In order to address the diffusion challenge posed by reactive elements and oxygen, it would be beneficial to promote the use of TBCs that do not contain bond coat in the near future. A unique TBC system was developed, comprising only the topcoat and the underlying substrate. Using selective laser melting processing, the addition of 1 wt.% CNTs to NiCrAlY eliminates the need for an additional bond coat material, resulting in improved structural clarity and increased properties. In addition, the growth of TGOx was also hindered even at a temperature of 1,800°C, which is higher than the typical operating temperature of a turbine engine. This also decreases extra weight, thus improving system performance. Therefore, in the near future, this creates opportunities for the development of TBCs without the need for bond coat materials [45].

The scientists have developed two basic methods to prevent coatings from separating or popping off. The first method is to develop a conventional TBC system and deposit a thick top coat layer to reduce the arrival rate of oxygen at the bond coat. Unfortunately, this method is not economical and also leads to increased weight. The second method is to develop a unique protective oxide like silicon-hafnium oxide [46]. In order to access its potential in the future, tests on an actual engine should be conducted. This would permit researchers to move forward and open new possibilities in TBCs.

Environmental barrier coatings play a vital role in safeguarding turbine engines from the detrimental effects of high temperatures. These coatings can safeguard vital components of turbines, especially those developed from silicon-based ceramics. These ceramic materials are highly appreciated by the research community for their contributions in enhancing turbine efficiency, thanks to their high-temperature properties and low-weight nature. However, without environmental barrier coatings (EBCs), these ceramic materials can degrade quickly under strenuous operating conditions. Scientists/materials engineers depend on EBCs to combat oxidation and corrosion of structural components inside a turbine engine. EBC materials, such as mullite and yttrium silicates, provide improved reliability and thermal stability.

1.6 HIGH-TEMPERATURE EROSION

Concerns have been for decades regarding the erosion of components at elevated temperatures [47,48]. When industries engraining in the power sector experience shutdowns as a result of erosion, it not only imposes a substantial financial burden on the fiscal well-being of a country but also interrupts the crucial function of supplying power to society [49]. A case of failure caused by erosion is documented during a visit to a local industry in Punjab, India. The burst tube is manufactured using SA210 GRA1 steel, which is classified as a ferritic steel. This particular steel exhibits excellent mechanical properties that make it well-suited for power generation applications. However, due to extended operations in areas with high temperatures, the steel tube will be consistently subjected to aging. Consequently, the mechanical performance and microstructure of the tube will inevitably deteriorate [6]. The boiler tube exhibited a window-type rupture on the parent metal, accompanied by secondary indications of erosion, after being in service for 1 year (inset Figure 1.10).

Various erosion mechanisms contribute to the deterioration of materials. The aforementioned mechanisms have the capability to function simultaneously and are

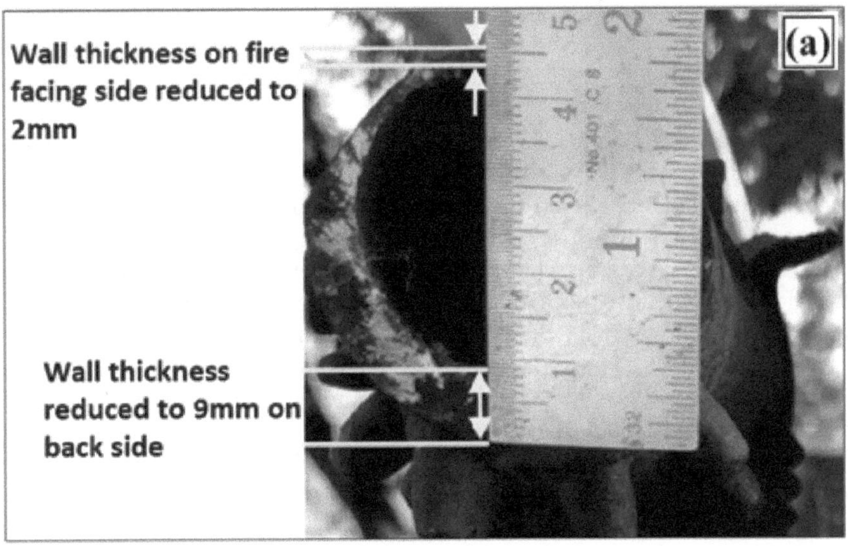

FIGURE 1.10 Failed boiler bed tube.

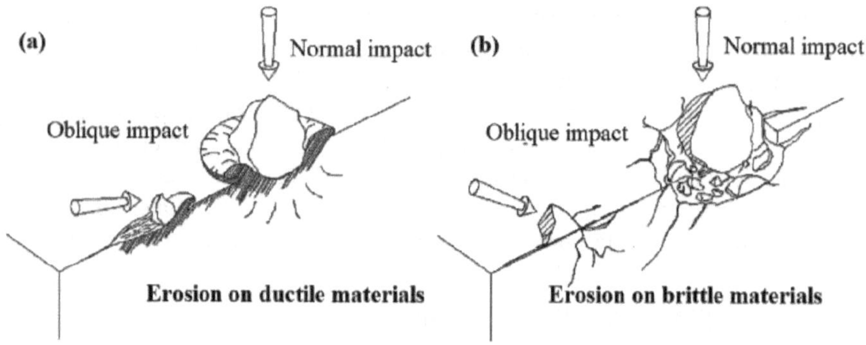

FIGURE 1.11 The erosion mechanism of (a) ductile and (b) brittle materials [52].

capable of interacting with one another. Therefore, determining the precise mechanism that contributes to erosion is challenging. As a result, the rate of erosion is calculated by considering the relative roles of these mechanisms [50]. Bitter proposed two fundamental erosion models for ductile and brittle materials [51]. The phenomenon is further complicated by the interaction of multiple parameters, including the characteristics of erosive particles, properties of the eroded material, working conditions, and distinct erosion mechanisms. Erosion is commonly attributed to two main processes, referred to as "cutting" and "deformation," regardless of the material's ductility or brittleness [51]. The cutting mechanism is characterized by the removal of material at low impact angles. However, the deformation mechanism occurs when particles strike the surface at a perpendicular angle, leading to plastic deformation or the formation of surface and subsurface cracks [52]. The schematic diagram illustrating the mechanisms for ductile and brittle materials is depicted in Figure 1.11. The primary mechanism for removing material in ductile materials such as metals and alloys is micro-cutting and micro-plowing. Failure in brittle materials, such as ceramics, is attributed to deformation and crack initiation, followed by crack propagation [53].

The coatings, which are composed of Ni-based superalloy powder, provide protection against surface degradation as a result of their high creep strength and excellent oxidation resistance. Nevertheless, the ductile nature of the material can lead to a decrease in hardness, which is a critical characteristic for withstanding erosion. The improvement of this can be achieved by integrating reinforcements. The utilization of reinforcement additives such as WC, Al_2O_3, Cr_3C_2, SiC, and TiO_2 has proven to be highly effective in improving the strength and durability of different materials [32].

In order to effectively withstand erosion in challenging environments, coatings must exhibit a high level of microhardness and improved fracture toughness. Bimodal composite coatings, composed of a mixture of micrometric-sized and nano-sized feedstock particles, have been shown to effectively enhance the mechanical properties, such as microhardness and fracture toughness, as well as improve the erosion resistance of the coating [54].

The bimodal composite (IN625-ALB in Figure 1.12) demonstrates superior erosive wear resistance compared to all other coatings regardless of testing angles.

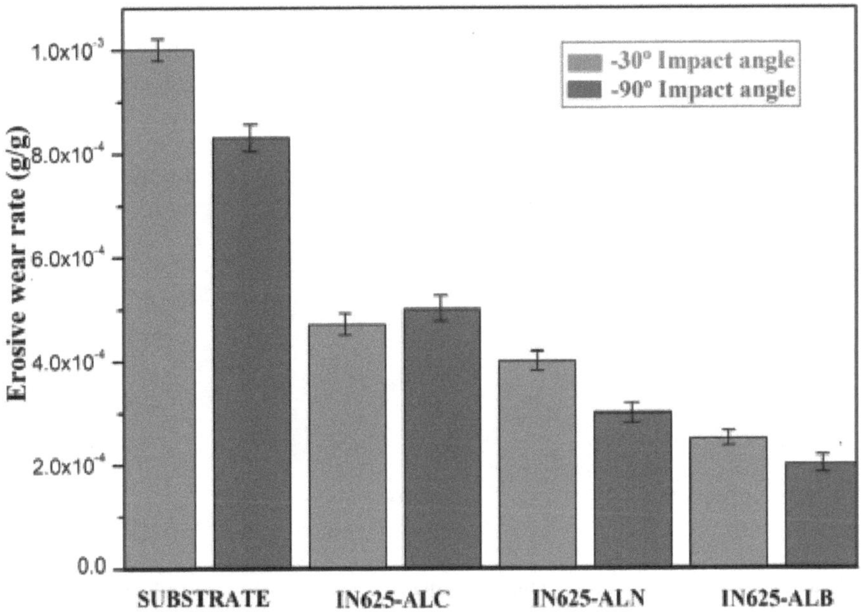

FIGURE 1.12 Erosive wear rates for all samples [54].

The erosive wear resistance of the composite structure is determined by the material's hardness and fracture toughness, as these factors influence the erosion behavior of the coating.

1.6.1 Erosion in Gas Turbine Blades

Erosion is a significant factor contributing to the failure of gas turbine blades. This phenomenon has a detrimental effect on the lifespan of the turbine blades [5]. The erosion of rotor blades due to exposure to sandy atmospheres results in thinning and rounding of the blade's leading edge, as illustrated in Figure 1.13. In order to prevent maintenance costs, it is necessary to have a comprehensive understanding of the mechanisms that are responsible for material loss.

The microstructure of the TBC significantly affects its erosion behavior. According to a study [56], the microstructure has been found to have a greater impact on erosion performance compared to the composition of the TBC. The study conducted by Nicholls et al. [57] demonstrated that columnar microstructured electron beam physical vapor deposition (EB-PVD) TBCs exhibit a tenfold increase in erosion resistance compared to the lamellar microstructured TBCs produced through APS. Multiple fracture mechanisms are involved in the erosion of TBCs.

There are two predominant mechanisms for EB-PVD TBCs: near-surface cracking and dendrite-initiated cracking [58]. The term "near-surface cracking" refers to cracking that occurs specifically within the top 20 μm of the coating. This type of cracking is visually represented in Figure 1.14a. The region in question is responsible for the bulk of material removal that occurs during the course of erosion. The cracks

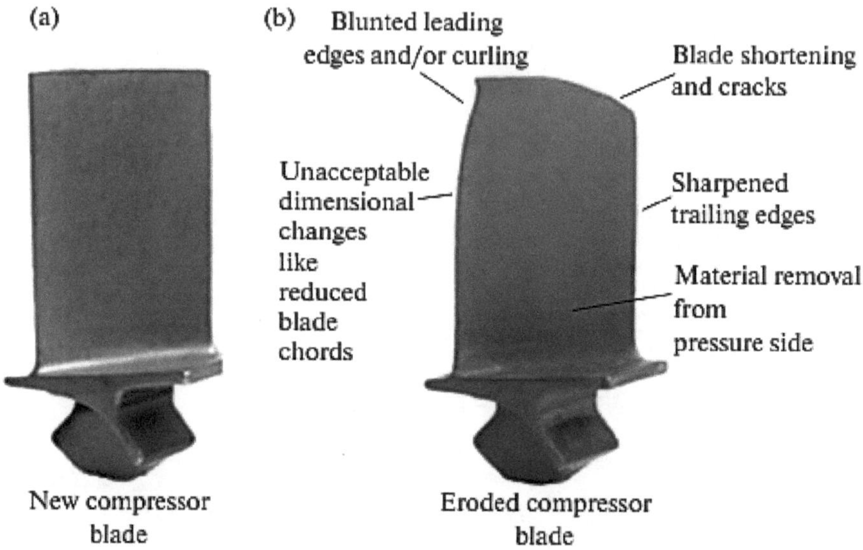

FIGURE 1.13 The gas turbine compressor blade is presented in (a) to showcase the new blade design, (b) displays the gas turbine blade after it has undergone erosion [55]. (*Under the CC-4.0 International License.*)

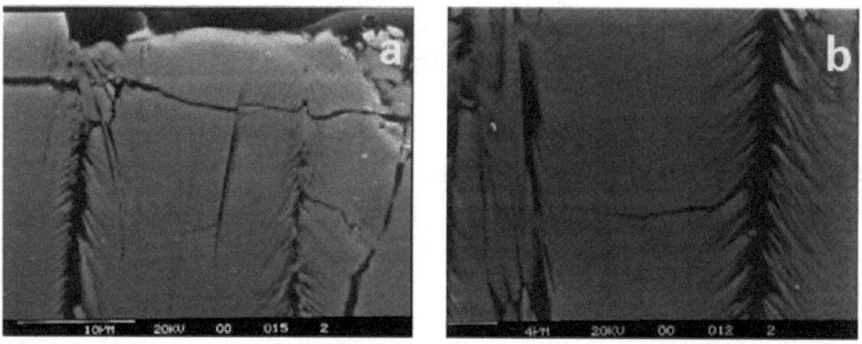

FIGURE 1.14 SEM micrographs highlighting (a) near-surface cracking and (b) crack initiated from dendrite [58]. SEM, Scanning electron microscope.

begin to develop in the near surface, and subsequent particle impact leads to the removal of tiny chunks of material. The crack propagation to the adjacent column is prevented by the presence of columnar boundaries, which effectively localizes the damage. Dendrite-initiated cracking is commonly observed at a depth greater than 20 µm within the coating, as depicted in Figure 1.14b. The columns exhibit a dendritic structure, characterized by a feathery edge. Cracks in the column originate from the edge structure, as opposed to the center, as stated in a study [58]. During impact conditions, it is possible for the columns to experience bending, and the presence of notches can exacerbate this effect by acting as stress concentrators [59]. The presence of these edge defects leads to the initiation of cracks.

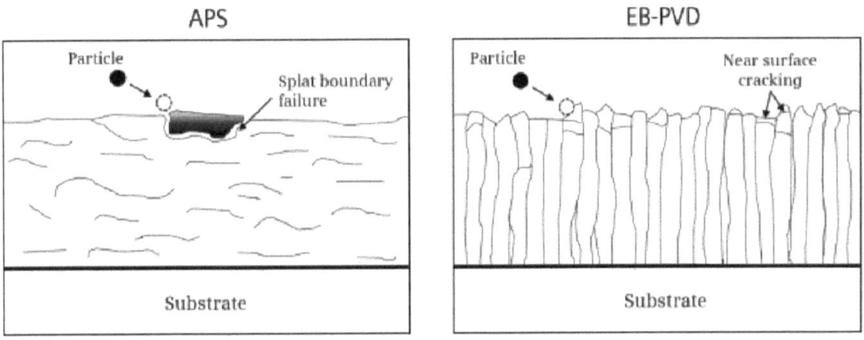

FIGURE 1.15 Failure modes for APS and EB-PVD TBCs [57]. APS, Air plasma spraying; EB-PVD, electron beam physical vapor deposition; TBCs, thermal barrier coatings.

The erosion mechanisms of APS TBCs are primarily caused by the weakly bonded lamellar splat boundaries [60]. The fundamental principles governing the process of material removal in the context of erosion of APS and EB-PVD coatings are depicted in Figure 1.15. When a particle comes into contact with the surface of an APS coating, it causes stress waves to be generated within the coating [61]. Cracks have the potential to propagate along splat boundaries due to the presence of a micro-crack network. This can result in the formation of pores and subsequent material loss [62]. The densified regions (columns) of EB-PVD TBC generally have fewer flaws compared to APS coatings, which results in a lower erosion rate for EB-PVD coatings [62]. It requires a greater number of successive impacts to connect the cracks together in comparison to flawed APS coatings.

Eaton and Novak [63] have identified three erosion mechanisms that are associated with different erosion rates of APS coatings. These mechanisms are illustrated in Figure 1.16. Local furrow plowing is a common occurrence during periods of low erosion rates. When the erosion rate is increased, the erodent particles induce fractures, and cracks propagate within and around individual splats, leading to the ejection of splats and the exposure of splat boundaries. This phenomenon is referred to as "secondary failure." Tunneling can occur during periods of higher erosion rates. This phenomenon is characterized by the interconnection of pores along the surface through fracture, resulting in the removal of a larger chip of material. Kumar et al. [40] also suggested that the presence of coarse porosity is undesirable due to its tendency to facilitate material removal through a tunneling mechanism (Figure 1.17).

The erosion rate has been demonstrated to be influenced by impact angles. The maximum erosion rates have been observed at an angle of 90° [57]. According to research [64], it has been observed that both APS and EB-PVD coatings experience lower erosion rates when subjected to lower impact angles. Lima et al. [65] discovered that the primary cause of failure when subjected to a 15° impact was the development of cracks around the boundaries of the splats in an APS 8YSZ TBC. Additionally, it was observed that the TBC material was most likely eliminated in clusters when subjected to a 90° impact. The main mechanism at a 90° impact angle was suggested to be tunneling via pores. The study conducted by Ramachandran

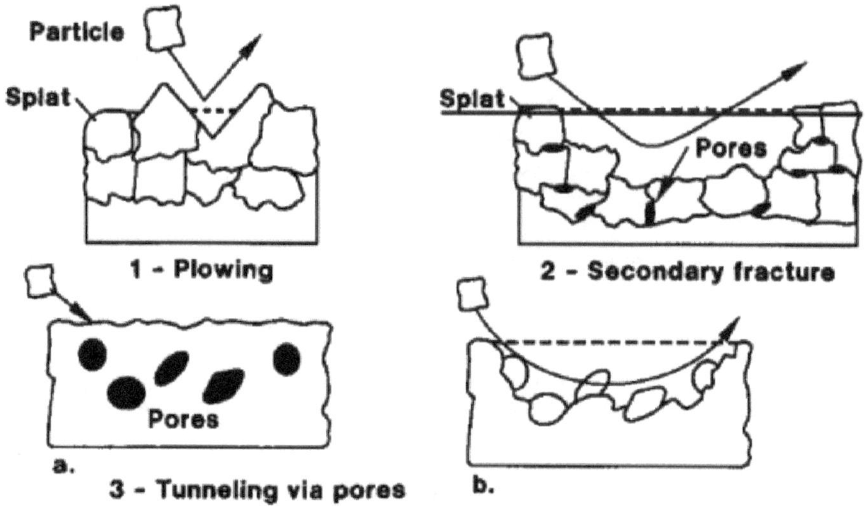

FIGURE 1.16 The schematic figure illustrates three failure mechanisms observed in APS TBCs: (1) plowing, (2) secondary fracture, and (3) tunneling via pores [63]. APS, Air plasma spraying; TBCs, thermal barrier coatings.

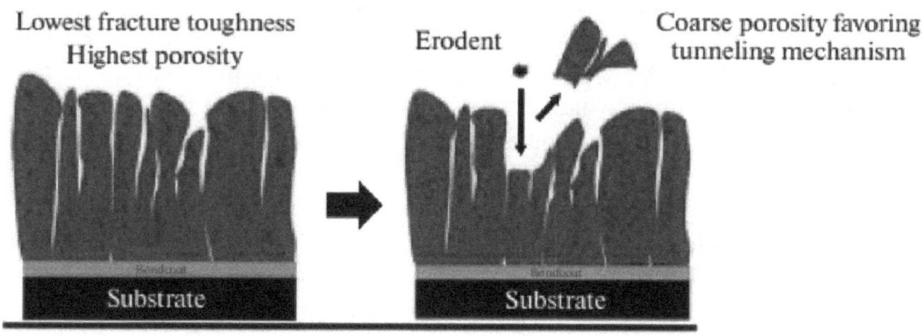

FIGURE 1.17 Proposed erosion mechanism of SPS TBCs [40]. SPS, Suspension plasma spray; TBCs, thermal barrier coatings. (*Under the CC-4.0 International License.*)

et al. [60] revealed that an impact angle of 30° primarily resulted in plowing and minimal brittle cracking for a thick 8YSZ APS coating with a porosity level of 4%.

The advancements in high-temperature corrosion and wear-resistant coatings have significantly enhanced the performance and lifespan of components in demanding environments, including power generation and aerospace applications [66–72]. Research on thermally sprayed coatings, such as Ni-based and Inconel-based systems, highlights their exceptional oxidation and erosion resistance, achieved through tailored microstructures and innovative processing techniques like microwave heating and plasma spraying [73–78]. Moreover, the integration of heat treatments has further refined their properties, ensuring superior protection against aggressive conditions [79,80]. These developments underscore the critical role of surface engineering in addressing high-temperature challenges.

1.7 CONCLUSION

High-temperature wear is considered to be the prime failure mechanism in power plants and gas turbine blades. In a reducing environment, the corrosion rate is much higher than in an oxidizing environment. A reducing environment is present as a result of the coal's incomplete combustion in the furnace. Preventing or detecting these failure mechanisms in initial stages is crucial to avoid accidents and economic loss. When biomass, bagasse, and wood waste are burned, they produce sulfur, alkaline metals, and chlorine, which can lead to severe corrosion in power plants. There is a challenge for engineers to research the severity of corrosion and find suitable materials and protection methods to minimize high-temperature corrosion and prevent premature failure of boiler tubes. Based on extensive research, it has been determined that thermally sprayed coatings are highly effective in safeguarding boiler tube materials from the harsh conditions of advanced coal-fired thermal power plants. It is possible to distinguish between Type I hot corrosion and Type II hot corrosion by analyzing various mechanisms. By implementing/depositing the right combination of TBCs, hot corrosion can be minimized. EBC materials like mullite and yttrium silicates provide improved reliability and thermal stability. The bimodal composite coating exhibited superior performance, displaying enhanced resistance to erosion and oxidation. The effectiveness of the bimodal coating in preventing oxidation and erosion was found to be superior to that of its unimodal counterparts. Achieving high-performance TBCs can be challenging due to the complexities of optimizing the columnar microstructure. Factors such as overall porosity and distribution of pore size play a significant role in this process. The impact of factors such as column density on the durability of TBCs is still not fully understood and will be the scope of study in the future. This chapter draws the strong connection between material science and engineering.

REFERENCES

1. Shahzad Baig, K. and Yousaf, M., 2017. Coal fired power plants: emission problems and controlling techniques. *Journal of Earth Science & Climatic Change*, *8*(404), p. 2.
2. Soeder, D.J. and Borglum, S.J., 2019. *The Fossil Fuel Revolution: Shale Gas and Tight Oil*. Elsevier.
3. Prashar, G. and Vasudev, H., 2020. Hot corrosion behavior of super alloys. *Materials Today: Proceedings*, *26*, pp. 1131–1135.
4. Mishra P. and Mishra, S.B., 2023. A review of high-temperature oxidation behaviour of thermally sprayed boiler tube materials in advanced coal-fired thermal power plants. *Nano World Journal*, *9*(S1), pp. S374–S379.
5. Prashar, G., Vasudev, H. and Thakur, L., 2023. A comprehensive review on the hot corrosion and erosion performance of thermal barrier coatings. *Protection of Metals and Physical Chemistry of Surfaces*, *59*(3), pp. 461–492.
6. Prashar, G., Vasudev, H. and Thakur, L., 2022. High-temperature oxidation and erosion resistance of Ni-based thermally-sprayed coatings used in power generation machinery: a review. *Surface Review and Letters*, *29*(03), p. 2230003.
7. Kumar, M., Singh, H. and Singh, N., 2013. Study of Ni-20Cr coatings for high temperature applications – a review. *Archives of Metallurgy and Materials*, *58*. DOI:10.2478/amm-2013-0030.

8. Radhika, N., Noble, N. and Adediran, A.A., 2024. Electrochemical and hot corrosion behaviour of annealed AlCoCrFeNi HEA coating over steel. *Scientific Reports*, *14*(1), p. 5652.
9. Nimmervoll, M., Schmid, A., Mori, G., Hönig, S. and Haubner, R., 2021. Surface sulphide formation on high-temperature corrosion resistant alloys in a H_2S-HCl-CO_2 mixed atmosphere. *Corrosion Science*, *181*, p. 109241.
10. Yan, Y.F., Xu, X.Q., Zhou, D.Q., Wang, H., Wu, Y., Liu, X.J. and Lu, Z.P., 2013. Hot corrosion behaviour and its mechanism of a new alumina-forming austenitic stainless steel in molten sodium sulphate. *Corrosion Science*, *77*, pp. 202–209.
11. Schilke, P.W., 2004. *Advanced Gas Turbine Materials and Coatings. GE Energy.* General Electric Company.
12. Viswanathan, R., 2001. An investigation of blade failures in combustion turbines. *Engineering Failure Analysis*, *8*(5), pp. 493–511.
13. Khajavi, M.R. and Shariat, M.H., 2004. Failure of first stage gas turbine blades. *Engineering Failure Analysis*, *11*(4), pp. 589–597.
14. Saunders, S.R.J., Hossain, M.K. and Ferguson, J.M., 1982. Comparison of hot-salt corrosion test procedures. In: Brunetaud, R., Coutsouradis, D., Gibbons, T.B., Lindblom, Y., Meadowcroft, D.B., & Stickler, R. (eds), *High Temperature Alloys for Gas Turbines 1982: Proceedings of a Conference held in Liège*, Belgium, 4–6 October 1982 (pp. 177–206). Springer Netherlands.
15. Singh, A., Sharma, V., Mittal, S., Pandey, G., Mudgal, D. and Gupta, P., 2018. An overview of problems and solutions for components subjected to fireside of boilers. *International Journal of Industrial Chemistry*, *9*, pp. 1–15.
16. Hocking, M.G., 1993. Coatings resistant to erosive/corrosive and severe environments. *Surface and Coatings Technology*, *62*(1–3), pp. 460–466.
17. Pawlowski, L., 2008. *The Science and Engineering of Thermal Spray Coatings.* John Wiley & Sons.
18. Heath, G.R., Heimgartner, P., Irons, G., Miller, R.D. and Gustafsson, S., 1997, October. An assessment of thermal spray coating technologies for high temperature corrosion protection. In *Materials Science Forum* (Vol. 251, pp. 809–816). Trans Tech Publications Ltd.
19. Talib, R.J., Saad, S., Toff, M.R.M. and Hashim, H., 2003. Thermal spray coating technology: a review. *Solid State Science and Technology*, *11*(1), pp. 109–117.
20. Singh, H., Kaur, M. and Prakash, S., 2016. High-temperature exposure studies of HVOF-sprayed Cr3C2-25 (NiCr)/(WC-Co) coating. *Journal of Thermal Spray Technology*, *25*, pp. 1192–1207.
21. Bala, N., Singh, H. and Prakash, S., 2009. High-temperature oxidation studies of cold-sprayed Ni–20Cr and Ni–50Cr coatings on SAE 213-T22 boiler steel. *Applied Surface Science*, *255*(15), pp. 6862–6869.
22. Singh, G., Bala, N. and Chawla, V., 2019. High temperature oxidation behaviour and characterization of NiCrAlY-B4C coatings deposited by HVOF. *Materials Research Express*, *6*(8), p. 086436.
23. Singh, G., Bala, N. and Chawla, V., 2020. Oxidation behaviour of HVOF sprayed NiCrAlY and NiCrAlY-20SiC coatings on T-91 boiler tube steel. *Protection of Metals and Physical Chemistry of Surfaces*, *56*, pp. 134–150.
24. Bala, N., Singh, H. and Prakash, S., 2011. Characterization and high-temperature oxidation behavior of cold-sprayed Ni-20Cr and Ni-50Cr coatings on boiler steels. *Metallurgical and Materials Transactions A*, *42*, pp. 3399–3416.
25. Mishra, S.B., Chandra, K. and Prakash, S., 2013. Erosion–corrosion performance of NiCrAlY coating produced by plasma spray process in a coal-fired thermal power plant. *Surface and Coatings Technology*, *216*, pp. 23–34.

26. Higuera, V., Belzunce, F.J. and Riba, J., 2006. Influence of the thermal-spray procedure on the properties of a CoNiCrAlY coating. *Surface and Coatings Technology, 200*(18–19), pp. 5550–5556.
27. Singh, G., Bala, N. and Chawla, V., 2017. High temperature oxidation behaviour of HVOF thermally sprayed NiCrAlY Coating on T-91 boiler tube steel. *Materials Today: Proceedings, 4*(4), pp. 5259–5265.
28. Bala, N., Singh, H. and Prakash, S., 2010. Accelerated hot corrosion studies of cold spray Ni–50Cr coating on boiler steels. *Materials & Design, 31*(1), pp. 244–253.
29. Kaushal, G., Singh, H. and Prakash, S., 2011. High-temperature erosion-corrosion performance of high-velocity oxy-fuel sprayed Ni-20 Cr coating in actual boiler environment. *Metallurgical and Materials Transactions A, 42*, pp. 1836–1846.
30. López, A.J., Proy, M., Utrilla, V., Otero, E. and Rams, J., 2014. High-temperature corrosion behavior of Ni–50Cr coating deposited by high velocity oxygen–fuel technique on low alloy ferritic steel. *Materials & Design, 59*, pp. 94–102.
31. Prashar, G. and Vasudev, H., 2022. Structure–property correlation of plasma-sprayed Inconel625-Al_2O_3 bimodal composite coatings for high-temperature oxidation protection. *Journal of Thermal Spray Technology, 31*(8), pp. 2385–2408.
32. Prashar, G. and Vasudev, H., 2023. High-temperature oxidation behavior of direct-aged bimodal Al_2O_3-reinforced Inconel625 plasma sprayed composite coatings. *Surface and Coatings Technology, 475*, p. 130156.
33. Vasudev, H., Thakur, L., Bansal, A., Singh, H. and Zafar, S., 2019. High temperature oxidation and erosion behaviour of HVOF sprayed bi-layer Alloy-718/NiCrAlY coating. *Surface and Coatings Technology, 362*, pp. 366–380.
34. Vasudev, H., Thakur, L., Singh, H. and Bansal, A., 2020. An investigation on oxidation behaviour of high velocity oxy-fuel sprayed Inconel718-Al_2O_3 composite coatings. *Surface and Coatings Technology, 393*, p. 125770.
35. Vasudev, H., Thakur, L., Singh, H. and Bansal, A., 2020. A study on processing and hot corrosion behaviour of HVOF sprayed Inconel718-nano Al_2O_3 coatings. *Materials Today Communications, 25*, p. 101626.
36. Binal, G., 2023. Isothermal oxidation and hot corrosion behavior of HVOF sprayed 80Ni-20Cr coatings at 750°C. *Surface and Coatings Technology, 454*, p. 129141.
37. Drexler, J.M., Chen, C.H., Gledhill, A.D., Shinoda, K., Sampath, S. and Padture, N.P., 2012. Plasma sprayed gadolinium zirconate thermal barrier coatings that are resistant to damage by molten Ca–Mg–Al–silicate glass. *Surface and Coatings Technology, 206*(19–20), pp. 3911–3916.
38. Bhatia, T., Ozturk, A., Xie, L., Jordan, E.H., Cetegen, B.M., Gell, M., Ma, X. and Padture, N.P., 2002. Mechanisms of ceramic coating deposition in solution-precursor plasma spray. *Journal of Materials Research, 17*(9), pp. 2363–2372.
39. Karaoglanli, A.C., Ogawa, K., Turk, A. and Ozdemir, I., 2013. Thermal shock and cycling behavior of thermal barrier coatings (TBCs) used in gas turbines. *Progress in Gas Turbine Performance, 237*, p. 260.
40. Kumar, N., Mahade, S., Ganvir, A. and Joshi, S., 2021. Understanding the influence of microstructure on hot corrosion and erosion behavior of suspension plasma sprayed thermal barrier coatings. *Surface and Coatings Technology, 419*, p. 127306.
41. Gupta, M., Li, X.H., Markocsan, N. and Kjellman, B., 2020. Design of high lifetime suspension plasma sprayed thermal barrier coatings. *Journal of the European Ceramic Society, 40*(3), pp. 768–779.
42. Rabiei, A.G.E.A. and Evans, A.G., 2000. Failure mechanisms associated with the thermally grown oxide in plasma-sprayed thermal barrier coatings. *Acta Materialia, 48*(15), pp. 3963–3976.

43. Ebach-Stahl, A., Schulz, U., Swadźba, R. and Munawar, A.U., 2021. Lifetime improvement of EB-PVD 7YSZ TBCs by doping of Hf or Zr in NiCoCrAlY bond coats. *Corrosion Science*, *181*, p. 109205.
44. Takahashi, R.J., Assis, J.M.K., Neto, F.P. and Reis, D.A.P., 2020. Heat treatment for TGO growth on NiCrAlY for TBC application. *Materials Research Express*, *6*(12), p. 126442.
45. Choudhary, S., Islam, A., Mukherjee, B., Richter, J., Arold, T., Niendorf, T. and Keshri, A.K., 2021. Plasma sprayed Lanthanum zirconate coating over additively manufactured carbon nanotube reinforced Ni-based composite: unique performance of thermal barrier coating system without bondcoat. *Applied Surface Science*, *550*, p. 149397.
46. Deijkers, J.A. and Wadley, H.N., 2021. A duplex bond coat approach to environmental barrier coating systems. *Acta Materialia*, *217*, p. 117167.
47. Praveen, A.S., Sarangan, J., Suresh, S. and Subramanian, J.S., 2015. Erosion wear behaviour of plasma sprayed NiCrSiB/Al_2O_3 composite coating. *International Journal of Refractory Metals and Hard Materials*, *52*, pp. 209–218.
48. Hidalgo, V.H., Varela, F.B., Menéndez, A.C. and Martınez, S.P., 2001. A comparative study of high-temperature erosion wear of plasma-sprayed NiCrBSiFe and WC–NiCrBSiFe coatings under simulated coal-fired boiler conditions. *Tribology International*, *34*(3), pp. 161–169.
49. Prashar, G. and Vasudev, H., 2021. High temperature erosion behavior of plasma sprayed Al_2O_3 coating on AISI-304 stainless steel. *World Journal of Engineering, 18*, pp. 760–766.
50. Islam, M.A. and Farhat, Z.N., 2014. Effect of impact angle and velocity on erosion of API X42 pipeline steel under high abrasive feed rate. *Wear, 311*(1–2), pp. 180–190.
51. Bitter, J.G.A., 1963. A study of erosion phenomena part I. *wear*, *6*(1), pp. 5–21.
52. Wang, Y.F. and Yang, Z.G., 2008. Finite element model of erosive wear on ductile and brittle materials. *Wear*, *265*(5–6), pp. 871–878.
53. Stachowiak, G. and Batchelor, A.W., 2013. *Engineering Tribology*. Butterworth-Heinemann.
54. Prashar, G. and Vasudev, H., 2022. Structure-property correlation and high-temperature erosion performance of Inconel625-Al_2O_3 plasma-sprayed bimodal composite coatings. *Surface and Coatings Technology*, *439*, p. 128450.
55. Alqallaf, J., Ali, N., Teixeira, J.A. and Addali, A., 2020. Solid particle erosion behaviour and protective coatings for gas turbine compressor blades—a review. *Processes*, *8*(8), p. 984.
56. Mahade, S., Ruelle, C., Curry, N., Holmberg, J., Björklund, S., Markocsan, N. and Nylén, P., 2019. Understanding the effect of material composition and microstructural design on the erosion behavior of plasma sprayed thermal barrier coatings. *Applied Surface Science*, *488*, pp. 170–184.
57. Nicholls, J.R., Deakin, M.J. and Rickerby, D.S., 1999. A comparison between the erosion behaviour of thermal spray and electron beam physical vapour deposition thermal barrier coatings. *Wear*, 233, pp. 352–361.
58. Wellman, R.G. and Nicholls, J.R., 2000. Some observations on erosion mechanisms of EB PVD TBCS. *Wear, 242*(1–2), pp. 89–96.
59. Steinberg, L., Naraparaju, R., Heckert, M., Mikulla, C., Schulz, U. and Leyens, C., 2018. Erosion behavior of EB-PVD 7YSZ coatings under corrosion/erosion regime: effect of TBC microstructure and the CMAS chemistry. *Journal of the European Ceramic Society, 38*(15), pp. 5101–5112.
60. Ramachandran, C.S., Balasubramanian, V. and Ananthapadmanabhan, P.V., 2013. Erosion of atmospheric plasma sprayed rare earth oxide coatings under air suspended corundum particles. *Ceramics International, 39*(1), pp. 649–672.
61. Zeng, J. and Kim, T.J., 1996. An erosion model of polycrystalline ceramics in abrasive waterjet cutting. *Wear, 193*(2), pp. 207–217.

62. Schmitt, M.P., Harder, B.J. and Wolfe, D.E., 2016. Process-structure-property relations for the erosion durability of plasma spray-physical vapor deposition (PS-PVD) thermal barrier coatings. *Surface and Coatings Technology, 297*, pp. 11–18.
63. Eaton, H.E. and Novak, R.C., 1987. Particulate erosion of plasma-sprayed porous ceramic. *Surface and Coatings Technology, 30*(1), pp. 41–50.
64. Tabakoff, W., 1989. Investigation of coatings at high temperature for use in turbomachinery. *Surface and Coatings Technology, 39*, pp. 97–115.
65. Lima, R.S., Guerreiro, B.M. and Aghasibeig, M., 2019. Microstructural characterization and room-temperature erosion behavior of as-deposited SPS, EB-PVD and APS YSZ-based TBCs. *Journal of Thermal Spray Technology, 28*, pp. 223–232.
66. Prashar, G., Vasudev, H. and Thakur, L., 2022. High-temperature oxidation and erosion resistance of Ni-based thermally-sprayed coatings used in power generation machinery: a review. *Surface Review and Letters, 29*(03), p. 2230003.
67. Singh, G., Vasudev, H., Bansal, A. and Vardhan, S., 2021. Influence of heat treatment on the microstructure and corrosion properties of the Inconel-625 clad deposited by microwave heating. *Surface Topography: Metrology and Properties, 9*(2), p. 025019.
68. Prashar, G. and Vasudev, H., 2020. Hot corrosion behavior of super alloys. *Materials Today: Proceedings, 26*, pp. 1131–1135.
69. Vasudev, H., Prashar, G., Thakur, L. and Bansal, A., 2022. Electrochemical corrosion behavior and microstructural characterization of HVOF sprayed inconel718-Al_2O_3 composite coatings. *Surface Review and Letters, 29*(02), p. 2250017.
70. Singh, P., Bansal, A., Vasudev, H. and Singh, P., 2021. In situ surface modification of stainless steel with hydroxyapatite using microwave heating. *Surface Topography: Metrology and Properties, 9*(3), p. 035053.
71. Vasudev, H., Singh, P., Thakur, L. and Bansal, A., 2020. Mechanical and microstructural characterization of microwave post processed Alloy-718 coating. *Materials Research Express, 6*(12), p. 1265f5.
72. Prashar, G. and Vasudev, H., 2022. Structure–property correlation of plasma-sprayed Inconel625-Al_2O_3 bimodal composite coatings for high-temperature oxidation protection. *Journal of Thermal Spray Technology, 31*(8), pp. 2385–2408.
73. Prashar, G. and Vasudev, H., 2022. Surface topology analysis of plasma sprayed Inconel625-Al_2O_3 composite coating. *Materials Today: Proceedings, 50*, pp. 607–611.
74. Prashar, G. and Vasudev, H., 2022. A review on the influence of process parameters and heat treatment on the corrosion performance of Ni-based thermal spray coatings. *Surface Review and Letters, 29*(01), p. 2230001.
75. Yedida, V.S. and Vasudev, H., 2022. A review on the development of thermal barrier coatings by using thermal spray techniques. *Materials Today: Proceedings, 50*, pp. 1458–1464.
76. Majji, B.G.R., Vasudev, H. and Bansal, A., 2022. A review on the oxidation and wear behavior of the thermally sprayed high-entropy alloys. *Materials Today: Proceedings, 50*, pp. 1447–1451.
77. Sunitha, K. and Vasudev, H., 2022. A short note on the various thermal spray coating processes and effect of post-treatment on Ni-based coatings. *Materials Today: Proceedings, 50*, pp. 1452–1457.
78. Vasudev, H., 2020. Wear characteristics of Ni-WC powder deposited by using a microwave route on mild steel: microwave cladding of Ni-WC. *International Journal of Surface Engineering and Interdisciplinary Materials Science (IJSEIMS), 8*(1), pp. 44–54.
79. Prashar, G., Vasudev, H., Thakur, L. and Bansal, A., 2023. Performance of thermally sprayed hydroxyapatite coatings for biomedical implants: a comprehensive review. *Surface Review and Letters, 30*(01), p. 2241001.
80. Parkash, J., Saggu, H.S. and Vasudev, H., 2022. A short review on the performance of high velocity oxy-fuel coatings in boiler steel applications. *Materials Today: Proceedings, 50*, pp. 1442–1446.

2 Corrosion and Wear Mechanisms in Surface Engineering

Kazi Sabiruddin and Setu Suman

2.1 INTRODUCTION

In our daily lives, we observe that metallic components and apparatus are exposed to an open atmosphere, which results in deterioration over time. Following a period of time, we see that they have sustained complete destruction. Then, we tried to figure out why the components and equipment were broken and found that rust was the main reason. In light of this, we need to explore corrosion and its mechanisms. Safety concerns, the economy, the environment, and conservation efforts are further compelling reasons for studying corrosion. Failure of buildings, bridges, vehicle components, ships, and towers due to corrosion may have a direct influence on human life, resulting in harm or even death. Failure of an operating machine or equipment can have the same disastrous results. Corrosion is one of the most critical aspects that contribute to the degradation of metallic materials in various industries, such as the chemical industry, the gas industry, the windmill rotor industry, the shipping industry, the automotive industry, and the turbine industry. Corrosion is pervasive, affecting all engineered materials, from significant industrial components to commonplace objects in our daily lives, as seen in Figure 2.1a–e.

Wear accounts for approximately 80% of mechanical moving system failures [1]. The combined action of corrosion and wear can result in significantly more significant overall material losses than the cumulative impact of either process alone. In the end, the gradual degradation of metallic surfaces used in huge industrial facilities due to corrosion and wear finally results in a loss of plant efficiency. Corrosion and wear unequivocally compromise materials directly and indirectly, resulting in substantial economic losses. Every day, we read in the papers about something that happens a lot, like bridges falling. Corrosion and wear lead to a wide variety of problems, including the sinking of ships, the crash of airplanes, the failure of vehicle components, and many more.

In this chapter, we have discussed the fundamental aspects of corrosion and wear in relation to surface engineering. We have provided readers with a better understanding through the use of schematic diagrams and authentic images depicting the phenomena of corrosion and wear. Both corrosion and wear are surface and subsurface phenomena that are crucial in terms of economics and the potential

DOI: 10.1201/9781003542957-2

FIGURE 2.1 Examples of corrosion that occur in everyday life include the following: (a) corrosion of a metallic plate when it is uniform; (b) corrosion that happens at the metallic part of a window; (c) corrosion of a bicycle part; (d) corrosion of a door hinge; and (e) corrosion of a nut that is attached to a bolt.

impact on human safety, which could even result in fatalities. It is essential to understand the specific phenomenon in each case and attribute the response to the protective mechanism.

2.2 ROLE OF CORROSION IN SURFACE ENGINEERING

The natural process of corrosion is a phenomenon that has an effect on materials, particularly metals, and leads to the depreciation and deterioration of such materials. To put it simply, corrosion is caused by the chemical or electrochemical processes that take place between a substance and the environment in which it is located [2]. Taking this into consideration, it is necessary to protect against the deterioration of the material caused by corrosion assaults. There are many different kinds of corrosion protection coatings that are applied to the components in order to prevent corrosion. These coatings include organic, inorganic, metallic, and ceramic corrosion protection. The next part contains additional in-depth talks on the primary mechanism of corrosion, the many types of corrosion, and the procedures used to test for corrosion.

2.2.1 Basic Mechanism of Corrosion

Corrosion is the process by which metals are broken down into oxides and sulfides by the chemical deposition of the metal into its constituent parts through the interaction of other chemical or electrochemical processes [3]. Corrosion has become one of the most challenging areas of science and engineering due to the interdisciplinary approach [3]. Of all the several types of corrosion processes, the most prevalent one is corrosion in aqueous solutions, which includes water, saltwater, and other operations in industry that create an aqueous medium. Corrosion is the process by which a metal reverts to its thermodynamic state. This involves forming the oxide or sulfide from which most materials originated when extracted from the earth and polished into engineering materials [4].

2.2.2 Classification of Corrosion

Fontana and Green have classified the most prevalent types of corrosion into the following eight categories. Below, the eight categories of corrosion are discussed in terms of their mechanisms, characteristics, and preventive measures.

1. General or uniform corrosion
2. Pitting corrosion
3. Crevice corrosion
4. Galvanic corrosion
5. Intergranular corrosion
6. Selective corrosion
7. Stress corrosion cracking (SCC), including corrosion fatigue and hydrogen embrittlement
8. Erosion corrosion, impingement corrosion, cavitation corrosion, and fretting corrosion

2.2.2.1 General or Uniform Corrosion

As the name implies, uniform corrosion is the most popular type of corrosion, occurring throughout the entire exposed metallic surface in a uniform way. The uniform distribution of the anodic and cathodic reactions throughout the surface is the cause of this. The metallic components react chemically and dissolve into metallic ions as a result of the resulting rate of mass loss. Rusting of steel in the environment is the most typical example of uniform corrosion. The corrosion of a metallic plate is an example of general corrosion, which is depicted in Figure 2.1a. Under these circumstances, the corrosion product is insoluble; however, it is porous and has weak adhesion, and as a result, it does not provide any protection [5]. Uniform corrosions are predictable and can be readily controlled. However, most other types of corrosion are much more difficult to predict.

2.2.2.2 Pitting Corrosion

This type of corrosion is the most intense and localized form of corrosion, resulting in the formation of holes on the surface of the metal. As a result of the fact that holes with tiny diameters are frequently difficult to detect over the surface, such holes pose a greater threat than uniform corrosion. In addition to this, it is challenging to acquire quantitative measurements, and it is frequently challenging to simulate in laboratory tests. Pitting is a phenomenon that typically takes place in passivated metals and alloys, and it is frequently linked to surroundings that are saline environments [3]. There are many different types of corrosion, but pitting corrosion is one of the most dangerous types for marine construction [6]. The pitting corrosion is typically accelerated by a variety of factors, including the type of metal, salinity, pH value, temperature of the aqueous electrolyte, and dissolved oxygen [6].

2.2.2.3 Crevice Corrosion

Crevice corrosion follows a similar mechanism to pitting corrosion, which is a localized form of corrosion. Typically, crevice corrosion occurs when stagnant solutions are present in confined spaces, such as underlap joints, washers, gaskets, and

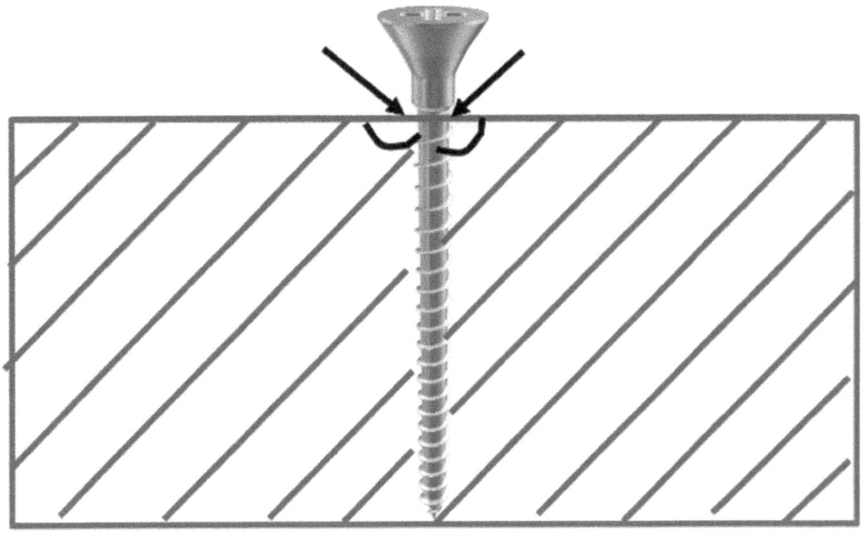

FIGURE 2.2 Schematic diagram of crevice corrosion.

openings. For example, between bolts and nuts (Figure 2.2), all metallic materials are prone to crevice corrosion. Stainless steel and aluminum, for example, can develop an oxide coating that makes them susceptible to crevice corrosion, especially in salty environments [2]. Stainless steels are prone to crevice corrosion due to the depletion of oxygen within the crevice, which prevents the maintenance of the thin passive layer of Cr_2O_3.

2.2.2.4 Galvanic Corrosion

Galvanic corrosion, also known as two-metal corrosion, takes place between two metals as a result of the potential difference that occurs when both metals are submerged in a solution that is corrosive [7]. Galvanic corrosion causes failure in numerous industries, including electrical and domestic plumbing components. A classic example of the detrimental effects is the perforation of the copper skin of the Statue of Liberty [3]. As shown in Figure 2.3, corrosion around the battery terminal is a type of galvanic corrosion. It happens when two different metals, like the battery terminal and the connecting cable, come into contact with each other in the presence of an electrolyte.

2.2.2.5 Intergranular

This type of corrosion has happened along the grain boundary. All metals are basically made up of grains and crystals. Grain borders are formed when one grain is next to another grain. This form of corrosion has been observed to occur most frequently in stainless steel when heated within a critical range of 723K–1,073K. During this time, the Cr_2O_3 reacts with $Cr_{23}O_6$, primarily at the grain boundaries, resulting in a depletion of chromium and an inability to form the protective passive thin layer of Cr_2O_3 [3]. For instance, this particular type of corrosion is most likely occurring in the heat-affected zone of stainless steel welds.

FIGURE 2.3 Example of galvanic corrosion (battery terminal).

2.2.2.6 Selective Corrosion

Selective corrosion is referred to as dealloying. This type of corrosion occurs when an element is removed from the alloys. Most of the time, this corrosion is something that cannot be seen by the human eye. The most active component of an alloy will be preferentially introduced into the solution during this process. In contrast, the remaining components will continue to exist as a porous mass that has been mechanically weakened [8]. For example, the dezincification of brass happens, particularly under dirt deposits. There are two forms of dezincification: (i) uniform-layer type dezincification and (ii) localized-plug type dezincification.

2.2.2.7 Stress Corrosion Cracking

The propagation of stress corrosion cracks is accelerated by corrosive media and tensile stress. SCC was the cause of the collapse of the Silver Bridge, which was located over the Ohio River, on December 15, 1967 [9]. The key elements influencing SCC are metallic material structure and chemical composition, the nature of the corrosive solution, temperature, and stress [2].

2.2.2.8 Erosion Corrosion, Impingement Corrosion, Cavitation Corrosion, and Fretting Corrosion

Erosion corrosion occurs when a metal surface moves quickly in contact with a corrosive fluid, speeding up the entire corrosion process. Cavitation corrosion is a type of corrosion characterized by the production and impulsion of gas bubbles in a liquid near a metal surface. It may compromise a protective coating, resulting in surface degradation.

2.2.3 Corrosion Testing Methods

Numerous tests are available to evaluate the corrosion and dependability of protective coatings on metal substrates. There are already tests available along with their standards, and some are being developed further. Organizations that contribute to the development and standardization of corrosion tests include ASTM, API, NACE International, SAE, NCCS, and ISO. The following is a concise overview of the most frequently employed corrosion test methodologies in the context of surface engineering applications:

- Salt spray tests
- Electrochemical tests

2.2.3.1 Salt Spray (Fog) Tests

The most widely used specified salt spray test to check the corrosivity consistency of the coating is described in ASTM B 117 standards [10]. Salt spray tests are considered to be the most useful approach for assessing the behavior of closely similar materials in marine atmospheres because they produce a saline environment within a closed chamber. Under these conditions, it simulated the basic conditions with some acceleration caused by moisture and temperature. The concentration of salt, the pH value of the solution, the air pressure, the temperature of the chamber, and the rate at which salt is deposited are some of the typical parameters. In this particular test, sodium chloride salts are typically utilized to create a saline environment. The saline solution will be formulated by dissolving 5% sodium chloride (NaCl) in 95% water. The chamber temperature influences the pH of a salt solution derived from water saturated with carbon dioxide at ambient temperature. The pH of a salt solution can be changed by raising the temperature of the water used to make the solution to at least 35°C (95°F) and then changing the pH number. Using the same parameters discussed before, salt spray tests are carried out on the plasma-sprayed Al_2O_3 coating that is applied to low-carbon steel for 24 hours. Following the test, the sample is depicted in Figure 2.4. The following experiments assess the corrosive conditions within a salt spray chamber and determine the mass losses over a specified time duration. This might be done in a single day, once a month, or even more regularly to guarantee that the operation remains consistent throughout time.

2.2.3.2 Electrochemical Tests

Electrochemical testing methods are widely used for studying metal or alloy corrosion behavior by measuring its electrochemical behavior in a corrosive environment. These techniques involve measuring the electrical properties of a material immersed in a corrosive solution. This is typically done by immersing the sample in a solution and measuring the potential and current flow between the sample and a reference electrode placed in the same solution. The electrochemical tests are carried out using an electrochemical workstation with a conventional three-electrode test setup, as depicted in Figure 2.5. Numerous investigators have used DC and AC electrochemical methods to study the performance and quality of protective coatings, including passive films on metallic substrates, and to measure the effectiveness of pretreatment surfaces. Two standard methods that come under electrochemical tests are (i) potentiodynamic polarization (ASTM F2129) and (ii) electrochemical impedance spectroscopy.

The primary purpose of corrosion testing is to assess and understand the behavior of materials in corrosive environments. The results obtained from corrosion testing

FIGURE 2.4 Coating sample after the salt sprayed for 24 hours in mist NaCl solution.

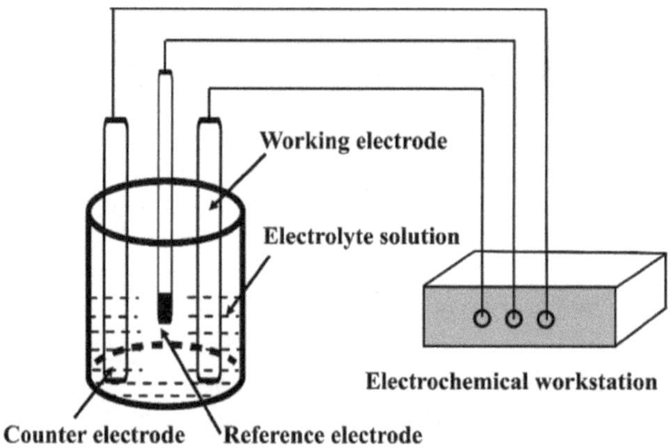

FIGURE 2.5 Schematic diagram of the electrochemical test setup.

are used for material selection, quality control, design optimization or modification, change of environment, intervention in the reactions application of coatings [11], and the development of adequate corrosion protection and mitigation strategies. By conducting corrosion testing, industries can ensure the longevity and reliability of their assets, reduce maintenance costs, and improve safety concerns. So, corrosion testing plays a vital role in managing and mitigating the damaging effects of corrosion, enabling industries to make informed decisions and take protective action against infrastructure, components, and equipment.

2.3 ROLE OF WEAR IN SURFACE ENGINEERING

Surface engineering is the process of improving the surface qualities of a component in order to extend its lifespan and improve its overall performance. Enhancement in the areas of aesthetics, optical characteristics, and wear resistance can be achieved through the employing of coatings.

2.3.1 Principle of Wear Mechanism

Wear is a surface or subsurface phenomenon that gradually deteriorates to a solid surface with material loss caused by relative motion between that surface and counter substances. Based on the mechanism, wear can be broadly classified into four categories. In the context of material surfaces, wear is defined as the damage that occurs as a result of the mechanical action of a surface that is in contact with moving objects.

2.3.2 Classification of Wear Mechanism

Based on the mechanism, wear can be broadly divided into four forms. All of these forms can remove surface material. Figure 2.6 illustrates the classification of wear mechanisms. The following is a comprehensive explanation of the mechanisms that encourage wear:

2.3.2.1 Abrasion

In abrasion wear, the mechanism of material removal occurs through the mechanical action of an abrasive. Abrasive materials are usually more complex than the surface they are abrading. Although metallic particles can also induce abrasion, non-metallic materials are the most common culprits that cause this type of wear

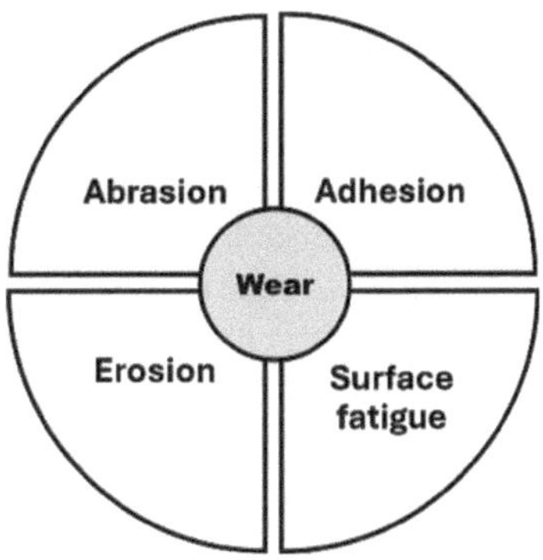

FIGURE 2.6 Classification of wear.

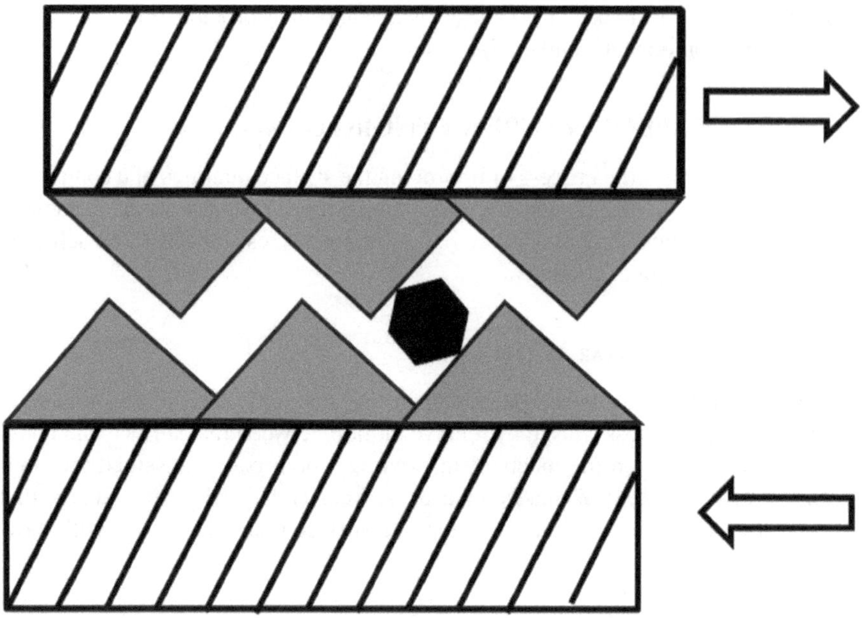

FIGURE 2.7 Schematic diagram of abrasive wear.

in metals. Figure 2.7 illustrates abrasion that occurs due to the relative motion of entrained solid particles between two sliding surfaces. Abrasive wear may originate from two surfaces rubbing together, with the harder surface grinding away at the softer surface. Coatings that are thick and firmly adherent provide good resistance to wear from abrasion [12].

2.3.2.2 Adhesion

This type of wear is the result of the transfer of material from one surface to another during relative motion under load. During adhesive wear, friction at the ends of opposing asperities in rubbing surfaces produces high localized temperatures [12]. Because of the localized temperatures, these tips are capable of going through deformation and possessing solid-state welding together. Figure 2.8 is provided for your reference. Adhesive wear can be classified as chemical wear due to the nature of the joining or welding processes. The particles that are taken from one surface during this process are either momentarily or permanently affixed to the other surface. Adhesive wear can occur between metallic materials, ceramics, polymers, or a mixture of these materials.

2.3.2.3 Erosion

Erosion is the process by which discrete small solid particles (alumina grits, steel grits, and steel shots) strike the surface with high impact, producing damage or material loss to the surface. This procedure causes problems for various components, including those that are responsible for propulsion, turbine blades, pipelines,

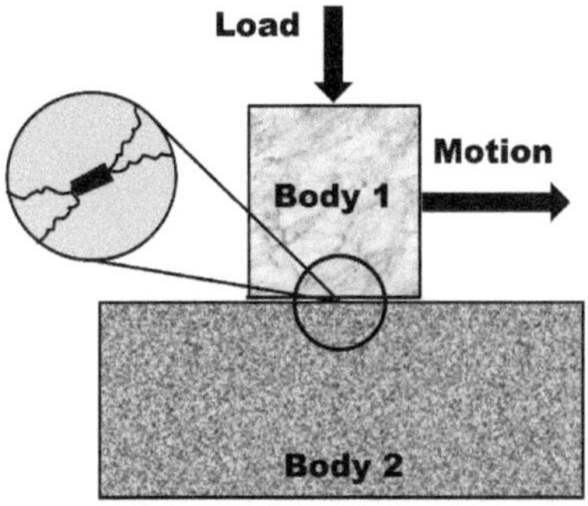

FIGURE 2.8 Schematic image of adhesive wear.

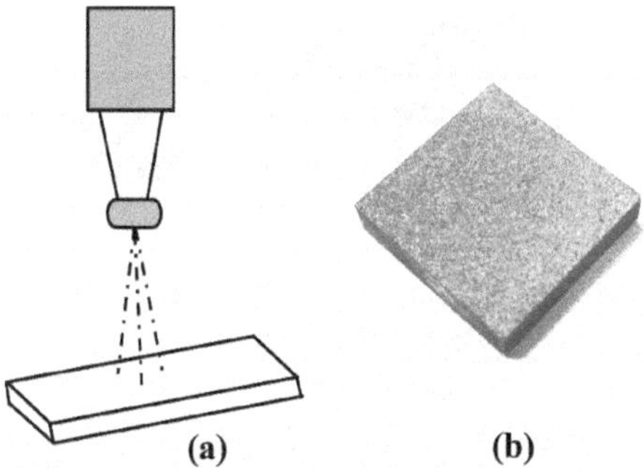

FIGURE 2.9 (a) Schematic image of solid particles impinges over the surface and (b) real image of grit-blasted surface.

and fluidized-bed combustion systems [1]. However, erosion has certain useful applications, particularly for cleaning and surface preparation prior to coatings and painting, utilizing the grit blasting method. Figure 2.9a shows a schematic image of the erosion process. In this process, very fast solid particles flow over the surface through the nozzle. Also, to get a better idea of the worn-down surface, see the real picture of the grit-blasted surface, shown in Figure 2.9b. Basically, erosion can be divided into three categories based on the type of material that is being worn away [13].

1. Cavitation erosion
2. Liquid-driven erosion
3. Solid particle erosion

2.3.2.4 Surface Fatigue

The cyclic forces created by continuous sliding on a solid surface cause material to fracture in the surface fatigue wear phase. In this scenario, material removal occurred owing to surface or subsurface cracking. Initially, a pit will be formed as

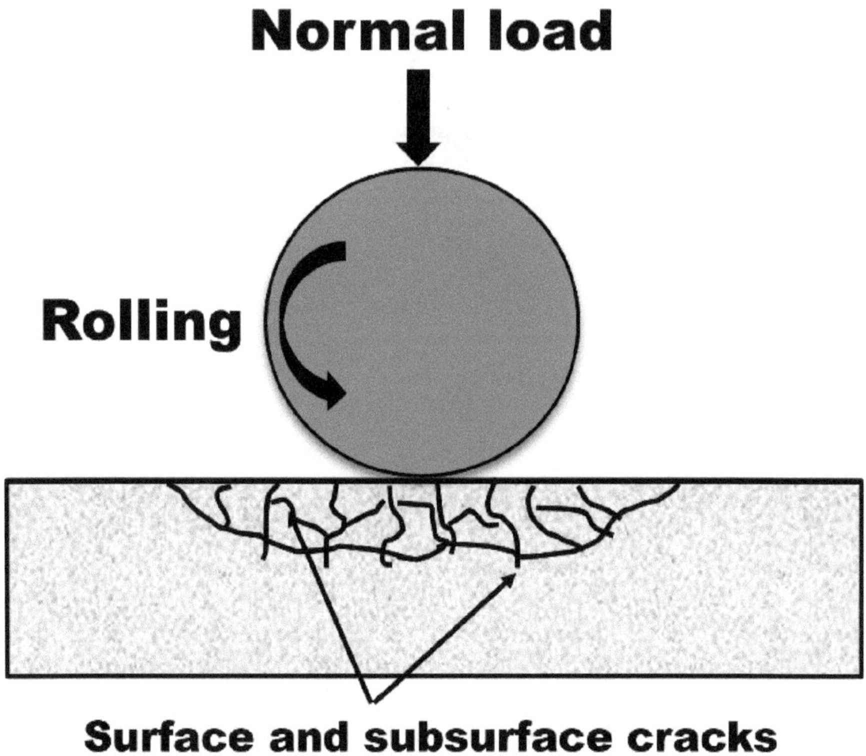

FIGURE 2.10 Schematic diagram of surface fatigue wear.

the subsurface cracking approaches the surface. Figure 2.10 illustrates a schematic diagram of surface fatigue wear. Fatigue wear results from elastic deformation and plastic deformation of the surface [14]. As a result, the overall mechanism is similar to abrasive wear, which occurs frequently. This particular mode of wear is the most challenging to predict.

2.4 WEAR TESTING METHODS

To evaluate the tribological properties of the materials, wear experiments have been performed in both linear reciprocating and rotating modes. Wear rate can be measured using a variety of ways once the test is completed [15]. These include:

- Weight loss method using the formula [16]:
  Net weight = Weight before wear − Weight after wear
- Wear rate is also calculated using the formula
  Wear rate = Net weight/time (g/s)
- Change in linear dimension of samples
- Optical measurement of the size of wear scars
- Profilometric measurement of wear volume

TABLE 2.1
Parameters of Ball-on-Disc Wear Test

Parameters	Values
Counter body	WC-6 wt.% Co (5 mm Dia.)
Load (N)	10, 20, 30
Frequency (Hz)	5, 10, 15
Time (minutes)	15
Temperature (°C)	25

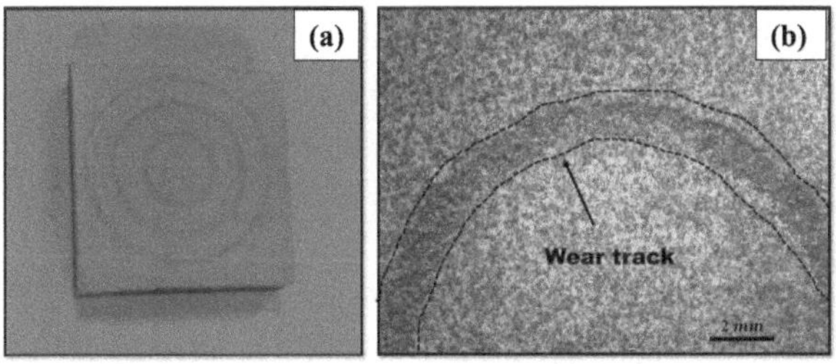

FIGURE 2.11 (a) Image of full wear track and (b) microscopic image of wear track on plasma-sprayed Al_2O_3 coating.

2.4.1 Ball-on-Disc

Ball-on-disc is the most frequently employed wear test method [17]. The two-body rotary wear can be determined with the help of these tests, which are preferred. The ball-on-disc test comprises a stationary ball hooked up to a metallic holder and then pressed against a rotating disc while the test is being performed. Friction is calculated as the force required to keep the ball from moving in the opposite direction. Additional information regarding the wear test conducted on the coated sample under a ball-on-disc tribometer is provided.

Ball-on-disc wear tests have been carried out on the polished surface of the plasma-sprayed Al_2O_3 coating using a (*Ducom TR-20LE*) ball-on-disk tribometer. This was performed in light of the fact that the coating was polished. Table 2.1 illustrates the test parameters for ball-on-disk. In accordance with the ASTM standards and the relevant literature, the parameters that were chosen were chosen appropriately. Following the completion of wear testing, the wear mechanisms of the coating were observed using an optical microscope (*Leica*), and the resulting wear track is depicted in Figure 2.11b. Furthermore, Figure 2.11a shows the macroscopic image of the whole wear track that was conducted on the coated samples with different values.

2.5 SURFACE ENGINEERING SOLUTIONS FOR CORROSION AND WEAR PROTECTION

In daily life, our minds automatically associate anything we observe with its outside look. It may be concluded that the surface's look has a major influence on the functionality of the component. Corrosion and wear are surface or subsurface phenomena that are observed on the surface, but they are occasionally challenging to observe with the naked eye. However, both of these things pose a threat to the life of the components, and they can either directly or indirectly influence human life. Taking this into consideration, corrosion and wear issues continue to be of significant importance in a wide variety of industrial applications and different kinds of products. The design of a component that is totally resistant to corrosion and wear is one direct strategy that may be taken to defend against the phenomenon of corrosion and wear; however, this technique is quite expensive. Due to the fact that corrosion and wear are surface phenomena, the coating technique is most suitable in terms of cost-effectiveness and maximizing the performance of components over their lifetime. There are several coating methods for applying the components. According to Bunshah, there are four general coating ways that cover most hard coating uses for protecting against corrosion and wear [12]:

1. Thermal Spray Coating
2. Chemical Vapor Deposition
3. Physical Vapor Deposition
4. Electroless/Electro-Plating Coating

Over the past few years, there has been a remarkable advancement in the research that focuses on corrosion and wear-resistant coatings. In order to develop coatings that have outstanding wear resistance properties and provide excellent corrosion resistance, researchers are actively exploring the design and synthesis methods of novel coating materials. Nowadays, advanced engineering materials, including ceramics, metal alloys, composite mixtures, and reinforced composites, including whisker and fiber reinforcements, are employed to mitigate the effects of corrosion and wear. Additionally, nanoparticles [18] and anticorrosive agents [19] were incorporated to enhance the resistance and protection of the coating against corrosive media. Numerous protective coatings have been developed to prevent corrosion and wear. Table 2.2 lists ceramic-based coatings that are resistant to corrosion when exposed to various corrosive conditions. However, Table 2.3 provides an illustration of the summary of the wear-resistant coatings, the manner of wear mechanism, and the wear tests.

TABLE 2.2
List of Ceramic-Based Corrosion-Resistant Coatings in Different Corrosive Medium

Coating Materials	Coating Process	Corrosive Medium and Duration	References
Al_2O_3-TiO_3	D-gun spray	5.2% of NaCl, 24 hours	[20]
$La_2Ce_2O_7$/YSZ	APS	$Na_2SO_4+V_2O_5$, 30 hours	[21]
Cr_2O_3	APS	3.5% of NaCl	[22]
Al_2O_3-13TiO_2	APS	6% of Na_2SO_4	[23]
YSZ	APS	V_2O_5 and $Na_2SO_4+MgSO_4$	[24]
Al_2O_3, Al_2O_3+TiO_2	APS	1 N H_2SO_4	[25]
YSZ, YSZ+Al_2O_3, YSZ/Al_2O_3	APS	55% V_2O_5 and 45% Na_2SO_4	[26]

TABLE 2.3
Summary of Wear Mechanism, Wear Test, and Ceramic-Based Wear-Resistant Coatings

Type of Coating	Mode of Wear Mechanism	Type of Wear Test	References
Al_2O_3-based coatings	Abrasive and fatigue wear	Relative humidity 10% ± 75%	[27]
TiO_2 coatings	Abrasive and fatigue wear	Ball-on-disk (ZrO_2 ball used)	[28]
Al_2O_3-Cr_2O_3 coatings	Abrasive wear	Linear sliding (WC + 6% Co)	[29]
Cr_3C_2-NiCr coatings	Adhesive and fatigue wear	Ball-on-disk	[30]
TiC-Co coating	Adhesive, abrasive, and fatigue wear	Ball-on-disk (Si_3N_4 ball used)	[31]
TiN, TiC, and TiB_2 coating	TiN and TiB_2: fatigue, adhesion, TiC: fatigue	Ball-on-disk (WC-Co ball)	[32]
Al_2O_3-SiC coating	Abrasive wear	Ball-on-disk (WC-6 wt.% Co)	[33]

Furthermore, the modification and optimization of the coating interface have become the focus of research to improve the compatibility and durability between the coating and the substrate, which strengthens the quality of coatings. Protective coatings are widely utilized to prevent corrosion and wear resistance on the surface of steel components used in various industrial applications. Thermal spraying is one of the best methods to fabricate protective coatings. Nowadays, thermal spraying techniques have gained much interest because they can deposit a wide range of materials to fabricate coatings. The thermal spraying process can deposit protective coatings like Al_2O_3, Cr_2O_3, yttria-stabilized zirconia (YSZ), TiO_2, SiC, TiN, TiC, TiO_2, TiB_2, and many others to improve the corrosion and wear resistance and extend the service life of components.

The rapid demand for thermal spray techniques in various industrial applications and the fabrication of protective coatings have helped to drive this demand. The thermal spray process can deposit metallic, ceramic, cermet, and some polymeric materials in the form of wire, rod, or powder used as feedstock material. Electric/gas heating is used as a source of melting the feedstock material; molten particles accelerate with high velocity and strike over the substrate, where they solidify and form a solid layer called a coating. There are several types of thermal spray techniques most commonly used in industrial applications [34], including:

a. Plasma spray
b. Flame spray
c. Electric arc spray
d. Warm spray
e. Detonation gun (D-gun)
f. Cold spraying
g. High-velocity oxy-fuel spraying (HVOF)
h. High-velocity air-fuel spraying (HVAF)

In recent years, plasma spray, flame spray, and electric arc spray have become the most popular techniques for applying corrosion-resistant coatings to steel substrates [35,36]. The following section provides a comprehensive discussion of the various coating materials.

Al_2O_3 Coatings: Corrosion-resistant coatings produced by the plasma spray process exhibited good benefits in certain conditions. It is well known that Al_2O_3 coatings are resistant to specific corrosive media and are primarily used in industry [25]. Al_2O_3 is one of the most influential and commonly employed coatings due to the formation of a passive layer over the coatings to protect from corrosion attacks [25,37]. The corrosion behavior of plasma-sprayed coatings depends on factors such as (i) corrosive media, (ii) environmental temperature, and (iii) composition, structure, porosity, adhesive strength, and other properties of coatings.

YSZ Coatings: Recently, it has been observed that thermal-sprayed YSZ-based coatings could provide resistance to the steel substrate of the superheater tubes against the molten salt corrosive attack at elevated temperatures [38,39]. A detailed review of the utilization of YSZ-based coatings for protection of the thermal barrier coatings against hot corrosive attacks was reported in [38]. YSZ-based coatings showed successful resistance against corrosive attack at elevated temperatures. Plasma-sprayed YSZ coatings performed successfully against molten salt attack (LiCl + KCl) at 600°C [40].

Cr_2O_3 Coatings: Chromium oxide coatings are considered one of the best corrosion-resistant materials to protect against the possible attack of chemicals on metallic components in the case of alkaline and acidic environments. The corrosion behavior of Cr_2O_3 coating has been found to be more effective, especially at higher temperatures [41]. Plasma-deposited Cr_2O_3 has not shown good wear-resistant properties at ambient temperature. This could be due to the high residual stress induced in the coating during the thermal spraying process [42]. Most of the research published for thermally sprayed Cr-based coatings suggested that Cr_2O_3 and Cr_3C_2-NiCr are corrosion-resistant coatings and maintain good mechanical properties [41,43,44].

2.6 CURRENT TRENDS AND FUTURE SCOPE

Corrosion and wear pose significant challenges for industries, resulting in both short-term and long-term damages. It is possible to reduce financial losses with early prediction and appropriate monitoring. The conventional methods that are used to predict and detect corrosion are processes that are not only time-consuming but also difficult to carry out in inaccessible regions. As a result of these factors, the majority of academics have turned their focus toward methodologies that are based on artificial intelligence, machine learning, and deep learning [45–49]. Currently, this is the most widely used tool among researchers for the early detection and prediction of corrosion. In the current era of digitalization and automation, more studies are being conducted in this field. However, more research is required to take into account and fully employ the latest breakthroughs in IR 4.0 technology in terms of modern sensors and artificial intelligence technologies. This is necessary in order to produce outstanding outcomes and excellent performance in the detection and prediction of corrosion.

In the future, work may be done on the development of duplex coatings, which are both wear- and corrosion-resistant coatings. Most of the studies have been conducted to deposit thick coating by thermal spray followed by plasma spraying. However, an alternatively novel technique, such as plasma spray-physical vapor deposition, has been utilized in order to decrease wear rates while producing successful resistance against corrosive attack [50,51].

2.7 SUMMARY

Corrosion and wear are the biggest concerns in industries as well as in our daily lives. This chapter summarizes the fundamentals of the corrosion and wear phenomena, various forms of corrosion and wear modes, and different testing methods for corrosion and wear. The field of surface engineering encompasses thermal spray coating. Thermally sprayed coatings have been widely used to enhance surface properties such as wear and corrosion resistance and increase the components' service life. Coating materials are essential in fabricating coatings for corrosion and wear resistance. Some of the common coating materials have been described in this chapter.

APS, air plasma spraying; YSZ, yttria-stabilized zirconia.

REFERENCES

1. Wu, J., Chen, Y. and Zhu, H., 2022. A review on the tribological performances of high-entropy alloys. *Advanced Engineering Materials*, *24*(8), p. 2101548.
2. Ibrahimi, B.E., Nardeli, J.V. and Guo, L., 2021. An overview of corrosion. In: Hussain, C.M. and Verma, C. (eds.), *Sustainable Corrosion Inhibitors I: Fundamentals, Methodologies, and Industrial Applications* (pp. 1–19). American Chemical Society.
3. Hansson, C.M., 2011. The impact of corrosion on society. *Metallurgical and Materials Transactions A*, *42*, pp. 2952–2962.
4. Schweitzer, P.A., 2009. *Fundamentals of Corrosion: Mechanisms, Causes, and Preventative Methods*. CRC Press.
5. Graedel, T.E. and Frankenthal, R.P., 1990. Corrosion mechanisms for iron and low alloy steels exposed to the atmosphere. *Journal of the Electrochemical Society*, *137*(8), p. 2385.
6. Wei, S. and Su, H., 2019. Mechanistic aspects of factors affecting pitting corrosion of metallic materials for marine application: a review paper. *International Journal of Electrochemical Science*, *14*(4), pp. 3949–3967.
7. Goyal, M., Kumar, S., Bahadur, I., Verma, C. and Ebenso, E.E., 2018. Organic corrosion inhibitors for industrial cleaning of ferrous and non-ferrous metals in acidic solutions: a review. *Journal of Molecular Liquids*, *256*, pp. 565–573.
8. Revie, R.W., 2008. *Corrosion and Corrosion Control: An Introduction to Corrosion Science and Engineering*. John Wiley & Sons.
9. Lichtenstein, A.G., 1993. The silver bridge collapse recounted. *Journal of Performance of Constructed Facilities*, *7*(4), pp. 249–261.
10. ASTM Committee G-1 on Corrosion of Metals, 2011. *Standard Practice for Operating Salt Spray (Fog) Apparatus*. ASTM International.
11. Aljibori, H.S., Alamiery, A. and Kadhum, A.A.H., 2023. Advances in corrosion protection coatings: a comprehensive review. *International Journal of Corrosion and Scale Inhibition*, *12*(4), pp. 1476–1520.

12. Bunshah, R.F., 1994. *Handbook of Deposition Technologies for Films and Coatings: Science, Applications and Technology.* William Andrew.
13. Mellor, B.G. (ed.), 2006. *Surface Coatings for Protection against Wear.* Woodhead Publishing.
14. Shepherd, D.E.T. and Dearn, K.D., 2012. Wear processes in polymer implants. In: Jenkins, M. and Stamboulis, A. (eds.), *Durability and Reliability of Medical Polymers* (pp. 143–163). Woodhead Publishing.
15. Gee, M.G. and Owen-Jones, S., 1998. Wear testing methods and their relevance to industrial wear problems. *NPL Report. CMMT(A)92.*
16. Suman, S. and Sabiruddin, K., 2024. Role of FSP in surface engineering. In: Rathee, S., Srivastava, M., and Davim, J.P. (eds.), *Friction Stir Welding and Processing: Fundamentals to Advancements* (pp. 55–66). John Wiley & Sons, Inc.
17. Ayerdi, J.J., Aginagalde, A., Llavori, I., Bonse, J., Spaltmann, D. and Zabala, A., 2021. Ball-on-flat linear reciprocating tests: critical assessment of wear volume determination methods and suggested improvements for ASTM D7755 standard. *Wear, 470*, p. 203620.
18. Xu, Z., Xie, Y., Ebrahimnia, M. and Dang, H., 2021. Effect of B4C nanoparticles on microstructure and properties of laser cladded IN625 coating. *Surface and Coatings Technology, 416*, p. 127154.
19. Wang, P., He, B., Du, Y., Wang, B., Gao, J., Liu, S., Ye, Q. and Zhou, F., 2022. Functionalized Ti3C2Tx-based nanocomposite coatings for anticorrosion and antifouling applications. *Chemical Engineering Journal, 448*, p. 137668.
20. Surya, A., Prakash, R., Senthil Kumar, P. and Bharath Balji, G., 2023. Development of alumina-titania composite layers on stainless steel through the detonation spray method and investigation of salt spray corrosion behavior along with surface examination. *International Journal of Chemical Engineering, 2023*(1), p. 1445360.
21. Lashmi, P.G., Ananthapadmanabhan, P.V., Chakravarthy, Y., Unnikrishnan, G., Balaji, N. and Aruna, S.T., 2017. Hot corrosion studies on plasma sprayed bi-layered YSZ/La2Ce2O7 thermal barrier coating fabricated from synthesized powders. *Journal of Alloys and Compounds, 711*, pp. 355–364.
22. Leivo, E.M., Vippola, M.S., Sorsa, P.P.A., Vuoristo, P.M.J. and Mäntylä, T.A., 1997. Wear and corrosion properties of plasma sprayed AI_2O_3 and Cr_2O_3 coatings sealed by aluminum phosphates. *Journal of Thermal spray technology, 6*, pp. 205–210.
23. Wang, Y., Tian, W., Zhang, T. and Yang, Y., 2009. Microstructure, spallation and corrosion of plasma sprayed Al_2O_3–13% TiO_2 coatings. *Corrosion Science, 51*(12), pp. 2924–2931.
24. Marple, B.R., Voyer, J., Thibodeau, M., Nagy, D.R. and Vassen, R., 2006. Hot corrosion of lanthanum zirconate and partially stabilized zirconia thermal barrier coatings. *Journal of Engineering for Gas Turbines and Power, 128*(1), pp. 144–152.
25. Celik, E.R.D.A.L., Ozdemir, I., Avci, E. and Tsunekawa, Y., 2005. Corrosion behaviour of plasma sprayed coatings. *Surface and Coatings Technology, 193*(1–3), pp. 297–302.
26. Afrasiabi, A., Saremi, M. and Kobayashi, A., 2008. A comparative study on hot corrosion resistance of three types of thermal barrier coatings: YSZ, YSZ+ Al_2O_3 and YSZ/Al_2O_3. *Materials Science and Engineering: A, 478*(1–2), pp. 264–269.
27. Bolelli, G., Bonferroni, B., Cannillo, V., Gadow, R., Killinger, A., Lusvarghi, L., Rauch, J. and Stiegler, N., 2010. Wear behaviour of high velocity suspension flame sprayed (HVSFS) Al_2O_3 coatings produced using micron-and nano-sized powder suspensions. *Surface and Coatings Technology, 204*(16–17), pp. 2657–2668.
28. Wang, H.D., He, P.F., Ma, G.Z., Xu, B.S., Xing, Z.G., Chen, S.Y., Liu, Z. and Wang, Y.W., 2018. Tribological behavior of plasma sprayed carbon nanotubes reinforced TiO_2 coatings. *Journal of the European Ceramic Society, 38*(10), pp. 3660–3672.

29. Dhakar, B., Chatterjee, S. and Sabiruddin, K., 2017. Linear reciprocating wear behaviour of plasma-sprayed Al_2O_3–Cr_2O_3 coatings at different loading and sliding conditions. *Sādhanā*, *42*, pp. 1763–1772.
30. Chhabra, P. and Kaur, M., 2020. Wear and friction characteristics of atmospheric plasma sprayed Cr3C2–NiCr coatings. *Tribology-Materials, Surfaces & Interfaces*, *14*(3), pp. 177–192.
31. Du, Y., Chen, H., Yang, G., Liu, B., Gao, Y. and Luo, H., 2018. Effect of cobalt content on high-temperature tribological properties of TiC-Co coatings. *Ceramics International*, *44*(12), pp. 14186–14194.
32. Hong, D., Niu, Y., Li, H., Zhong, X., Tu, W., Zheng, X. and Sun, J., 2019. Comparison of microstructure and tribological properties of plasma-sprayed TiN, TiC and TiB2 coatings. *Surface and Coatings Technology*, *374*, pp. 181–188.
33. Sharma, V. and Kazi, S., 2019. An investigation on D-gun sprayed Al_2O_3-SiC coatings. *Surface and Coatings Technology*, *375*, pp. 303–314.
34. Oksa, M., Tuurna, S. and Varis, T., 2013. Increased lifetime for biomass and waste to energy power plant boilers with HVOF coatings: high temperature corrosion testing under chlorine-containing molten salt. *Journal of Thermal Spray Technology*, *22*, pp. 783–796.
35. Fauchais, P. and Vardelle, A., 2012. Thermal sprayed coatings used against corrosion and corrosive wear. *Advanced Plasma Spray Applications*, *10*, p. 34448.
36. Pardo, A., Casajús, P., Mohedano, M., Coy, A.E., Viejo, F., Torres, B. and Matykina, E., 2009. Corrosion protection of Mg/Al alloys by thermal sprayed aluminium coatings. *Applied Surface Science*, *255*(15), pp. 6968–6977.
37. Habib, K.A., Saura, J.J., Ferrer, C., Damra, M.S., Giménez, E. and Cabedo, L., 2006. Comparison of flame sprayed Al_2O_3/TiO_2 coatings: their microstructure, mechanical properties and tribology behavior. *Surface and Coatings Technology*, *201*(3–4), pp. 1436–1443.
38. Jones, R.L., 1997. Some aspects of the hot corrosion of thermal barrier coatings. *Journal of Thermal Spray Technology*, *6*, pp. 77–84.
39. Lu, G.X., Hao, L.J., Liu, C. and Ye, F.X., 2014. Thermal analysis and failure behaviour of YSZ thermal barrier coatings on low heat rejection diesel engine piston. *Materials Science and Technology*, *30*(11), pp. 1273–1281.
40. Shankar, A.R., Mudali, U.K., Sole, R., Khatak, H.S. and Raj, B., 2008. Plasma-sprayed yttria-stabilized zirconia coatings on type 316L stainless steel for pyrochemical reprocessing plant. *Journal of Nuclear Materials*, *372*(2–3), pp. 226–232.
41. Suegama, P.H., Fugivara, C.S., Benedetti, A.V., Guilemany, J.M., Fernández, J. and Delgado, J., 2004. The influence of gun transverse speed on electrochemical behaviour of thermally sprayed Cr_3C_2–NiCr coatings in 0.5 M H_2SO_4 solution. *Electrochimica Acta*, *49*(4), pp. 627–634.
42. Jarosinski, W.J.C. and Temples, L.B., Praxair ST Technology Inc, 2003. Corrosion resistant powder and coating. U.S. Patent 6,503,290.
43. Bram, M., Ahmad-Khanlou, A., Buchkremer, H.P. and Stöver, D., 2002. Vacuum plasma spraying of NiTi protection layers. *Materials Letters*, *57*(3), pp. 647–651.
44. Sidhu, B.S. and Prakash, S., 2003. Evaluation of the corrosion behaviour of plasma-sprayed Ni3Al coatings on steel in oxidation and molten salt environments at 900 C. *Surface and Coatings Technology*, *166*(1), pp. 89–100.
45. LeCun, Y., Bottou, L., Bengio, Y. and Haffner, P., 1998. Gradient-based learning applied to document recognition. *Proceedings of the IEEE*, *86*(11), pp. 2278–2324.
46. Li, Y., Wang, G., Nie, L., Wang, Q. and Tan, W., 2018. Distance metric optimization driven convolutional neural network for age invariant face recognition. *Pattern Recognition*, *75*, pp. 51–62.

47. Jaderberg, M., Vedaldi, A. and Zisserman, A., 2014. Deep features for text spotting. In: Fleet, D., Pajdla, T., Schiele, B., Tuytelaars, T. (eds.). *Computer Vision–ECCV 2014* (ECCV Lecture Notes in Computer Science, vol 8692). Cham: Springer International Publishing. https://doi.org/10.1007/978-3-319-10593-2_34
48. Aghdam, H.H., Heravi, E.J. and Puig, D., 2016. A practical approach for detection and classification of traffic signs using convolutional neural networks. *Robotics and Autonomous Systems*, *84*, pp. 97–112.
49. Yammen, S. and Muneesawang, P., 2014. An advanced vision system for the automatic inspection of corrosions on pole tips in hard disk drives. *IEEE Transactions on Components, Packaging and Manufacturing Technology*, *4*(9), pp. 1523–1533.
50. Picas, J.A., Menargues, S., Martin, E., Colominas, C. and Baile, M.T., 2017. Characterization of duplex coating system (HVOF+PVD) on light alloy substrates. *Surface and Coatings Technology*, *318*, pp. 326–331.
51. Shao, F., Zhao, H., Liu, C., Zhong, X., Zhuang, Y., Ni, J. and Tao, S., 2017. Dense yttria-stabilized zirconia coatings fabricated by plasma spray-physical vapor deposition. *Ceramics International*, *43*(2), pp. 2305–2313.

3 High-Temperature Tribological Behavior of Polymer-Based Composite Coating

Divya Gurkar Somashekar, Keshavamurthy Ramaiah, Anand Ashok Kakhandaki, and Yogesh Bhalerao

3.1 INTRODUCTION

3.1.1 Overview of Polymer-Based Composite Coatings

Coating technology is considered an economic and efficient solution. Polymer-based composite coatings are special materials or substances that incorporate polymers and reinforcing fibers or fillers with reasonable mechanical, thermal, and chemical performance. Polymers are recognized as excellent coating materials due to their enhanced barrier properties, easier adjustability, and compatibility for large-scale production [1], with unique combined properties suggesting these composite coatings are suited for a wide range of applications. These coatings are versatile and can be designed as required in terms of durability, cost-effectiveness, abrasion resistance, corrosion protection (chemical), impact resistance, or UV radiation [2].

Because of that lightweight, they find themselves in industries like automotive and aerospace, where such a thing is important. In industries such as aerospace, these coatings are used to protect an aircraft's structure from corrosion, high temperatures, and mechanical stress. In the automotive sector, they are often used to coat different components for better protection and less wear. They may be applied in numerous finishes and colors for improved surface appearance as well as simplicity of application [3,4]. In construction, they prolong the lifespan of buildings and infrastructure [5,6]. The anti-corrosive features are favorable to some marine applications where saltwater would corrode the material; however, biocompatibility is useful for medical applications. Marine applications utilize polymer-based coatings reinforced with graphene, providing outstanding mechanical strength and corrosion resistance as well as excellent barrier properties that benefit performance reliability in severe marine conditions [7,8]. In addition, they are used as active agents in food packaging to preserve the shelf life [9,10], electronics for moisture protection, and textiles to repel water or stain [11]. These coatings have also been employed in superhydrophobic and self-healing applications [12,13], demonstrating the vast versatility of coated textiles. Typical polymers in these coatings include thermosetting resins,

DOI: 10.1201/9781003542957-3

such as epoxy or polyurethane and phenolic, which are well known for their excellent adhesion properties, mechanical-physical integrity, and chemical resistance. These coatings exhibit significantly improved performance over traditional polymer-based coating materials when action traces are reinforced with carbon fibers, glass fibers, or ceramics.

3.1.2 Importance in High-Temperature Applications

Polymer-based composite coatings show huge potential for high-temperature applications due to their excellent thermal stability and protective capabilities. These coatings that combine polymers with reinforcing elements such as fibers or nanoparticles are made to withstand high-temperature levels without undergoing much material degradation. This is in sharp contrast to the traditional coatings, which begin losing adhesion as thermal breakdown occurs at elevated temperatures and cannot withstand a maximum failure temperature—thus polymer-based composite coatings are ideal for protecting high-heat components like turbine blades or exhaust systems. Some materials have been used up to 900°C, such as aluminum oxide and chromium(III) oxide [14], though silicon dioxide has performed well up to about 1,800°C [15]. These oxide films provide protection that must be porous-free, well-sticking, and dense, as well as thermal shock-stable materials within the high-temperature ambience. Their capacity to maintain functionality and protect underlying materials is essential in places where extreme heat is a regular hazard [16].

In the aerospace sector, polymer-based composite coatings are very important for spacecraft and aircraft safety [17]. An example is NASA's Space Shuttle, which used silica fiber polymer composites in its thermal protection system to safeguard the shuttle during re-entry into Earth's atmosphere, where it can face temperatures over 1,500°C. These coatings were crucially important for the shuttle to return safely and successfully complete its mission because they had a good ability to resist such extreme heat [18,19]. Polymeric coatings in the automotive industry also provide heat protection, covering engine components and exhaust systems when exposed to temperatures exceeding 800°C, thereby protecting critical parts while preventing thermal distortion that could adversely impact performance [20].

Meanwhile, the energy industry profits greatly from polymer-based composite coatings, especially at elevated temperatures utilized in power generation. Gas turbine engines are exposed to temperatures beyond 1,250°C, and ceramic-matrix polymer-based composites are vital to ensure thermal protection. By preventing the degradation caused by excessive heat and oxidation to the turbine blades, such coatings enhance the efficiency and performance while also eliminating maintenance requirements [21].

3.2 POLYMER MATRIX MATERIALS

3.2.1 Widely Used Polymers

The right choice of polymer matrix materials is vital for the proper functioning, performance, and longevity of composite coatings in high-temperature tribological applications. It depends on various properties and requirements for a specific application, such as temperature range, mechanical load, environmental impact, and price. In high-temperature tribological coatings, polyetheretherketone (PEEK) can work continuously at temperatures up to 250°C, polytetrafluoroethylene (PTFE) can operate at temperatures up to 260°C because of its excellent thermal properties, and polyimides, epoxy resins, silicones, and polyphenylene sulfide (PPS) have high thermal stability, ranging from 200°C to 240°C [22]. Figure 3.1 shows how polymer matrix composites can be grouped. They are mainly classified into two main types: thermosets and thermoplastics. This depends on the polymer matrix used. Thermosets include certain resins such as epoxy, polyester, and bismaleimide, while thermoplastics include polypropylene, polyamide, and PEEK. These polymers are commonly reinforced with fibers like carbon, glass, and organic materials to enhance the mechanical properties and broaden the applications of the composites.

Epoxy resins represent the most common matrix materials in composite coatings because of their excellent mechanical properties and high chemical resistance, which is combined with an excellent adhesion level. They also exhibit good thermal stability, with a typical temperature range of 150°C–200°C, which can be pushed even higher with specialized formulations. One more added advantage of epoxies is their very low shrinkage during curing, critical to maintaining the coating integrity under high-temperature exposures. The limitation is normally based on T_g; hence, they are not suitable for very high-temperature applications above [23]. Polyimides represent one of the most thermally stable polymers. They can bear continuous use up to 300°C and short-term exposure up to 500°C. They exhibit excellent mechanical strength combined with good chemical resistance and strong tribological properties, making them suitable for high-temperature applications. High T_g of the polyimides ensures that the structural integrity is retained under extreme conditions. The major drawback of polyimides involves the difficulty and increased cost associated with the processing when compared

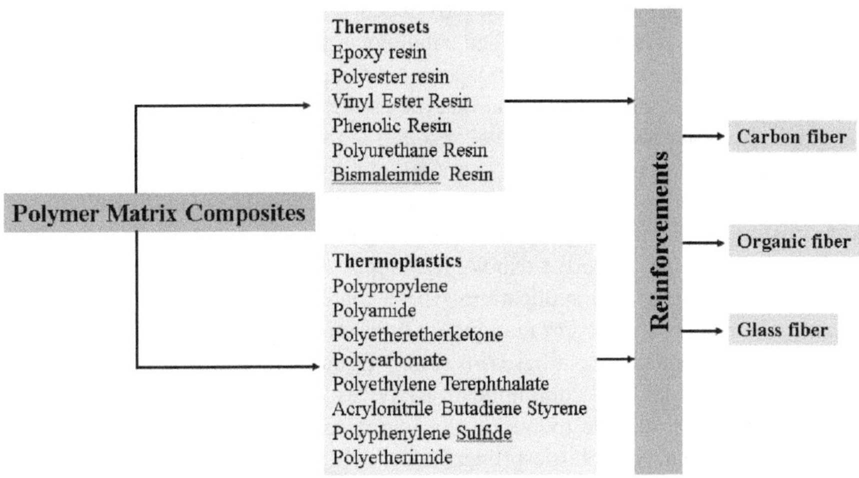

FIGURE 3.1 Classification of polymer matrix composites.

with other polymers [24]. Silicones possess excellent flexibility and stability over a wide range of temperatures ranging from −60°C to +300°C. They exhibit tremendous oxidation and heat resistance; hence, silicones have been good materials for temperature-varying applications. Silicones also exhibit very good resistance to environmental factors such as UV radiation, ozone, and moisture. Their mechanical strength is, however, comparably low with respect to other polymers, and they will likely be reinforced to improve their load-carrying potential in tribological applications [25,26].

3.2.2 Properties and Selection Criteria

When selecting a polymer matrix for high-temperature tribological applications, it is crucial to consider several factors. The polymer must have excellent thermal stability to endure high temperatures and thermal cycling without degrading. It should also exhibit strong tribological performance with low friction and minimal wear to enhance coating lifespan. Adhesion to substrates and reinforcing materials is essential for durability under thermal and mechanical stress. The polymer must resist chemical attacks from corrosive environments and be practical to process and cost-effective for large-scale use [27].

3.3 ROLE OF REINFORCEMENTS AND FILLERS

3.3.1 Reinforcements and Fillers

In high-temperature tribological coatings, selecting the right reinforcements and fillers is essential for boosting performance and durability. Materials like alumina, boron nitride (BN), and silicon carbide are favored for their superior hardness, thermal stability, and wear resistance. They enhance the composite's mechanical strength and its ability to endure high temperatures and abrasive conditions [28,29]. Titanium oxide-graphene oxide hybrids can be introduced into epoxy to greatly enhance its corrosion-resistant properties [30]. The tribological testing of aluminum alloy samples with oxide coatings (Al_2O_3) demonstrated sufficient durability when paired with polymer composites. The study found that the addition of fillers such as cotton fibers and graphite to the polymer composites resulted in the lowest coefficients of friction, ensuring stable performance over a 15 km sliding distance under a load of 0.5 MPa. This research provides valuable insights into the frictional behavior of aluminum oxide coatings in combination with various polymer composite fillers, highlighting their potential for effective tribological applications [31]. They serve multiple purposes, including optical applications where coatings with tuned particle surface plasmon resonances are used to manipulate light for plasmonic effects. In magnetic materials, high-frequency coatings with cut-off frequencies above 1 GHz are crucial for advanced electronic and communication devices [32]. The study investigates how adding nano silicon dioxide to carbon fiber-reinforced epoxy composites enhances wear resistance and viscoelastic properties, which are essential for high-temperature tribological applications. It also shows improved thermal stability, with notable increases in storage modulus and glass transition temperature as the filler content increases [33]. Ceramic fillers contribute to maintaining structural integrity and prolonging the coating's lifespan by providing a hard, protective barrier against wear and thermal degradation.

The addition of metals such as aluminum, chromium, and nickel further improves the mechanical properties and thermal conductivity of the composites. Metals add toughness and resistance to impact on the coating, with heat dissipation being an important factor in applications at high temperatures. Medical coating has been one of the most critical contributions toward the development of the biocompatibility and bio-functionality of biomedical devices to further improve such devices for clinical needs. Composite coatings have been developed by incorporating polymers with inorganic material, improving the device's biocompatibility and bioactivity. It is a huge promise for biomedical applications. The incorporation of 1 wt.% surface-functionalized silver nanoparticles (F-AgNPs) synthesized using *Couroupita guianensis* leaves extract into DGEBA-based epoxy resin significantly enhanced the corrosion resistance, hardness, and T_g of the coatings [34]. Researchers demonstrated that adding aluminum and zinc nanoparticles to carbon fabric-reinforced epoxy composites significantly improves their mechanical and adhesive wear properties, which are essential for high-temperature applications. The study finds that the composite with 0.5 wt.% Zn shows superior wear resistance [35]. Furthermore, they examined how aluminum and zinc nanofillers improve the abrasive wear resistance of carbon fabric-reinforced epoxy composites, a key factor in high-temperature tribological applications. Surface analysis highlights the role of fiber and filler failure modes in determining the composites' wear performance [36].

Graphite, graphene oxides (GOs), and MWCNTs are reinforcements with very good thermal conductivity, low friction, and strength with a high weight ratio. These reinforcements will greatly enhance the performance by reducing friction and wear while improving thermal stability in conditions from a tribological point of view; such materials contribute to making the composite lightweight, a desirable trait in use. In this work, multi-walled carbon nanotubes (CNTs) were dispersed homogeneously into an urushiol-formaldehyde polymer. Therefore, the best content included 0.5 wt.% GO and 1.0 wt.% MWCNTs, giving a remarkable enhancement in hardness, adhesion, and corrosion resistance with a 30-minute drying time. The better performance of such composite coatings was found, especially concerning the corrosion rate reduction and protection efficiency increase, thus making them highly appropriate for application to offshore steelworks [37].

3.3.2 Impact on Coating Durability and Performance

The use of these reinforcements and fillers affects the coating's durability and performance. Ceramics boost hardness and wear resistance, metals enhance toughness and heat dissipation, and carbon-based materials lower friction and improve thermal conductivity. The optimal combination of these materials can greatly improve the composite coating's performance, ensuring it meets the rigorous demands of high-temperature tribological conditions. Proper selection of reinforcements and fillers is crucial for achieving the best balance between mechanical strength, thermal stability, and tribological efficiency [38,39].

3.4 FUNDAMENTALS OF TRIBOLOGY

Tribology, in general, is the interaction of surfaces in motion, embracing friction, wear, and lubrication. The major terms involved here are coefficient of friction (COF), wear, and lubrication, with their influence on the performance and service life of materials. Undoubtedly, optimization in these three aspects improves the efficiency, durability, and reliability of components and mechanical systems of all varieties.

3.4.1 Principles of Tribology

Tribology is the study of the interaction between surfaces in relative motion concerning friction, wear, and lubrication. Friction refers to the resistance of sliding surfaces; it is measured by the COF, which is a variable depending on the materials and the conditions. Wear means material erosion due to mechanical forces that might cause deterioration and loss in efficiency. Application of lubrication involves substances such as oil or grease, which act to lessen direct contact between surfaces and consequently reduce friction and wear. These principles are very vital in machinery design and its regular maintenance, as they improve the performance and extend the lifespan of the involved components through efficient surface treatment and materials [40,41].

When applied to polymer-based composite coatings, tribology addresses how these coatings interact with surfaces in motion. Friction (μ) in these composites is influenced by the coating's surface texture and polymer matrix, often reinforced with nanoparticles or fibers to reduce resistance and improve smoothness. The value of μ varies depending on whether the motion has just begun or is already in progress. Initially, static friction $\mu_s = F/N$ applies, describing the resistance to the start of motion. Once the specimen is in motion, dynamic or kinematic friction, given by $\mu_d = F/N$, represents the ratio of the force required to maintain a constant speed to the applied force [42]. Wear involves the gradual degradation of the coating from mechanical contact, which can affect protective properties and cause damage; therefore, material selection and bonding are key. Lubrication may involve inherent properties of the coating or additional lubricants to further minimize friction and wear. Mastering these tribological aspects is vital for optimizing the performance and durability of mechanical systems with polymer-based composite coatings.

3.4.2 Influence of High Temperatures on Tribological Behavior

High temperatures can significantly impact the tribological behavior of polymer-based composite coatings, affecting friction, wear, and lubrication. In automotive applications where coatings are exposed to engine heat, the polymer matrix in composites like those reinforced with carbon fibers or nanoparticles can soften, increasing friction due to reduced hardness and greater deformation. This softening can also lead to higher wear rates, as seen in studies involving high-temperature tribological testing of epoxy-based composites, where thermal degradation resulted in accelerated material loss and compromised protection [43]. At higher frictional speeds, the COF increases and wear resistance decreases because PIF fibers are crushed and dispersed faster, leading to more abrasive wear. Additionally, increased heat softens the polymer matrix, reducing its load-bearing capacity and making the composite more vulnerable to plowing and scratching [44]. High temperatures can diminish the effectiveness of lubricants, as observed in experiments with silicone-based lubricants in bearings, where elevated temperatures caused lubricant evaporation and reduced viscosity, leading to increased friction and wear [45].

The high-temperature-resisting particles in polymers significantly improve the thermal stability and performance at elevated temperatures. Silica (SiO_2) enhances thermal stability and strength in mechanical aspects [46]. Al_2O_3 has a melting point of approximately 2,050°C, which enhances the wear property and hardness, making it suitable for high-temperature applications up to extreme temperatures [47]. BN depicts excellent thermal conductivity and stability, thus remaining effective up to about 1,000°C [48]. molybdenum sulfate (MoS_2) offers high-temperature lubrication with performance up to 1,200°C, contributing to reducing friction and wear [49]. Graphene and GO deliver excellent thermal conductivity, keeping their properties up to around 2,500°C. Titanium dioxide enhances thermal stability and UV resistance. Its degradation occurs at temperatures above 1,000°C. Ceramic particles like zirconia and titanium carbide have an excellent high-temperature performance with melting points of ~2,700°C and ~3,150°C, respectively [50]. These particles are highly crucial for enhancing high-temperature durability in polymer-based materials so the materials remain effective in extreme applications. Similarly, in polymer coatings within aerospace components, there is a need to exhibit thermal stability with enhanced filler-material bonding to retain performance. The test results of high-temperature testing showed that polymer matrix composites reinforced with ceramic particles demonstrate better resistance to wear and lower friction coefficients. This again outlines the proper selection of materials and design strategies capable of ensuring sustainable efficiency under thermal stresses.

3.5 THERMAL STABILITY AND MECHANICAL PROPERTY CHANGES

It is necessary to understand the changes in thermal stability and mechanical properties to enable good evaluation for high-temperature tribological applications of polymer-based composite coatings. The factors that may influence the performance at elevated temperatures are the thermal stability of the material, which refers to the degree to which a material resists degradation, maintaining its properties, coefficient of thermal expansion (CTE), which covers dimensional stability due to large or small temperature swings, while mechanical property changes pertain to changes in hardness and wear resistance, affecting coating effectiveness concerning friction and thermal stresses.

3.5.1 Arrhenius Equation and Decomposition Rates

The Arrhenius equation is essential for assessing the thermal stability and decomposition rates of materials, especially for polymer-based composite coatings subjected to high temperatures. Equation (3.1) is expressed as:

$$k = Ae^{-\frac{E_a}{RT}} \tag{3.1}$$

where

k represents the reaction or decomposition rate constant,
A is the pre-exponential factor, indicating the frequency of collisions or interactions in the reaction,
E_a is the activation energy, the energy barrier that must be surpassed for the reaction to proceed,
R is the gas constant (8.314 J/mol·K), and
T is the absolute temperature in Kelvin (K)

According to this equation, the rate of a reaction or decomposition accelerates exponentially with temperature. As temperature increases, the term $e^{-\frac{E_a}{RT}}$ decreases rapidly, leading to a higher rate constant k. This means that at higher temperatures, materials decompose more quickly due to increased thermal energy overcoming the activation energy barrier.

For polymer-based composite coatings, this means that high temperatures can accelerate the degradation of both the polymer matrix and its reinforcing elements. The Arrhenius equation allows the determination of the rate at which a material will degrade for any particular thermal condition, hence giving useful information on thermal stability and durability under high-temperature service. This is of prime importance in designing coatings that are intended to face long periods of high-temperature exposure without losing either performance or structural integrity.

3.5.2 Coefficient of Thermal Expansion

The CTE is a key factor in assessing the high-temperature performance of polymer-based composite coatings [51]. CTE quantifies how much a material expands or contracts with temperature changes, calculated using Equation (3.2).

$$\text{CTE} = \frac{1}{L_o}\frac{\Delta L}{\Delta T} \tag{3.2}$$

where

CTE represents the coefficient of thermal expansion,
L_o is the initial length of the material,
ΔL is the change in length, and
ΔT is the temperature change.

CTE in composite polymer-based coatings influences changes in coating size and mechanical properties according to changes in temperature. A coating with a higher CTE will greatly expand upon heating; this may result in problems such as delamination or cracking, provided the thermal expansion is not managed uniformly within the coating and substrate. Conversely, a low CTE is related to minor expansion, enhanced dimensional stability, and improved adhesion of the coating. For instance, in aerospace applications, coatings must resist extreme temperature variations without degradation. Reinforcement in polymer-based coatings by the addition of materials like CNTs or ceramics reduces CTE, hence increasing the resilience of the coating to thermal stresses by maintaining integrity [52].

3.5.3 Thermal and Mechanical Property Assessments

The thermal properties assessments are some of the important tests in establishing the efficiency in the performance of the polymer-based composite coatings in high temperatures. Techniques such as TGA measure the weight loss of the coating while it undergoes heating; this was used to identify the temperature at which significant degradation occurs, which is of value for applications exposed to high temperatures over time. This ensures that the coating is stable and functional. Thermal conductivity is determined by methods such as Laser Flash Analysis. It reveals the ability of the coating to conduct heat without hazardous heat buildup that may influence its capability for damage or failure. Differential scanning calorimetry (DSC) characterizes the thermal properties in glass transition and melting temperature of the polymer-based composite coatings. This information becomes important in estimating the performance of coatings at high temperatures, which in turn dictates their thermal stability, wear resistance, and overall effectiveness in tribological environments. CTE refers to the amount the coating expands or contracts with temperature changes. A smaller CTE is preferred, as this means less dimensional change, and this minimizes the possibility of different kinds of problems such as delamination, cracking, or loss of adherence due to thermal cycling. This will maintain the coating's protective ability and structural integrity during variable thermal conditions [53–55].

The mechanical property analysis gives an idea about the performance of the polymer-based composite coatings under high temperatures and mechanical stresses. Hardness tests, either Vickers or Rockwell, determine the resistance of a coating to indentation or abrasion, thus ensuring protection for the critical component from wear during high-speed operations, such as in aerospace. Scratch tests or peel tests measure adhesion strength, which must be sufficiently strong to keep the coating bonded onto the substrate, especially under continuous thermal expansions and contractions in high-temperature applications such as automotive engines [56,57]. The combination will, therefore, yield comprehensive data on coatings' mechanical durability and provide an assurance of maintaining protective and functional properties under extreme operating conditions that, in turn, create a foundation for the extension of the life and reliability of components.

The interaction in tribology involves mutual determination of thermal and mechanical properties, exactly as is done for friction, wear, and lubrication performance. Thermal conductivity describes the property of conducting heat in a material; the mechanical property of hardness and strength, thus, may be affected by the thermal conductivity, as those with high thermal conductivity will dissipate friction-generated heat efficiently and preserve their mechanical properties at higher temperatures. Thermosetting polymers possess better strength, stiffness, weight, and thermal resistance compared with thermoplastics. Their performance is based on thermal expansion and chemical structure. The basic properties that are influenced by polymer chemistry and crosslinking include thermal stability and behavior [58,59].

Similarly, thermal expansion affects dimensional stability and component fit, and high temperatures can lead to thermal softening, reducing a material's strength. Thermal expansion is a key physical property of polymers when exposed to temperature changes, influenced by their structure, chemical composition, and bond strength. Polymers typically have a CTE ranging from 30 to 100 ppm/K, which is significantly higher than metals and ceramics, whose CTE values are between a few and 20 ppm/K [60]. This high thermal expansion can limit the industrial use of polymers. Studies show that polymers with flexible linkages and bent bonds exhibit higher CTE values compared to rod-like polymers with rigid segments. Thermosetting polymers, like epoxies, form dense three-dimensional networks upon curing, resulting in lower CTE in their glassy or solid state below the T_g. However, CTE increases in the rubbery state above T_g due to greater free volume, enhanced Brownian motion of the polymer chains, and reduced density [61]. Research has shown that adding small quantities of nanofillers—such as carbon black, calcium carbonate, nano metal oxides, nano-graphite, talc, nanoclay, CNTs, organically modified clays, montmorillonite, magnesia, mesoporous silica, and alumina—can significantly alter the thermal expansion and glass transition temperature (T_g) of thermoset polymers. These nanofillers enhance thermal stability and reduce residual stress in polymer composites due to their lower CTE compared to unfilled polymers. Additionally, incorporating graphite nano-platelets has been observed to further decrease both linear and axial CTE in these composites [62–64]. Researchers found that sliding wear and dynamic mechanical tests reveal that incorporating 3 wt.% nano silicon dioxide (SiO_2) into carbon epoxy composites increases the storage modulus and decreases both the loss modulus and COF. This indicates improved elasticity and hardness in the glassy region, contributing to reduced wear and heat dissipation. The higher T_g observed with 3 wt.% $nSiO_2$ suggests enhanced molecular interactions and a stronger cross-linking network, which further decreases the COF and wear. The addition of 3 wt.% $nSiO_2$ enhanced the mechanical and thermal properties of the composites. Ensuring that tribological materials retain their strength and performance under varying thermal conditions is crucial for effective system operation.

3.6 WEAR AND FRICTION TESTING METHODS

Wear and friction testing methods are inevitable for the characterization of high-temperature tribological properties of polymer-based composite coatings. Pin-on-disc, ball-on-disc, and reciprocating sliding tests simulate real-life conditions under definite loads and at controlled temperatures. High-temperature tribometers, generally used in such tests, manage the temperature accurately and monitor it continuously for the measurement of wear rates and friction coefficients with high precision. Generally, it is used in combination with these methods to test the thermal stability and degradation at high temperatures. Surface analyses using SEM and EDS after the test provide comprehensive information on the wear mechanisms of adhesive, abrasive, and thermal degradation. These basic insights form the understanding of the behavior of polymer-based composite coatings when exposed to high-temperature conditions, which are of vital importance in practical development toward enhanced performance in highly demanding applications.

3.6.1 Pin-on-Disc Tests

Among the different available techniques, the pin-on-disk test is one of the most commonly used to characterize the wear and friction properties of polymer-based composite coatings, especially at high temperatures. The principle involves pressing a pin, normally in a ball or cylindrical form, against a rotating disk coated with the material under test. This test method is an excellent simulation of the sliding contact between surfaces and is, therefore, highly applicable in coatings performance evaluation for real applications, in which temperature rise significantly influences friction and wear. The figure shows the pin-on-disc apparatus. Using Equations (3.3) and (3.4), wear volume and the specific wear rate were determined, respectively. Figure 3.2 is a schematic diagram depicting a pin-on-disc tribometer arrangement designed for testing material wear. The test specimen is fixed onto a steel disc, which is driven by a motor that manages its rotational movement. Displacement resulting from the specimen's wear is measured by a lever arm equipped with a linear variable differential transformer (LVDT). The load applied to the specimen is regulated by weights attached to a pulley system, and a damper is used to reduce vibrations during the testing process. The entire setup is supported by a base plate for stability.

$$\text{Wear volume is given by } (V) = \frac{m}{\rho} \, [\text{mm}^3] \tag{3.3}$$

where m is the weight loss in g and ρ is the actual density in g/mm^3.

$$\text{Specific wear rate is given by } (K_s) = \frac{V}{L \times D} \, [\text{m}^3/\text{Nm}] \tag{3.4}$$

where V is wear volume (m^3), L is load (N), and D is abrading distance (m).

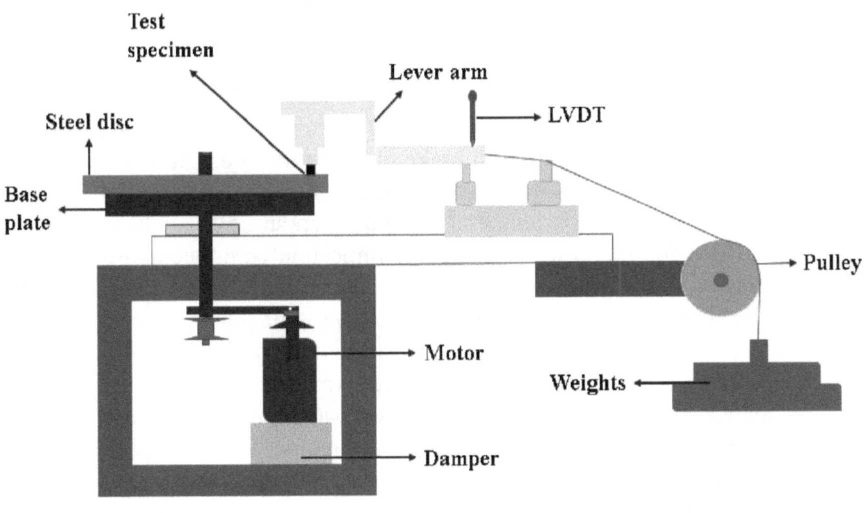

FIGURE 3.2 Schematic diagram of a pin-on-disc tribometer setup for wear testing.

For high-temperature scenarios, the pin-on-disk test is performed in a controlled setting where both the pin and disk are heated to the required temperature, often surpassing 200°C, depending on the specific application. This controlled environment allows for a thorough assessment of the coating's wear resistance, frictional behavior, and thermal stability under challenging conditions. Wood et al. conducted a study using a pin-on-disk tribometer equipped with LVDTs, temperature and friction transducers, and an electrostatic wear sensor. The test conditions were set at a sliding velocity of 5 m/s and a contact stress of 2.05 GPa to know the wear conditions of the samples at elevated temperatures [65]. The study found that the addition of Al_2O_3 and graphite to PA6 composites significantly reduced heat generation, with decreases of 7.07% for PA6 containing 30 wt.% Al_2O_3 and 12.3% for PA6 with 20 wt.% graphite compared to PA6 alone. Thermal stability analysis using TGA revealed that weight loss increased with filler loading, but notable weight loss occurred only at higher temperatures—450.9°C for PA6 with 30 wt.% Al_2O_3 and 466.7°C for PA6 with 20 wt.% graphite [66]. Throughout the test, key parameters like applied load, sliding speed, and temperature are meticulously monitored to gauge the coating's performance. The wear rate is typically determined by measuring the material loss from either the disk or the pin, while the friction coefficient is calculated based on the force needed to sustain the sliding movement [67].

The most valuable benefits of polymer-based composites are their low thermal and electrical conductivity, which makes them very effective as insulators of components that need to be protected from heat and electrical currents. In case the design requires parts with thermal conductivity, it would be possible to add thermally conductive materials to the composite to retain this property. For example, polyimide composites provide good thermal conductivity and low weight compared to metals. This composite has a high glass transition temperature; it can be tailored for different uses by replacing carbon fibers with those having low CTE [68]. Adhesive wear is also common in the automotive industry as some portions are prone to serve under severe friction and high temperatures, such as brake pads and the parts making up the engine body. The use of high-performance materials with superior wear resistance—high hardness tool steels and advanced polymers such as PTFE and PEEK—can reduce the problem [69]. Among the most common composite materials used in brake pads are those based on phenolic, as they provide a high friction coefficient and demonstrate wear reduction at room temperature and at increased temperatures up to about 250°C. Such materials also contribute to the increase of the life of important components due to their material transfer and degradation characteristics. Because of its high COF and the capability to retain reduced wear at both room temperature and up to about 250°C, the phenolic composites were used for automotive brake pads [70]. The applications for these polymer-based composites are in chassis, brake pads, hoods, bumpers, fenders, engine cradles, interior and exterior components, tires, and sunroofs. These composites provide a unique combination of lightweight and high strength, resistance to damage, stiffness, thermal insulation, and wear resistance at an affordable cost.

3.6.2 Block-on-Ring Tests

The block-on-ring test is an effective method for evaluating the tribological performance of polymer-based composite coatings, especially under high-temperature conditions. In this test, a coated rectangular block is pressed against a rotating ring, simulating sliding contact in various mechanical applications. This setup replicates different wear scenarios, such as sliding, rolling, or combined loading, making it highly relevant for studying the durability and reliability of these coatings in environments where friction and wear are critical. Figure 3.3 illustrates a block-on-ring tribometer setup designed to assess the friction and wear characteristics of materials. In this setup, the test specimen, secured by a block, is pressed against a rotating test ring that is immersed in test oil. The temperature within the system is regulated by a heater and refrigerant, while a thermocouple is used to track the temperature throughout the testing process, ensuring precise measurement of the thermal impact on wear and friction.

Polymer tribology is more complex and less understood compared to metals, as polymers do not adhere to established friction laws used for metals and ceramics. This complexity arises from polymers' relative softness, lower thermal conductivity, and lower melting points, which affect heat generation and wear behavior. If these factors are not properly considered, it can lead to significant issues in applications involving rolling, sliding, or bearing components made from polymers [71]. Approximately 23% of global energy consumption is attributable to tribological contacts. Of this, 20% is spent overcoming friction, while 3% is used for remanufacturing worn parts and replacing equipment affected by wear and related failures. Tong et al. analyzed the tribo-performance of carbon fabric-reinforced phenolic-based composites to know whether they are effective for self-lubricating bearings and gears using a block-on-ring test setup. Adding CNTs in MoS_2 hybrids

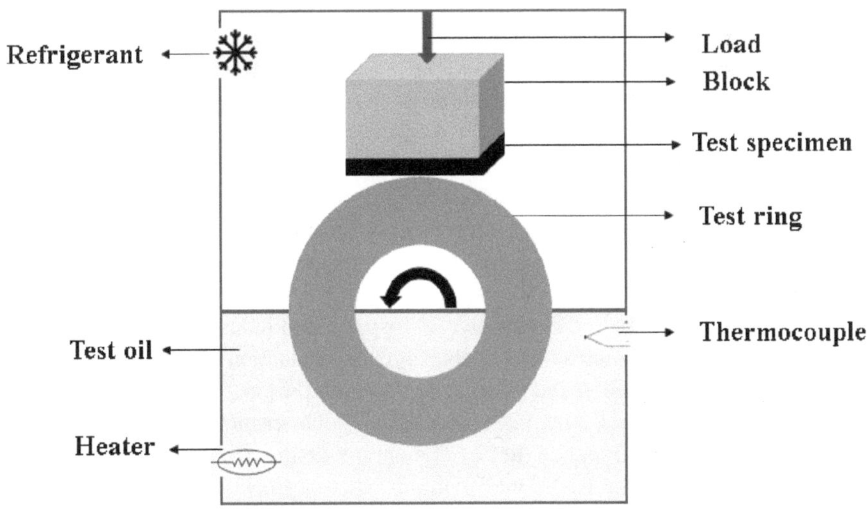

FIGURE 3.3 Schematic diagram of a block-on-ring tribometer setup.

improved their tribological properties, reducing friction as hybrid content increased. While 1 wt.% hybrids decreased wear resistance due to abrasive wear, 2 and 4 wt.% hybrids enhanced wear resistance. This improvement results from CNTs' mechanical properties, strong bonding with the matrix, and MoS_2 transfer film formation [72]. For high-temperature applications, the block-on-ring test is conducted in a controlled environment where both the block and ring are heated, often exceeding 200°C. The test apparatus can include heaters, temperature sensors, and an inert gas atmosphere to prevent material degradation. Key parameters such as applied load, sliding speed, and temperature are meticulously controlled to evaluate the coating's wear resistance, frictional behavior, and thermal stability. The versatility of this test makes it a crucial tool in developing and optimizing polymer-based composite coatings for high-temperature applications in industries like automotive and aerospace. The experiments conducted on glass/polyester composite coatings revealed that as the applied load increases, the film thickness decreases and weight loss rises due to increased contact and shear forces at the rubbing surfaces. Higher loads lead to more significant damage on the composite surface, including detachment of the outer polyester layer. The new film thickness measurement technique is found to be more reliable and easier than previous methods, and it is recommended for use in other tribological setups, with further research needed on its performance at varying rotational speeds [73]. High-performance polymers, often produced in developed countries, offer superior physical properties, high melting points, and excellent resistance to degradation, but they can be costly, reaching up to 10,170 INR/kg. These polymers include ultra-high-performance types like polyaramids (e.g., Kevlar) and the PAEK family (e.g., PEEK), which have melting points around 343°C and can operate up to 250°C. Polybenzimidazole, known for its exceptional temperature resistance with a glass transition temperature of 427°C, also falls into this category [74].

3.6.3 Performance under Simulated Operational Conditions

Grasping the mechanisms of friction and wear is essential for developing advanced polymers for tribological uses; yet traditional research methods often fall short in examining these processes at the micro- or mesoscopic levels. Molecular dynamics (MD) simulation has recently become valuable for its capability to explore atomic-scale interactions at contact interfaces [75]. Assessing performance under simulated operational conditions is crucial for understanding how polymer-based composite coatings behave at high temperatures. These coatings are engineered to endure harsh environments, including extreme temperatures, mechanical stresses, and continuous sliding contact, typical in sectors like automotive, aerospace, and industry. Simulating these conditions in a controlled lab setting enables precise evaluation of the coatings' wear resistance, frictional stability, and thermal degradation [76].

Awwad et al. analyzed the wear profile of polymer-metal contacts using a block-on-ring configuration and finite element modeling (FEM) combined with experimental data. Unlike traditional methods, this approach accounted for stress distribution, revealing that wear profiles have distinct normal and critical zones. The FEM model, integrated with Archard's wear equation and linked with Abaqus software, accurately predicted stress distributions and wear patterns. SEM analysis confirmed that the critical zone, characterized by high-stress concentrations, exhibited more severe damage compared to the normal zone, highlighting significant differences in wear behavior [77]. The research investigates the impact of nanocopper particles on improving wear resistance in carbon fiber-reinforced epoxy composites, crucial for high-temperature tribological applications. It also develops an ANN model for predicting wear behavior, enhancing the understanding of composite performance under different conditions [78]. In contrast to macroscopic tribological tests, MD simulation allows for a clear observation of the friction process. Utilizing various visualization tools, such as the Materials Studio 2024's visualization module, OVITO 3.10.6, and VMD 1.9.4, enables real-time monitoring of polymer deformation, adhesion, and abrasion during motion [79]. This approach aids researchers in examining atomic-level interactions at friction contact interfaces. The significant reduction in the wear rate of a composite was observed due to the adsorption of silica particles onto polymer chains by comparing the microscopic friction behavior of polyvinyl alcohol (PVA)/polyacrylamide (PAM) composites with and without silica nanoparticles [80]. Similarly, polymer friction in water-lubricated conditions can be examined. Song and Zhao studied the friction behavior of PTFE/copper under both dry friction and water lubrication conditions. Their observations revealed that, compared to dry friction, wear was significantly reduced under water lubrication because water molecules diminished the adhesion of copper to PTFE molecules [81]. In simulations, the coated materials are heated to temperatures above 200°C and subjected to sliding or rolling contact with other surfaces under varying loads. This controlled environment allows detailed monitoring of parameters such as wear rate, friction coefficient, and material property changes over time. Techniques like pin-on-disk or block-on-ring tests are used to recreate the real-world stresses these coatings encounter. The data obtained from these simulations are essential for detecting potential failure mechanisms, such as oxidative wear, adhesive wear, or thermal cracking, which might not be visible under standard testing conditions. Understanding how coatings perform under these simulated conditions helps manufacturers refine their formulations to improve durability, extend service life, and ensure reliable performance in challenging high-temperature applications.

3.7 APPLICATIONS IN VARIOUS INDUSTRIES

High-temperature tribological performance, provided by polymer-based composite coatings, acts decisively in areas such as aerospace, automotive, and industrial machinery. The extreme-condition performance of this class of materials increases component durability and enhances operational efficiency, further promoting overall system functionality. This is certainly the means by which the challenges posed to contemporary engineering applications will be overcome and the frontiers of technological advancement in many industries pushed further.

3.7.1 Aerospace

Reliability and efficiency in the high-temperature tribological performance of the various components are thus imparted to the aerospace industry by the polymer-based composite coatings. The requirement of a number of exquisite mechanical, physical, and chemical properties enforces the use of specially formulated polymers, whose quality and characteristics are further enhanced when used in nanocomposites. Nanocomposites offer improved strength and flexibility, with reduced weight and cost. This review outlines the benefit of using polymer-based nanocomposites, one of which involves fibers such as glass, and advanced applications involving propylene polycarbonate (PPC), which is an environmentally friendly material with very high thermal decomposition and tensile strength for aerospace applications [82]. Nanocomposites improve functionality, flexibility, and resistivity while reducing the weight and cost of the material. The following example identifies PPC as the most suitable material, due to the excellent properties it provides, such as high thermal decomposition and strength, in aerospace applications.

3.7.1.1 Advanced Uses of Polymeric Composites

Aerospace systems, including jet engines, turbines, and high-speed aircraft parts, operate under extreme temperatures and difficult conditions [83]. Because of the advanced thermal stability and wear resistance shown by the polymer-based composite materials, high value is derived in the usage at such extreme conditions. These are applied in critical components where high temperatures and friction, like in engine parts, bearings, and seals, are often undergone. Advanced composites, including carbon fiber or ceramic fillers, are applied to turbine blades and high-performance bearings to guarantee minimal friction while providing improved thermal resistance. This has been fundamental in the enhancement of efficiency and durability for aerospace components. Their ability to handle temperatures significantly greater than 600°C without degradation inspires great confidence in their performance and structural integrity under extremely extreme conditions of high-speed flight and engine performance. Composites are finding their place in a number of aerospace applications, such as airframe structures, rocket motor castings, antenna dishes, engine nacelles, and various kinds of control surfaces like horizontal and vertical stabilizers, landing gear doors, and turbine blades, many of which are safety-critical [84]. Boron fiber/epoxy and carbon fiber/epoxy composites have found applications in many aircraft components such as flaps, stabilizers, landing gear doors, wing box skins, and fuselages. Aramid fiber/epoxy is used in fairings, ducts, and interior components, while hybrid composites are applied in panels, fairings, and landing gear doors. In the case of helicopters, glass fiber/epoxy and carbon fiber/epoxy are used in rotor blades while carbon fiber/thermoplastic is used in tail fins. Carbon fiber/epoxy is being used in space shuttles for payload bay doors and in rockets for filament-wound cases [85].

3.7.2 Automotive

The high-temperature friction properties in the case of polymer-based composites are highly desirable to increase the life and performance in the automotive sector. Applications involving extremely high temperatures combined with considerable mechanical stress in automotive engines and brake systems raise the importance of advanced composite coatings. These composites are used in parts such as engine pistons, valves, and brake pads to improve wear resistance and reduce friction. Most of the brake linings contain friction-reducing and wear-reducing fillers like graphite or ceramics, which enhance the efficiency of braking by prolonging the life of parts.

Besides, such coatings provide good heat dissipation, which is crucial for maintaining proper running of the motor and avoiding overheating. Higher fuel efficiencies with attendant lower emissions to assist industry goals of sustainability and performance enhancement will result from the progress that will be made on the development of polymer composites resistant at high temperatures. Due to their advanced tribological properties, in particular self-lubricating and thermal stability, polymer composites find highly increasing applications in transportation systems [86].

Hybrid bushings play an important role in the reliable operation of many car components over an extended period. Such examples include, among others, windshield wiper motors, shock absorbers, steering shaft joints, and door hinges. A hybrid structure combining the superior tribological properties of plastics with the load-carrying and thermal conductivity of metals can be obtained by integrating a plastic material into a porous bronze sinter layer applied to a steel backing [87]. The tribological properties ensue when solid lubricants within the plastic matrix form a lubricating film between sliding surfaces, thereby enabling the operation of such bearings without oil or grease. In one of the studies on the tribological behavior of two PEEK compounds in a ring-on-plate test arrangement, nano-modified PEEK outperformed even commercial PTFE-based references for higher loads and temperatures [88]. This happens to be the case with hybrid composite coatings, which, in numerous researches, have shown better wear resistance. In the present work, a multimodal PPS composite with 3 wt.% MWCNT, 10 wt.% graphite, and 10 wt.% short carbon fibers gave the best performance concerning low friction and wear properties. Hence, the combination gives the best overall profile, more specifically electrical and tribological characteristics. The results of these tests showed that the application of optimized filler combinations opens perspectives for a multifunctional improvement of the property profile of PPS composites beyond the limits of single filler systems [89].

3.7.3 Meeting High-Temperature Demands

The tribological properties of polymers and polymer composites are intimately related to their viscoelastic and temperature-dependent properties. When two surfaces come into sliding contact, heating effects occur at the asperities, which increase the temperature at the frictional surface and influence their viscoelastic responses, including their stress, adhesion, and transfer characteristics [90]. This represents one of the major challenges when one addresses the high-temperature demands in the development of composite coatings based on polymers. Such coatings should assure mechanical and tribological properties at high temperatures for reliable operation in applications. To this effect, high-temperature polymer composites are developed to withstand such extreme thermal conditions with little degradation.

PPS is a high-temperature thermoplastic with relatively excellent processability, rigidity, heat resistance, impact resistance, creep properties, and chemical corrosion characteristics, suited to various industrial applications. While low wear resistance and high friction coefficient (>0.35) prevent it from being suitable in applications involving alterations to friction-related applications [91]. Then, these fillers can be ceramic-based or metal and are even carbon fibers to enhance the thermal stability of coatings as well as their wear resistance. The thorough testing thus done judges their performance in a high-temperature environment to be satisfactory to sustain aerospace, automotive, and industrial environment demands. Silicone rubber is used in aircraft due to its excellent performance over temperature, irradiation, chemicals, aging, and electrical insulation. Carbon nanoparticles such as graphene, CNTs, and carbon black improve the resistance to oxidation in air [92]. This is the exact reason why the development of advanced polymer composites with operations above 200°C is essential to meet modern engineering performance demands. In mitigating problems associated with high-temperature conditions, these coatings are increasingly playing a decisive role in technology advancements along with efficiency and reliability improvements in many sectors. Researchers studied the friction and wear behavior of the PTFE/MoS_2 composites at high temperatures. Thus, the friction coefficient of composites varies from 0.14 to 0.19, and the wear rate ranges from 4.18 to 13.38×10^{-4} mm^3/Nm in the testing temperature from 25°C to 250°C. Significantly, above 200°C, the PTFE/MoS_2 composites demonstrated superior performance than pure PTFE and more precisely with 2 and 5 wt.% MoS_2, with lower friction and wear rates. The wear mechanism changes from mild abrasive wear at low temperature (25°C–100°C) to fatigue and severe abrasive wear at high temperature (150°C–250°C) [93]. The researchers say that normal load, sliding speed, and temperature are parameters that greatly affect the tribological properties of polyamide, HDPE, and their composites. Hybrid kenaf/glass fiber-based composites increase both specific strength and rain erosion resistance of aircraft, whereas, in aircraft brake applications, carbon fiber-reinforced silicon carbide provides excellent temperature resistance to up to 1,200°C [94].

3.8 CONCLUSION

3.8.1 Summary of Key Findings

Polymer-based composite coatings demonstrated significant improvements in high-temperature thermal stability and tribological properties. This material is developed to be able to withstand high heat while still maintaining its mechanical strength. Their robustness is demonstrated, as they are excellent at the reduction of wear and friction in high-temperature environments.

Heat-resistant polymer composites are formed by adding metals, ceramics, or pure carbon fiber as filler. The fillers help in increasing the thermal stability and wear and friction performance of these composites so that they can be efficiently put to use across diverse challenging applications within aerospace, automotive, and metal-forming industries.

The fillers, such as graphite or ceramics, showed substantially better wear resistance and friction reduction in studies. A composite shows improved wear resistance at higher loadings of hybrid fillers (2–4 wt.%) and poor performance when the same individually contains 1% filler content, due to a coupled abrasive wear action. This is a curious case of how maximizing the concentration of filler particles may have an effect on real performance.

High-temperature polymer composites are very versatile and used in various applications across multiple sectors, catering to industries like aerospace, automotive, and industrial machinery. Known for providing an additional level of performance and long-life endurance that enhances deployment support, excellent tough applications involving temperature resistance can benefit from their use as well.

3.8.2 Future Directions in Polymer-Based Composite Coatings

Polymer-based composites with a high-temperature performance would be more widely used in engineering applications if advanced fillers and additives are developed. New advances in nanomaterials and hybrid fillers show increased potential to enhance thermal stability, wear resistance, and operations at high speeds (under light loads), as well as the ability to tailor material properties beyond what we have today.

The discovery of the implementation of advanced manufacturing methods like additive and precision molding promises to innovate in high-performance composite coating generation. The attractiveness of these methods is that while complex filler materials can be efficiently integrated, they allow for further refinement of the coatings' properties and final performance to a desired product-related property.

Extensive, long-term durability studies are essential to evaluate the ability of polymer-based composites to withstand high temperatures for extended periods. This embodies how perception-signifying research is crucial to obtaining better insight on the long-term performance and reliability of these materials in order for them to meet real-world necessities.

Nowadays, due to the rising environmental awareness, it is more meaningful than ever before to develop polymer composites that are sustainable and environmentally friendly. The potential as a novel type of composites for various applications is expected, and sustainability shall be pursued by developing biodegradable or recyclable materials through reduced environmental loads during their manufacturing processes.

This might enable the design of polymer-based composites whose properties can be customized for specific application requirements based on their material science advances. This includes the calibration of thermal, mechanical, and tribological properties aimed at specific industry requirements to achieve performance and efficiency.

REFERENCES

1. H.-Y. Li, D.-N. Huang, K.-F. Ren, and J. Ji, Inorganic-polymer composite coatings for biomedical devices, *Smart Mater. Med.*, vol. 2, pp. 1–14, 2021.
2. L. Wang et al., Multifunctional polymer composite coatings and adhesives by incorporating cellulose nanomaterials, *Matter*, vol. 6, no. 2, pp. 344–372, 2023.
3. V. Dogra et al., Polymer based composite coatings in engineering applications, in: A. Verma, H. S. Gupta, and S. K. Sethi (eds.), *Hybrid Composite Materials*, Singapore: Springer Nature Singapore, pp. 237–251, 2024.
4. A. Groza, Advances in polymer based composite coatings, *Polymers (Basel)*, vol. 13, no. 10, p. 1611, 2021.
5. S. Ganguly and S. Margel, Introduction to coatings and surface preparation, in: S. Thomas and J. S. George (eds.), *Polymer-Based Nanoscale Materials for Surface Coatings*, Amsterdam: Elsevier, pp. 1–18, 2023.

6. S. B. Ulaeto, R. P. Ravi, I. I. Udoh, G. M. Mathew, and T. P. D. Rajan, Polymer-based coating for steel protection, highlighting metal–organic framework as functional actives: A review, *Corros. Mater. Degrad.*, vol. 4, no. 2, pp. 284–316, 2023.
7. S. Verma, S. Mohanty, and S. K. Nayak, A review on protective polymeric coatings for marine applications, *J. Coat. Technol. Res.*, vol. 16, no. 2, pp. 307–338, 2019.
8. C. Yoshikawa, R. Takagi, T. Nakaji-Hirabayashi, T. Ochi, Y. Kawamura, and H. Thissen, Marine antifouling coatings based on durable bottlebrush polymers, *ACS Appl. Mater. Interfaces*, vol. 14, no. 28, pp. 32497–32509, 2022.
9. P. Cazón, G. Velazquez, J. A. Ramírez, and M. Vázquez, Polysaccharide-based films and coatings for food packaging: A review, *Food Hydrocoll.*, vol. 68, pp. 136–148, 2017.
10. V. Gupta, D. Biswas, and S. Roy, A comprehensive review of biodegradable polymer-based films and coatings and their food packaging applications, *Materials (Basel)*, vol. 15, no. 17, p. 5899, 2022.
11. J. J. Licari, *Coating Materials for Electronic Applications: Polymers, Processing, Reliability, Testing*. Norwich, NY: Noyes Publications, William Andrew Pub., pp. 1–531, 2003.
12. H. Zhang and Z. Guo, Recent advances in self-healing superhydrophobic coatings, *Nano Today*, vol. 51, no. 101933, p. 101933, 2023.
13. R. S. Sutar et al., Superhydrophobic Al_2O_3–polymer composite coating for self-cleaning applications, *Coatings*, vol. 11, no. 10, p. 1162, 2021.
14. G. Y. Lai, *High-Temperature Corrosion and Materials Applications*, vol. 2, Detroit, MI: ASM International, 2007.
15. M. Zielecka, A. Rabajczyk, K. Cygańczuk, Ł. Pastuszka, and L. Jurecki, Silicone resin-based intumescent paints, *Materials (Basel)*, vol. 13, no. 21, p. 4785, 2020.
16. I. Stojanović, B. Škrlec, M. Kurtela, V. Alar, and M. Odeljan, Physical and chemical properties of high-temperature silicone-based polymer coatings applied on different surface roughnesses, *Coatings*, vol. 13, no. 12, p. 2100, 2023.
17. R. Yadav, M. Tirumali, X. Wang, M. Naebe, and B. Kandasubramanian, Polymer composite for antistatic application in aerospace, *Def. Technol.*, vol. 16, no. 1, pp. 107–118, 2020.
18. D. Glass, Ceramic matrix composite (CMC) thermal protection systems (TPS) and hot structures for hypersonic vehicles, in: *15th AIAA International Space Planes and Hypersonic Systems and Technologies Conference*, Dayton, OH, 2008.
19. O. Uyanna and H. Najafi, Thermal protection systems for space vehicles: A review on technology development, current challenges and future prospects, *Acta Astronaut.*, vol. 176, pp. 341–356, 2020.
20. F. Khan et al., Advances of composite materials in automobile applications – A review, *J. Eng. Res.*, 2024. DOI:10.1016/j.jer.2024.02.017.
21. X. Ma, K. Rivellini, P. Ruggiero, and G. Wildridge, Novel thermal barrier coatings with phase composite structures for extreme environment applications: Concept, process, evaluation and performance, *Coatings*, vol. 13, no. 1, p. 210, 2023.
22. K. Friedrich, Polymer composites for tribological applications, *Adv. Ind. Eng. Polym. Res.*, vol. 1, no. 1, pp. 3–39, 2018.
23. Y. Wang, S. Chen, X. Chen, Y. Lu, M. Miao, and D. Zhang, Controllability of epoxy equivalent weight and performance of hyperbranched epoxy resins. *Comp. B Eng.*, vol. 160, pp. 615–625, 2019.
24. T. Ramakrishnan et al., Study of numerous resins used in polymer matrix composite materials, *Adv. Mater. Sci. Eng.*, vol. 2022, pp. 1–8, 2022.
25. H. J. Fabris and W. G. Knauss, Synthetic polymer adhesives, in: G. Allen (ed.), *Comprehensive Polymer Science and Supplements*, New York: Elsevier, pp. 131–177, 1989.

26. M. Englert, F. Minister, A. Moussaoui, and W. Pisula, Mechanical properties of thermo-oxidative aged silicone rubber thermally stabilized by titanium oxide based fillers, *Polym. Test.*, vol. 115, no. 107726, p. 107726, 2022.
27. L. Deleanu, M. Botan, and C. Georgescu, Tribological behavior of polymers and polymer composites, in: A. Patnaik, T. Singh, and V. Kukshal (eds.), *Tribology in Materials and Manufacturing - Wear, Friction and Lubrication*, IntechOpen, 2021.
28. M. D. Tyona and A. Itodo, Effects of surface modification (surface treatment) on friction and surface abrasion of ceramic composites, in: R. Jose and F.I. Ezema (eds.), *Surface Modification and Functionalization of Ceramic Composites*, Amsterdam: Elsevier, pp. 91–114, 2023.
29. L. Sawunyama, T. O. Ajiboye, O. Oyewo, and D. C. Onwudiwe, Ceramic-polymer composite membranes: Synthesis methods and environmental applications, *Ceram. Int.*, vol. 50, no. 3, pp. 5067–5079, 2024.
30. Z. Yu et al., Preparation of graphene oxide modified by titanium dioxide to enhance the anti-corrosion performance of epoxy coatings, *Surf. Coat. Technol.*, vol. 276, pp. 471–478, 2015.
31. J. Korzekwa, E. Bociąga, and D. Bochenek, Investigation of selected polymer composite-aluminum oxide coating tribological systems, *Materials (Basel)*, vol. 13, no. 23, p. 5491, 2020.
32. F. Faupel, V. Zaporojtchenko, T. Strunskus, and M. Elbahri, Metal-polymer nanocomposites for functional applications, *Adv. Eng. Mater.*, vol. 12, no. 12, pp. 1177–1190, 2010.
33. G. S. Divya, B. Suresha, H. M. Somashekar, and I. M. Jamadar, Dynamic mechanical analysis and optimization of hybrid carbon-epoxy composites wear using Taguchi method, *Tribol. Ind.*, vol. 43, no. 2, pp. 298–309, 2021.
34. R. Manjumeena, R. Venkatesan, D. Duraibabu, J. Sudha, N. Rajendran, and P. T. Kalaichelvan, Green nanosilver as reinforcing eco-friendly additive to epoxy coating for augmented anticorrosive and antimicrobial behavior, *Silicon*, vol. 8, no. 2, pp. 277–298, 2016.
35. G. S. Divya and B. Suresha, Role of metallic nanofillers on mechanical and tribological behaviour of carbon fabric reinforced epoxy composites, *Mater. Sci. Appl.*, vol. 09, no. 09, pp. 740–750, 2018.
36. G. S. Divya and B. Suresha, Metallic nanofillers effect in assessing the abrasive wear performance of carbon fabric reinforced epoxy composites, *AIP Conf. Proc.*, vol. 2057, p. 020050, 2019.
37. L. Zhang, H. Wu, Z. Zheng, H. He, M. Wei, and X. Huang, Fabrication of graphene oxide/multi-walled carbon nanotube/urushiol formaldehyde polymer composite coatings and evaluation of their physico-mechanical properties and corrosion resistance, *Prog. Org. Coat.*, vol. 127, pp. 131–139, 2019.
38. B. Jakab, I. Panaitescu, and N. Gamsjäger, The action of fillers in the enhancement of the tribological performance of epoxy composite coatings, *Polym. Test.*, vol. 100, no. 107243, p. 107243, 2021.
39. K. Bhavith, P. Pai, M. Sudheer, C. G. Ramachandra, M. Prashanth, and K. Kumar, The effect of metal filler on the mechanical performance of epoxy resin composites, *Eng. Proc.*, vol. 59, no. 1, p. 200, 2024.
40. I.-J. Kim (ed.), Basic principles of tribology, in: *Engineering Metrology for Pedestrian Falls Prevention and Protection*, Cham: Springer International Publishing, 2022, pp. 53–116.
41. S. Wen and P. Huang, *Principles of Tribology*, Hoboken, NJ: Wiley, 2017.
42. W. Brostow, V. Kovačević, D. Vrsaljko, and J. Whitworth, Tribology of polymers and polymer-based composites, *J. Mater. Edu.*, vol. 32, no. 5–6, 2010, pp. 273–290.

43. C. Tu, J. Cao, H. Huang, C. Yan, A. AL-Maqbol, and X. Wang, The investigation of microstructure and tribological properties of PTFE/PI-PAI composite coating added with VN, *Surf. Coat. Technol.*, vol. 432, no. 128092, p. 128092, 2022.
44. Y. Yan, B. Zhang, J. Wang, C. Cao, and F. Yan, MWCNT–polyimide fiber-reinforced composite for high-temperature tribological applications, *Coatings*, vol. 14, no. 2, p. 181, 2024.
45. S. M. Aouadi, H. Gao, A. Martini, T. W. Scharf, and C. Muratore, Lubricious oxide coatings for extreme temperature applications: A review, *Surf. Coat. Technol.*, vol. 257, pp. 266–277, 2014.
46. J. Raabe et al., Evaluation of reaction factors for deposition of silica (SiO_2) nanoparticles on cellulose fibers, *Carbohydr. Polym.*, vol. 114, pp. 424–431, 2014.
47. R. Komanduri and S. Iyengar, Conventional and super abrasive materials, in: R. Komanduri and S. Iyengar (Eds.), *Encyclopedia of Materials: Science and Technology*, Amsterdam: Elsevier, pp. 1629–1651, 2001.
48. T. L. Wong, C. Vallés, A. Nasser, and C. Abeykoon, Effects of boron-nitride-based nanomaterials on the thermal properties of composite organic phase change materials: A state-of-the-art review, *Renew. Sustain. Energy Rev.*, vol. 187, no. 113730, p. 113730, 2023.
49. P. Serles, K. Gaber, S. Pajovic, G. Colas, and T. Filleter, High temperature microtribological studies of MoS2 lubrication for low earth orbit, *Lubricants*, vol. 8, no. 4, p. 49, 2020.
50. P. Balaji, S. Rajan, K. Sathickbasha, B. Sethupathi, and D. Magadevan, The significance of low and high temperature solid lubricants for brake friction applications and their tribological investigation, *Tribol. Int.*, vol. 191, no. 109109, p. 109109, 2024.
51. R.-M. Wang, S.-R. Zheng, and Y.-P. Zheng (eds.), Other properties of polymer composites, in: *Polymer Matrix Composites and Technology*, Sawston, Cambridge: Elsevier, pp. 513–548, 2011.
52. K. F. Chan, M. H. M. Zaid, M. S. Mamat, S. Liza, M. Tanemura, and Y. Yaakob, Recent developments in carbon nanotubes-reinforced ceramic matrix composites: A review on dispersion and densification techniques, *Crystals (Basel)*, vol. 11, no. 5, p. 457, 2021.
53. D. J. Skrovanik and C. K. Schöff, Thermal mechanical analysis of organic coatings, *Prog. Org. Coat.*, vol. 16, no. 2, pp. 135–163, 1988.
54. L. Fedrizzi, A. Bergo, F. Deflorian, and L. Valentinelli, Assessment of protective properties of organic coatings by thermal cycling, *Prog. Org. Coat.*, vol. 48, no. 2–4, pp. 271–280, 2003.
55. H. M. Ng, N. M. Saidi, F. S. Omar, K. Ramesh, S. Ramesh, and S. Bashir, Thermogravimetric analysis of polymers, in: J.I. Kroschwitz and H. F. Mark (eds.), *Encyclopedia of Polymer Science and Technology*, Hoboken, NJ: Wiley, pp. 1–29, 2018.
56. K. Rahmani, G. Wheatley, A. Sadooghi, S. J. Hashemi, and J. Babazadeh, The experimental investigation of hardness and wear behaviors of inner surface of the resin tubes reinforced by fibers, *Results Eng.*, vol. 11, no. 100273, p. 100273, 2021.
57. N. Myshkin and A. Kovalev, Adhesion and surface forces in polymer tribology—A review, *Friction*, vol. 6, no. 2, pp. 143–155, 2018.
58. C. Li, G. A. Medvedev, E.-W. Lee, J. Kim, J. M. Caruthers, and A. Strachan, Molecular dynamics simulations and experimental studies of the thermomechanical response of an epoxy thermoset polymer, *Polymer (Guildf.)*, vol. 53, no. 19, pp. 4222–4230, 2012.
59. N. Saba and M. Jawaid, A review on thermomechanical properties of polymers and fibers reinforced polymer composites, *J. Ind. Eng. Chem.*, vol. 67, pp. 1–11, 2018.
60. K. Takenaka and M. Ichigo, Thermal expansion adjustable polymer matrix composites with giant negative thermal expansion filler, *Compos. Sci. Technol.*, vol. 104, pp. 47–51, 2014.

61. K. Makarian, S. Santhanam, and Z. N. Wing, Coefficient of thermal expansion of particulate composites with ceramic inclusions, *Ceram. Int.*, vol. 42, no. 15, pp. 17659–17665, 2016.
62. A. A. Asif, B. John, V. L. Rao, and K. N. Ninan, Surface morphology, thermomechanical and barrier properties of poly(ether sulfone)-toughened epoxy clay ternary nanocomposites, *Polym. Int.*, vol. 59, no. 7, pp. 986–997, 2010.
63. A. Warrier et al., The effect of adding carbon nanotubes to glass/epoxy composites in the fibre sizing and/or the matrix, *Compos. Part A Appl. Sci. Manuf.*, vol. 41, no. 4, pp. 532–538, 2010.
64. N. Suzuki, S. Kiba, Y. Kamachi, N. Miyamoto, and Y. Yamauchi, Mesoporous silica as smart inorganic filler: Preparation of robust silicone rubber with low thermal expansion property, *J. Mater. Chem.*, vol. 21, no. 14, p. 5338, 2011.
65. R. J. K. Wood et al., Electrostatic monitoring of the effects of carbon black on lubricated steel/steel sliding contacts, in: *Tribology and Interface Engineering Series*, Amsterdam: Elsevier, pp. 109–121, 2005.
66. S. Satheeskumar and G. Kanagaraj, Experimental investigation on tribological behaviours of PA6, PA6-reinforced Al_2O_3 and PA6-reinforced graphite polymer composites, *Bull. Mater. Sci. (India)*, vol. 39, no. 6, pp. 1467–1481, 2016.
67. K. Torkashvand, V. K. Selpol, M. Gupta, and S. Joshi, Influence of test conditions on sliding wear performance of high velocity air fuel-sprayed WC–CoCr coatings, *Materials (Basel)*, vol. 14, no. 11, p. 3074, 2021.
68. D.-J. Liaw, K.-L. Wang, Y.-C. Huang, K.-R. Lee, J.-Y. Lai, and C.-S. Ha, Advanced polyimide materials: Syntheses, physical properties and applications, *Prog. Polym. Sci.*, vol. 37, no. 7, pp. 907–974, 2012.
69. D. L. Burris and W. G. Sawyer, A low friction and ultra low wear rate PEEK/PTFE composite, *Wear*, vol. 261, no. 3–4, pp. 410–418, 2006.
70. B. Praveenkumar and S. Darius Gnanaraj, Case studies on the applications of phenolic resin-based composite materials for developing eco-friendly brake pads, *J. Inst. Eng. (India) Ser. D*, vol. 101, no. 2, pp. 327–334, 2020.
71. M. Fox, Polymer tribology, *Lube Mag.*, vol. 135, pp. 32–37, 2016.
72. Z. Tong, Q. Ma, Y. Ni, T. Mei, J. Li, and G. Dong, Tribological properties of carbon fabric reinforced phenolic-based composites containing CNTs@MoS2 hybrids, *J. Mater. Sci.*, vol. 54, no. 23, pp. 14354–14366, 2019.
73. B. F. Yousif, A. Al-Haddad, J. G. Alotaibi, T. F. Yusaf, Design of film thickness instrument for fibre polymer composites tribological experiments, *J. Eng. Sci. Technol.*, vol. 9, no. 2, pp. 154–166, 2014.
74. P. A. Steinerand and R. Sandor, Polybenzimidazoleprepreg: Improved elevated temperature properties with autoclave processability, *High Perform. Polym.*, vol. 3, pp. 139–150, 1991.
75. T. Yin et al., Molecular dynamics simulation on polymer tribology: A review, *Lubricants*, vol. 12, no. 6, p. 205, 2024.
76. D. M. C. McCarthy and S. B. Glavatskih, Assessment of polymer composites for hydrodynamic journal-bearing applications, *Lubr. Sci.*, vol. 21, no. 8, pp. 331–341, 2009.
77. K. Y. E. Awwad et al., Finite element analysis and experimental validation of polymer–metal contacts in block-on-ring configuration, *Friction*, vol. 12, no. 3, pp. 554–568, 2024.
78. G. S. Divya, R. Keshavamurthy, C. Siddaraju, and K. V. S. Murthy, Investigation on sliding wear properties of nano metallic particle reinforced hybrid composites through design of experiments and ANN, *J. Inst. Eng. (India) Ser. D*, vol. 105, pp. 1551–1562, 2023.

79. Y. Li, S. Wang, B. Arash, and Q. Wang, A study on tribology of nitrile-butadiene rubber composites by incorporation of carbon nanotubes: Molecular dynamics simulations," *Carbon N. Y.*, vol. 100, pp. 145–150, 2016.
80. Q. Ei et al., Evaluating the effects of nanosilica on mechanical and tribological properties of polyvinyl alcohol/polyacrylamide polymer composites for artificial cartilage from an atomic level, *Polymers*, vol. 11, p. 76, 2019.
81. J. Song and G. Zhao, A molecular dynamics study on water lubrication of PTFE sliding against copper, *Tribol. Int.*, vol. 136, pp. 234–239, 2019.
82. A. Rajput, Upma, S. K. Shukla, N. Thakur, A. Debnath, and B. Mangla, Advanced polymeric materials for aerospace applications, in: Inamuddin, T. Altalhi, and S.M. Adnan (Eds.), *Aerospace Polymeric Materials*, River Street Hoboken, NJ: Wiley, pp. 117–136, 2022.
83. R. Anderson, Polymeric materials used in aerospace applications, *J. Mater. Sci. Eng.*, vol. 12, no. 1, 1–2.
84. S. W. Ghori, R. Siakeng, M. Rasheed, N. Saba, and M. Jawaid, The role of advanced polymer materials in aerospace, in: M. Jawaid and M. Thariq (eds.), *Sustainable Composites for Aerospace Applications*, Sawston, Cambridge: Elsevier, pp. 19–34, 2018.
85. W. W. Wright, Polymers in aerospace applications, *Mater. Eng.*, vol. 12, no. 4, pp. 222–227, 1991.
86. S. V. Panin et al., Material design methodology for optimized wear-resistant thermoplastic–matrix composites based on polyetheretherketone and polyphenylenesulfide, *Materials (Basel)*, vol. 13, no. 3, p. 524, 2020.
87. F. Oster, F. Haupert, K. Friedrich, W. Bickle, and M. Müller, Tribologischeho chleistungsbeschichtungenausneuartigenpolyetheretherketon (PEEK)-compounds, *Tribol. Schmier.*, vol. 51, pp. 17–24, 2004.
88. F. Oster, F. Haupert, K. Friedrich, M. Müller, and W. Bickle, Neuartige Polyetheretherketon (PEEK)-Beschichtungen für hohe tribologische Beanspruchungen, *Mater. Werkst.*, vol. 35, pp. 690–695, 2004.
89. A. Noll, K. Friedrich, T. Burkhart, and U. Breuer, Effective multifunctionality of poly(p-phenylenesulfide) nanocomposites filled with different amounts of carbon nanotubes, graphite and short carbon fibers, *Polym. Compos.*, vol. 34, pp. 1405–1412, 2013.
90. A. Mimaroglu, H. Unal, and T. Arda, Friction and wear performance of pure and glass fibre reinforced poly-ether-imide on polymer and steel counterface materials, *Wear*, vol. 262, no. 11–12, pp. 1407–1413, 2007.
91. C. J. Schwartz and S. Bahadur, The role of filler deformability, filler-polymer boding, and counterface material on the tribological behaviour of polyphenylene sulphide (PPS), *Wear*, vol. 250, pp. 1532–1540, 2001.
92. S. Nambiar and J. T. W. Yeow, Polymer-composite materials for radiation protection, *ACS Appl. Mater. Interfaces*, vol. 4, no. 11, pp. 5717–5726, 2012.
93. J. Zhen et al., High temperature friction and wear behavior of PTFE/MoS_2 composites, *Lubricants*, vol. 11, no. 8, p. 312, 2023.
94. D. K. Rajak, D. D. Pagar, P. L. Menezes, and E. Linul, Fiber-reinforced polymer composites: Manufacturing, properties, and applications, *Polymers (Basel)*, vol. 11, no. 10, p. 1667, 2019.

4 An Overview of the Effects of Severe Plastic Deformation Processes on High-Temperature Erosion of Thermally Sprayed Coatings

Parvinkal Singh and Pardeep Kumar

4.1 INTRODUCTION

The demand for engineering materials with wear resistance and higher strength is increasing with advancements in technology. Research on improving productivity and efficiency in the field of material science is needed. The present-day components in various fields of industrial application have been exposed to stringent conditions. These metals and alloys still constitute the most important group of engineering materials. The degradation of materials under severe conditions prevailing in the industrial environment has long emerged as a prime area of research. Material degradation is found to be the major cause of failure of vital industrial components [1,2]. Degradation of material occurs due to exposure to heat, oxidation, or some mechanical actions [3,4]. Mechanical actions such as erosive and abrasive wear, melting wear, adhesive wear, diffusive wear, and fretting wear may lead to the degradation of the material. Exposure to ionizing radiation; corrosion in aqueous media, sulfur, and other halogens; dissolution in alkalis, liquid metals, or salts; and oxidation with oxygen also result in loss of material [5]. Rapid material degradation occurs due to severe conditions, such as high velocity of the mating parts, very high temperatures, and highly erosive environments [6–8]. As per the opinion of the authors, it would not be surprising that the annual cost due to degradation of materials reaches the figure of trillion dollars [9]. In another study, an annual depreciation of 5% was estimated due to material degradation [10]. Researchers have carried out studies to develop new materials and prevent degradation under stringent conditions.

These inflated power demands have led to advancements in the technology used in thermal power generating plants. A stringent environment and aggressive service conditions often result in premature failure of materials due to degradation [11,12].

DOI: 10.1201/9781003542957-4

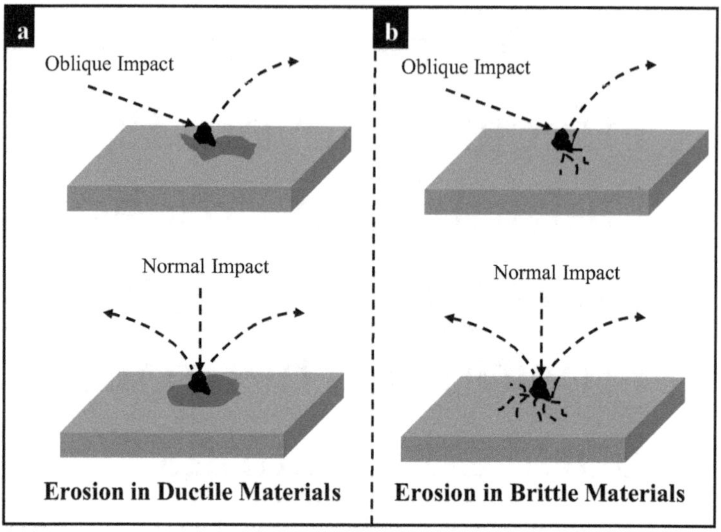

FIGURE 4.1 Erosion behavior of the (a) ductile and (b) brittle materials.

Research related to improving the life of components prevailing in such installations would be of great interest to all researchers. The main cause of industrial component failure has been identified as solid particle erosion (SPE) [13,14]. SPE is described as the gradual loss of material caused by repetitive solid particle impacts on the surface of metals or alloys [15,16]. According to Stein [17], the SPE resulted in a loss of US$1,500 per year. Various erosion mechanisms occur simultaneously and interact with each other. SPE of material is often classified as brittle erosion or ductile erosion, as shown in Figure 4.1. According to Hutchings [18], in brittle erosion, wear is induced by crack initiation and propagation, whereas in ductile erosion, wear is caused by plastic deformation (plowing or cutting). In a study by Islam and Farhat [19], the erosion mechanism was studied at different impact angles. The study concluded that the microplowing mechanism was responsible for erosion at an oblique impact angle, whereas at a high impact angle, the metal was removed by fracture. The removal of material from ductile metals occurs through cutting at low impact angles [20]. Erosion occurs through several mechanisms, such as plastic deformation, fatigue, plowing, microcutting, and particle fragmentation [10].

The material degradation at elevated temperatures results from blunt impact in addition to sliding, plowing, and cutting. The mechanism involves the loss of material from the debris generated by impacting particles that penetrate deeply into the target surface under external impacts [21]. According to the investigations of Zhang et al. [22], based on the impact angle, the material removal mechanisms can be characterized by forging, extrusion, cracking, cutting, scratching, rubbing, adhesion, and sputtering. The cracking and cutting mechanism resulted in the maximum wear rate when the impact angle was between 20° and 30°. A graphical representation of the different material removal mechanisms at different impact angles is presented in Figure 4.2, along with the associated acoustic emission energies [22].

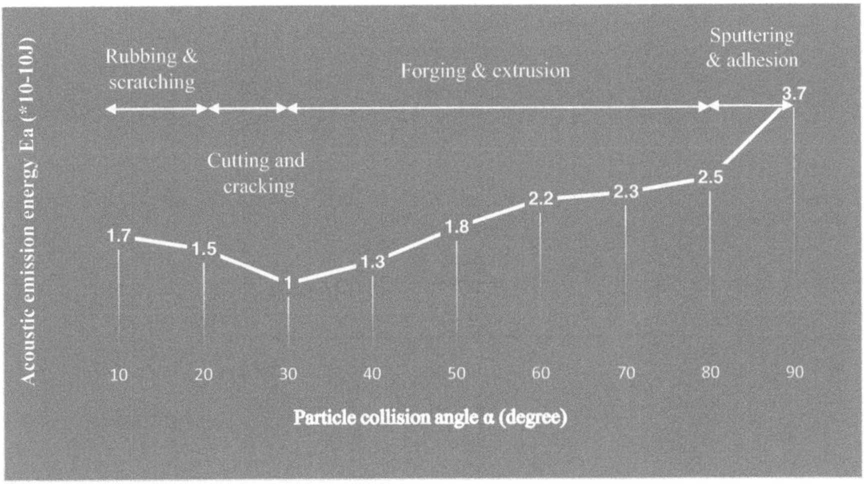

FIGURE 4.2 Various erosion mechanisms at different bedding angles [22].

A drastic change in this erosion trend was observed by Hutchings [23] and Roy and Sundarajan [24] for different shapes of erodent particles. The change in erosion behavior with respect to the change in impact angle also differs for erosion performed at different velocities and temperatures. In a review article, Roy [11] reported that 9Cr-Mo steel eroded maximally at a normal impact angle and at a temperature of 850°C when impacted at 30–70 m/s with round alumina particles (130 μm). However, when erosion testing was performed at a velocity of 20 m/s on the same material under similar conditions, the maximum erosion rate was reached at the oblique impact angle. Similarly, for cobalt, Singh et al. [25] reported that high erosion occurs at a 60° impact angle when wear experiments are performed at 680°C in the velocity range of 70–140 m/s. However, erosion experiments were performed on the same material under similar conditions but at a different velocity of 140 m/s and a temperature of 500°C, with the high erosion rate moved to an impact angle of 30° [26]. An SPE at high temperature is the main reason for downtime in energy generation systems [27]. At high temperatures, oxidation is also responsible for degradation and erosive wear [24]. According to Wellman and Nicholls [28], the combination of both oxidation and erosion causes high material loss in fluidized bed combustors (FBCs) and gas turbine engines. Figure 4.3 shows the effects of the different factors on SPE.

In thermal power plants, replacing damaged tubes is extremely expensive; it accounts for an estimated 56% of overall production expenses [29]. Tillmann and Hagen [30] reported that a higher temperature demand for the production of more energy resulted in a decreased lifetime of austenitic steels. Therefore, preventive measures must be taken against failure and material loss to assure reliability and safety in the operation of materials. Consequently, it is essential to understand the complexity of each type of loss of material. Many components of thermal power plants, such as pulverized coal burner nozzles (PCBNs) and FBCs, reheaters, superheaters, boiler heat exchangers, economizer tube banks, lock hopper

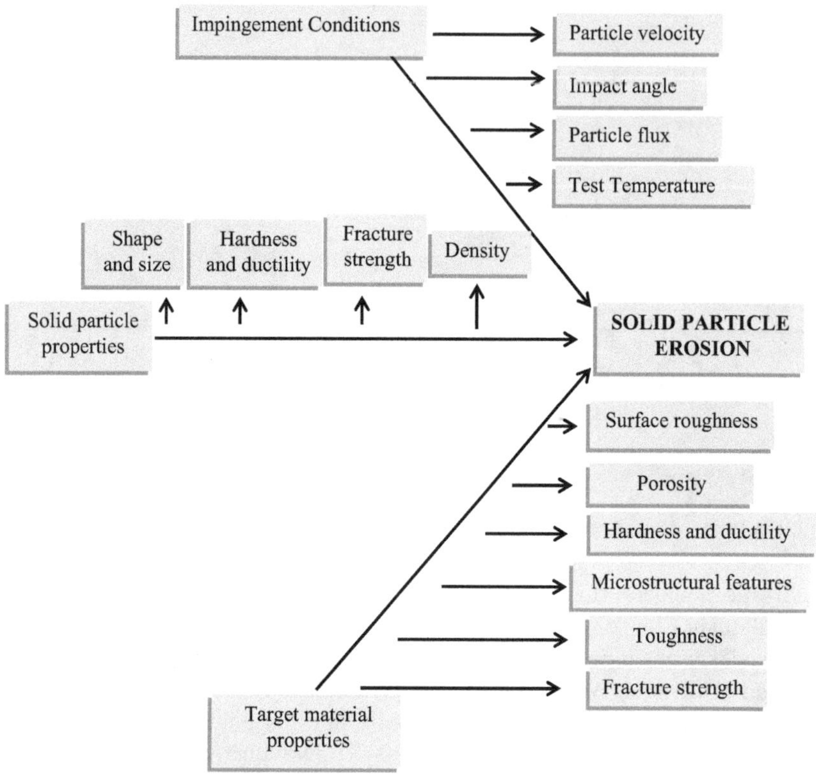

FIGURE 4.3 Factors influencing the erosion behavior in SPE. SPE, solid particle erosion.

valves, steam throttle valves, and steam turbines, are exposed to erosive wear [31–33]. The overall increase in production cost due to rapid degradation is a serious matter of concern.

4.2 PULVERIZED COAL BURNER NOZZLE

Extensive work has been conducted on the high-temperature erosion behavior of erodent particles at low to medium velocities, focused mainly on the protection of boiler tubes. According to Kumar and Sidhu [34], a PCBN is one of the major components and is used to feed pulverized coal-air mixtures to boiler furnaces, as demonstrated in Figure 4.4 and exposed to higher erosion rate than boiler tubes. The inclusion of a significant number of ash particles (approximately 50%) in coal containing hard quartz abrasive material (up to 1.5%) increases the likelihood of erosion of the coal used in thermal power plants [35,36]. In coal-fired thermal plants, a network of pulverized fuel pipes is used to transport pulverized fuel to the boiler [25]. Certain components, such as the burner assembly, pulverized fuel bends and elbows, orifice, and multiple pot outlets, are exposed to high erosive wear. According to Kumar and Sidhu [34], a wear life of 5,000–15,000 hours is observed for these components.

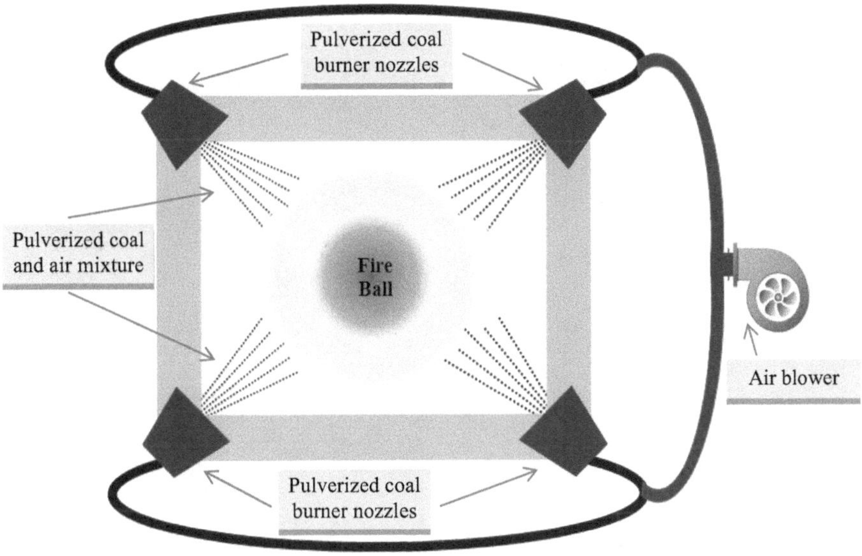

FIGURE 4.4 Plan of the boiler furnace showing the arrangement of the PCBNs. PCBNs, pulverized coal burner nozzles.

Pulverized coal passes through PCBNs with a very high velocity and high mass flux. Both of these factors make the erosion problem more prominent, leading to rapid degradation of the PCBNs.

The useful life of the outer portions of the burner nozzle tips is often destroyed by cracking, burning, and melting due to the heat-affected zones adjacent to the welds. Eroded burner nozzle tips (PCBNs) cause poor fuel distribution at the outlet of the nozzle tip, which results in flame variations [37]. The purely damaged PCBNs at different power generation plants are shown in Figure 4.5. The scanning electron microscopy (SEM) morphology of the eroded SS-304 steel (commonly used material for PCBNs) is shown in Figure 4.6. The penetrating particles initiate the material removal process, and these sites continuously degrade with a striking amount of eroded particles [38,39]. A large number of deep and wide voids are observed in the SEM micrograph of the SS-304 specimen (see Figure 4.6).

4.3 EROSION RESISTANCE

The development of new erosion-resistant materials at elevated temperatures is needed. The proper alloy selection, design aspect, chemical additive, shielding, fuel selection, and protective coatings can be utilized to avoid high-temperature erosion. A study by Chinnadurai and Bahadur [40] revealed that the erosion rate can be inhibited by using coating materials with high hardnesses, low porosities, and complex phase compositions. Currently, the application of many types of protective coatings and enhancements in their performances has been reported at different levels [41,42].

FIGURE 4.5 Damaged pulverized coal burner nozzles at different power generation plants. (*Under CC-4.0 License*).

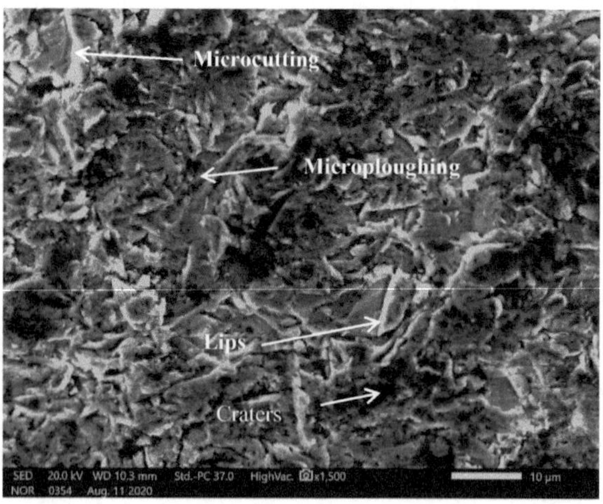

FIGURE 4.6 SEM morphology of eroded SS-304. SEM, scanning electron microscopy.

The coatings must exhibit effective protection against wear to provide economical and effective service behavior [43,44]. The consolidated information from studies related to the SPE behavior of coatings formulated on metallic substrates is discussed in Table 4.1.

A variety of protective coatings can be effectively developed using thermal spraying to reduce the harmful effects of wear in industrial components [45,46]. The basic idea behind the formulation of thermal spray coatings is demonstrated in Figure 4.7. A heat source creates molten or softened droplets of certain materials (metal/polymer/ceramic), which are accelerated and transported by a gas stream to reach the properly prepared surface of the substrate with great kinetic energy. Upon striking the substrate surface, the accelerated droplets became flattened and constructed a mechanical bond. A continuously striking stream of such droplets, which are uniform at different locations on the substrate surface, formulates a lamellar coating layer.

Various metals, ceramics, and cermets can be deposited via thermal spraying, which has been applied in different industries. The selection of an appropriate thermal spray process depends upon a number of factors, such as the type of coating material, the level of coating performance needed, the economics, the size of the component to be coated, and the portability [70]. Chen and Zhou [71] reported that the rapidly developing high-velocity oxy-fuel (HVOF) coating process is an effective method for satisfying the needs of protecting industrial components. HVOF spray processes differ from conventional flame spray processes because combustion occurs inside the nozzle, and the gases are initially at much higher pressures than those used in the flame spray process. These high pressures allow the HVOF process to spray at supersonic rates with improved powder particle heating and melting characteristics [5,72–74].

4.4 COATING DEFECTS

Thermal spray coatings' effectiveness is restricted by the existence of unmelted particles, nonhomogeneous microstructure, splats, and splat boundaries [75–77]. Some of the thermally sprayed coating defects are illustrated in Figure 4.8. The presence of unmelted or partially melted feedstock powder grains is visible in the cross-sectional SEM image of the cermet coating (Figure 4.9a). The thermal spray coatings exhibit pores and voids along with cracks (Figure 4.9b). The porosity present at splat boundaries initiates an erosive attack [77–81]. Thermal spray coatings have limited usefulness due to the interconnected porosity of their structure.

According to investigations performed by Kamal and Jayaganthan [83], the voids and pores found at splat boundaries originate from the spraying process. According to Padmavathi and Sarada [84], different mechanisms introduce two types of pore geometry in coatings. First, lamellar porosity originates from intermittent contact at lamellar or splat boundaries. The other variety of particles is spherical in shape and is the result of the expansion of trapped gases [85,86]. To reduce or minimize coating defects, some work has been conducted on postcoating treatments of HVOF and other thermal spray coatings [77,81,87–90].

TABLE 4.1
Studies Related to the Solid Particle Erosion of Different Coatings

Year	Base Material	Feedstock Material	Coating Technique	Results
2017	Mild steel	Ni-WC	Cold spray	• The 10.5% volume of WC component was insufficient to reinforce Ni for erosion resistance at an oblique impact angle • Smaller WC particles were used to achieve high WC retention in the coating • The coating exhibits ductile erosion behavior and shows better performance at normal impact angle [47]
2013	SS-304	WC-Co-Cr	HVOF	• Compared to standard cermet coating, the erosion resistance of fine-grain WC coating was higher • For lowering the degree of decarburization, the inflight particle temperature was decreased • In comparison to nanostructure WC, conventional cermet coating has inferior hardness and toughness [48]
2021	SS-304	CoCrFeNiSi	Laser cladding	• Single FCC structure was compared to laser cladding CoCrFeNiSi • Co-based alloy coating showed microhardness of 485 Hv, and CoCrFeNiSi coating exhibited 580 Hv • The erosion rate was higher for Co-based alloy coating compared to CoCrFeNiSi coating at all test conditions [49]
2015	AISI-304	NiCrSiB/Al_2O_3	Plasma spraying	• The coating exhibits hardness of 674 Hv and porosity of 2%–4% • The erosion resistance of the coating was found to be 2.5 times better at a 30° impact angle • At both a 30° and 90° impact angle, the coating showed higher erosion resistance [50]
2017	CA6NM	WC-10Co-4Cr	HVAF	• The coating showed less decarburizing of WC due to low porosity and well-bounded homogeneous structure • At 70 m/s impact velocity, both coated and substrate steel demonstrated larger cumulative volume loss than at 35 m/s • The cermet coating exhibits a mixed mode of erosion, whereas the substrate shows ductile behavior during erosion [51]

(Continued)

TABLE 4.1 (*Continued*)
Studies Related to the Solid Particle Erosion of Different Coatings

Year	Base Material	Feedstock Material	Coating Technique	Results
2014	Titanium alloy	[Cr_3C_2-25(Ni20Cr)]	HVOF	• The coating exhibits less semimelted powder particles and lower porosity • Oxidation resistance was followed by formation of nickel and chromium oxides • Both coatings provide the highest oxidation resistance at 800°C • Both coatings show parabolic behavior due to slow oxidation kinetics [52]
2020	SS-316L	WC-Cr_3C_2-Ni	Atmospheric plasma spray (APS) and HVOF	• The erosion rate of SS-316L was two to three times higher compared to APS and HVOF coating • At elevated temperatures, less erosion was due to low degree of decarburization, low porosity, and high intersplat sintering • HVOF consists of dense morphological structure [53]
2017	Mild steel	Yttria-stabilized zirconia (YSZ)-NiCrBSi	Atmospheric plasma spraying	• Higher amounts of NiCrBSi resulted in denser microstructure • The lower porosity obtained in the coating improved erosion resistance consequently • Lower porosity was due to low adhesion of NiCrBSi [54]
2020	Martensitic stainless steel FV520B	Cr_3C_2-NiCr	Shroud plasma spraying and HVOF	• HVOF has a denser coating than SPS • HVOF exhibits better erosion resistance • The SPS coating showed extensive fatigue fracture spalling and big cracks • Maximum erosion occurred at 60° and 75° impingement angles [55]
2015	17-4 PH steel	Nanolayered CrAlTiN and multilayered CrAlTiN-AlTiN	Cathodic arc coating	• Multilayered CrAlTiN-AlTiN and nanolayered CrAlTiN coatings were developed • The CrN coating showed high erosion rate compared to both coatings • CrAlTiN, which has a shorter bilayered period, outperformed its counterparts at a 30° impact angle [56]

(*Continued*)

TABLE 4.1 (*Continued*)
Studies Related to the Solid Particle Erosion of Different Coatings

Year	Base Material	Feedstock Material	Coating Technique	Results
2018	AISI 304	SiC-WC-Cr_3C_2	HVOF	• The coating exhibits average hardness of 1,058 Hv • The porosity was observed to be 3%–4% after formulation of coating • The erosion rate was observed to be increased with impact time • The coating showed better erosion resistance at all test temperatures [57]
2017	Nickel-based superalloy	Cr_3C_2-NiCr	HVOF	• The increase in air pressure leads to gradual increase in volume erosion rate • Volume wear rate increases with decrease in coating thickness • The rate of erosion was higher at 30° impact angle than at 90°, and the highest erosion was seen at 60° impact angle [58]
2021	Mild steel	FeAlCr	Detonation gun	• The coating exhibits brilliant microstructural integrity • Erosion rate of mild steel was decreased four times after coating with iron aluminide • The results conclude that coating can be utilized against erosive wear at 400°C • Spray deposition frequency significantly affects the coating structure [59]
2018	IN718	Thermal barrier coating (TBC) 7 wt.% YSZ	Air plasma spray	• Two sets of different microstructures of coating (19.5% ± 1.2% and 12.9% ± 0.5% porosity) were analyzed at elevated temperature erosion • The high porosity on top layer of coating results in higher erosion rate • The major reason for erosion on porous TBC is the detachment of lamellar structure [60]
2016	2Cr13 steel	Ni60-NiCrMoY	Supersonic atmospheric plasma spraying	• At various impingement angles, the influence of vacuum remelting of as-sprayed coatings on erosion behavior was investigated • Range of impact angles was broadened with vacuum remelting • The bond strength of coated specimen was improved with vacuum remelting of as-sprayed coatings [61]

(*Continued*)

TABLE 4.1 (*Continued*)
Studies Related to the Solid Particle Erosion of Different Coatings

Year	Base Material	Feedstock Material	Coating Technique	Results
2014	13Cr-4Ni	Stellite 6	Laser cladding	• At laser energy densities of 32–52 J/mm^2, stellite cladding was created • Lowest dilution was observed at 32 J/mm^2 • Maximum rise in erosion resistance was observed at 32 J/mm^2 [62]
2020	M2 High speed steel	TiAlSiN/Cr	Direct current and high power impulse magneto sputtering (DCMS) (Hi PIMS)	• The coating has both an amorphous and crystalline phase of TiN • The Cr layer thickness of 21 nm in TiAlSiN/Cr coating showed brilliant erosion resistance with high hardness and toughness • The coating exhibits optimal comprehensive properties [63]
2013	Stainless steel $Cr_{18}Ni_9Ti$	Cr_3C_2-NiCr	HVOF	• The nanocrystalline grains and amorphous phase were observed by transmission electron microscopy • The coating exhibits high microhardness with low porosity and uniform microstructure • Micropores, microcutting, craters, lips, and cracks were observed on the eroded surface [64]
2016	Stainless steel	NiCrAlY	HVOF	• Electrical properties of eroded and as-sprayed samples were measured by impedance spectroscopy • The erosion mainly occurs due to spalling of sprayed splats • With increasing particle mass, the constant phase elements corresponding to microcracks increase, respectively [65]
2015	Ti6242 alloy	Ti_2AlC MAX phase coating	Magnetron sputtering coater	• The MAX phase was formed by annealing the specimens in a vacuum at 800°C for 1 hour • The erosion resistance of Ti6242 can be improved by development of MAX phase • The dominant wear mechanisms were observed to be the plowing and cutting [66]

(*Continued*)

TABLE 4.1 (*Continued*)
Studies Related to the Solid Particle Erosion of Different Coatings

Year	Base Material	Feedstock Material	Coating Technique	Results
2017	MDN321	NiCrAl/WC-Co cenosphere coating	Plasma spraying	• The coating exhibits 71% lower erosion rate than substrate steel • The erosion rate was reduced due to the creation of an oxide layer and the high thermal stability of components • The morphology of eroded surface shows a brittle mode of erosion [67]
2014	TiAl	MCrAlY/ ZrO_2-7 wt.% Y_2O_3	Plasma spraying	• The coating exhibits high porosity with typical lamellar stacking characteristics • After laser remelting, the compactness of coating was improved with decrease in surface lamellar defects • The ductile erosion characteristics with microcutting marks were evident on the coating surface • Erosion failure was mainly attributed to crushing of brittle ceramic coating and spalling of sprayed splats [68]
2016	23-8-N Nitronic steel	WC-10Co-4Cr	Thermal spray	• At two different impact angles, the erosion behavior of both coated steels was determined • The coated steel showed high fracture toughness, less decarburization, and homogeneous structure • The less wear rate was due to optimum substrate properties and well-bounded structure [69]

HVOF, high-velocity oxy-fuel; FCC, face-centered cubic; HVAF, high-velocity air-fuel; SPS, suspension plasma spray.

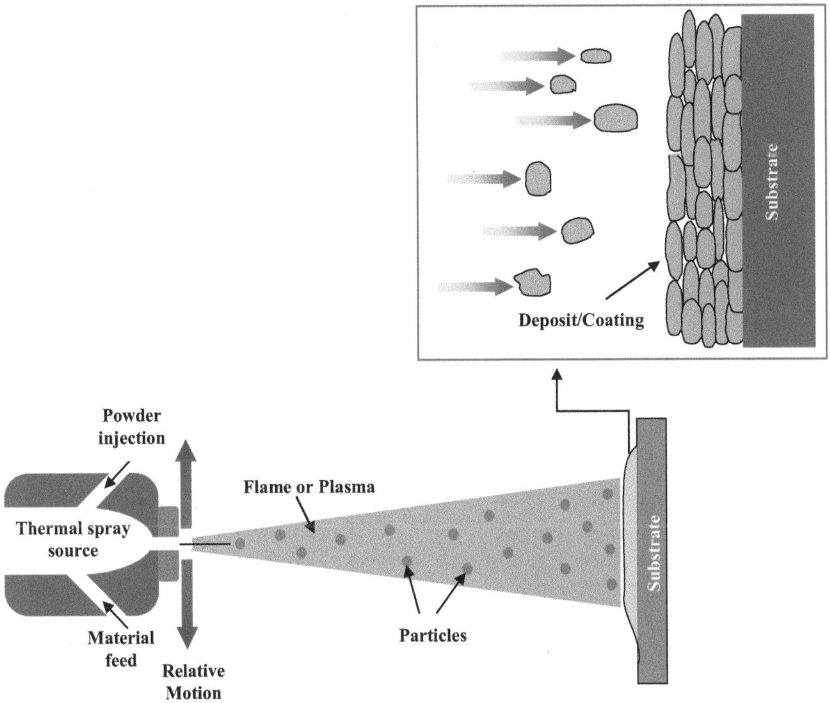

FIGURE 4.7 Schematic diagram of the thermal spraying process.

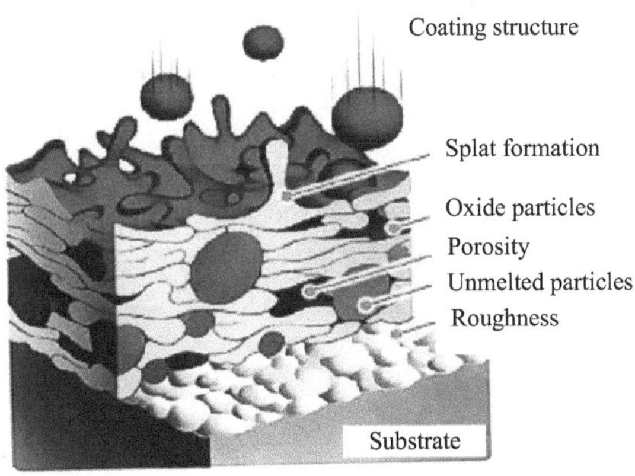

FIGURE 4.8 Defects in thermal spray coatings [82].

4.5 SEVERE PLASTIC DEFORMATION PROCESSING

Both the physical and mechanical states (radial and tangential stresses, microstructure, and microhardness) and the microgeometry (roughness and porosity) have decisive importance in determining the performance of components [92–95]. The

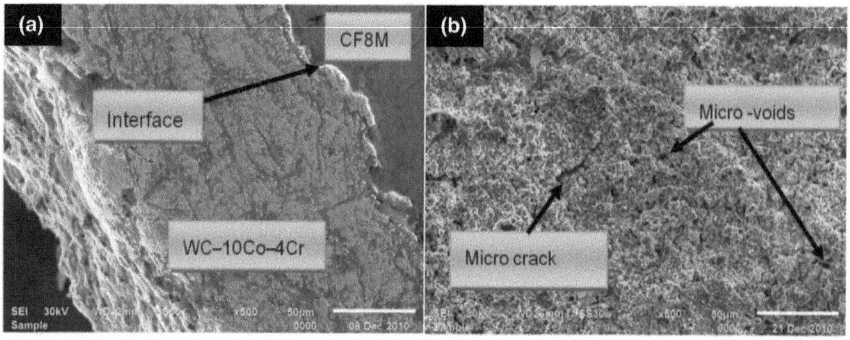

FIGURE 4.9 SEM image of the thermal spray cermet (WC-10Co-4Cr) coating: (a) cross-section and (b) surface [91]. SEM, scanning electron microscopy.

creation of a residual compressive stress field in the superficial layer significantly extends the wear life of components [94,95].

The different experimental techniques of metal forming that may be used to plastically deform materials are referred to as SPD processing. One special characteristic of SPD processing is the high strain applied without causing a noticeable alteration in the workpiece's overall size. Another characteristic that limits the material's free flow and creates a large hydrostatic pressure is the form maintained by employing specific tool geometries. This hydrostatic pressure is necessary to produce significant stresses and introduce high concentrations of lattice defects [96].

The production of high-strength and lightweight parts with environmental harmony is the main objective of SPD processing [97,98]. However, there is still a large gap between the application of SPD processes in large-scale production and on the laboratory scale for fundamental research. Little research has been reported on the enhancement of surface integrity while avoiding time-consuming and expensive postcoating treatment processes.

Numerous SPD procedures have been created to subject bulk metals to extraordinarily high strains without causing them to change form. The typical SPD processes include the following:

- cyclic extrusion compression
- equal channel angular pressing [99]
- super short multipass rolling [100]
- high-pressure torsion [100]
- repetitive corrugation and straightening [101]
- accumulative roll-bonding
- severe torsion straining [102]
- cyclic closed-die forging [103]

By using these procedures, the material's desired qualities can be obtained by managing the grain size, a crucial microstructural component affecting the metallic work materials' mechanical behavior and chemical reaction to their surroundings [104]. However, the conventional forming operations such as rolling, unidirectional extrusion, uniaxial compression, and tension are excluded from SPD processing.

4.6 FRICTION STIR PROCESSING

Friction stir processing (FSP) uses strong, localized plastic deformation to alter a metal's characteristics. Morisada [105] performed FSP on a thermally sprayed cemented carbide (WCCrC-Ni) layer. This process resulted in an approximately 1.5-fold increase in the coating hardness. Furthermore, the bonding between the coating and substrate improved due to the enhanced diffusion of metallic elements during processing. Hodder [106] homogenized alumina powder in an Al matrix produced by cold-gas dynamic spraying employing FSP as a postprocessing technique. Peat et al. [107] assessed how well cold spray-deposited metal-matrix composite coatings performed in terms of erosion following FSP treatment. According to reports, the procedure reduced the volume loss by about 40%.

4.7 BURNISHING PROCESS

Burnishing is an SPD process utilized as a postprocessing surface treatment process that can effectively modify the surface properties of components [108–110]. The burnishing process exerts pressure through a specially designed burnishing tool to plastically deform the surface irregularities on the component surface [111–114]. Burnishing is an SPD process that can generate high strain and strain rates [115]. Skalski et al. [116] reported that the material's yield strength is undoubtedly exceeded by the pressure used during the burnishing process. Consequently, plastic deformation suppresses the surface imperfections and fills the surrounding valleys. Burnishing is performed in different ways by different researchers [45,117–120]. A schematic diagram of the burnishing operation with a hard ball and roller is shown in Figure 4.10. Lin and Wang [121] reported that refining of the microstructure by fragmentation of splat boundaries can be achieved after burnishing. Another useful technique for adding beneficial compressive residual stresses to the surface area is burnishing, which

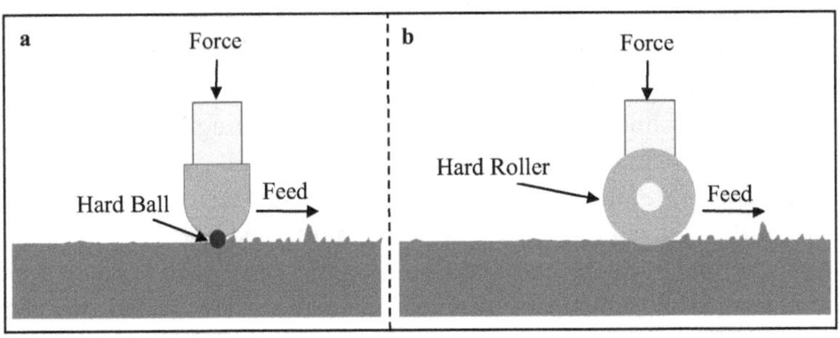

FIGURE 4.10 Schematic illustration of the burnishing mechanism: (a) ball burn and (b) roller burn.

extends the component's wear life [122]. One such technique that produces good performance qualities in the top layer of machined objects is surface plastic formation by burnishing [123]. The research conducted in this field encompasses not only the technology itself but also the issues regarding contact mechanics [124].

It has been observed that the treated surface layers of different metals and alloys exhibit significant grain refinement. Selecting the appropriate set of values for the process parameters of any production activity has always been difficult for production engineers. The selection of the appropriate process parameters has remained a challenge for production engineers. Determining these parameters is crucial for the process to be completed with the intended result in terms of production cost and cycle time.

The burnishing operation has many parameters, the values of which must be chosen precisely. The quality of the burnished surface is governed by many parameters and they include the burnishing force, workpiece rotation speed, feed at which the burnishing operation is carried out, and lubricant use during burnishing. The burnishing process is also influenced by other parameters related to the burnishing tool, such as tool diameter, tool material, and number of tool passes.

Prevey and Cammett [125] observed that compressive residual stresses are generated at sufficient depths, which enhances the life of the components. The shape of the burnishing tool, type of material, and working parameters define the extent of deformation after the burnishing operation. An altered geometric structure, a compressive state of internal stress, and increased hardness are achieved by the burnishing process. Moreover, the depth of the plastic region characterizes the range of these deformations. Hiegemann and Weddeling [126] reported that surface plastic formation by burnishing results in favorable performance in the upper layer of machined parts. Nikitin et al. [127] conducted a study on Ti-6Al-4V and AISI 304 alloys and reported that nanocrystalline regions are stable and are induced during short-term annealing and deep rolling. The nanocrystalline surface layer region was not altered by isothermal fatigue in the low-cycle fatigue regime at high-stress amplitudes up to 600°C. In another study using proper processing conditions [128], burnishing was found to reduce the wear rate by 44% for St-37 steel and 38% for copper. The roller burnishing tool applies pressure through the ball, which results in plastic deformation of the surfaces. Hamadache et al. [129] found improvements of up to 9% in hardness and 89% in wear when compared to a sample of structural Rb40 steel that had been turned after burnishing. In a study, Radziejewska et al. [130] revealed that laser alloying carbon steels with cobalt stellite reduced the wear rate from burnishing by 40%. The surface irregularities are leveled by moving the burnishing tool. Selected control of the rolling parameters is required to achieve a specific surface quality. A micrograph of the thermally sprayed cermet-coated and burnished specimens is shown in Figure 4.11. Compared to the as-coated sample, the burnished surface appears to be rather smooth with relatively few pores. Luca et al. [131] stated that the material being worked on, the rolling parameters (feed and rolling force), and the initial roughness all affect the surface conditions of a workpiece during burnishing. To adjust the surface roughness to particular needs, there are various methods for adjusting the burnishing parameters. Studies from Zhuang and Halford as well

FIGURE 4.11 Thermally sprayed cermet coating surfaces: (a) as-sprayed and (b) burnished [149].

as Hassan and Al-Bsharat [132] demonstrated that the burnishing process creates compressive residual stresses in the surface area, which may lessen the impact of surface cracks.

4.8 FUTURE PERSPECTIVES AND THE LATEST APPROACH FOR COMBATING EROSION FACTORS

4.8.1 Burnishing with Novel Technique

The conventional burnishing operation in most of the studies utilizes a roller burnishing tool. Avilés and Albizuri [133] reported that the cold operation of metallic surfaces ultimately improved the surface finish, increased the microhardness, and increased the wear resistance. The burnishing operation with a rotating burnishing tool can be utilized efficiently because it can provide many benefits over roller burnishing. The rotating burnishing tool generates greater radial and tangential forces on the surface of the coating. The higher forces in turn generate higher compressive residual stresses with increasing temperature on the coating surface [134–136]. The surface roughness and porosity, on the other hand, can be greatly reduced. The splat boundaries will be disrupted and combined together due to the rotation of the burnishing tool (see Figure 4.11), ultimately resulting in an increase in the microhardness of the coating surface. The burnishing forces and speed significantly affect the performance of burnishing tools [137–140]. Instead of a rolling tool, Yang and Dillon [141] used a fixed roller tool for burnishing the circumferential surface of a Co-Cr-Mo alloy. In contrast to roller burnishing, large shear stresses and strains develop on the surface of the workpiece, resulting in improved surface integrity. Such shear stresses may also be helpful if burnishing of the HVOF coating surface is performed in a similar manner. The shear stresses thus produced on the upper uneven coated surface may help in breaking the peaks and filling up the nearby pores. Similar results were also reported by Singh and Kumar [142], who improved the surface integrity of cermet coatings.

The literature reveals that the surface roughness and porosity are reduced after burnishing HVOF thermally sprayed coatings. Furthermore, increases in microhardness and compressive residual stresses have also been reported after the burnishing process. Cooling during the burnishing process is expected to improve the compressive residual stresses and microhardness. Cryogenic burnishing has also been carried out by different researchers [89,120,143–147].

4.8.2 Cryogenic Burnishing

The heat generated during the burnishing operation results in the growth of refined grains after deformation. The use of liquid nitrogen during the burnishing process removes the heat generated during the burnishing process [148,149]. Grain growth after dynamic recrystallization is restricted during cryogenic burnishing [141,150]. The cryogenic burnishing process results in thicker surface layers than normal or elevated temperature processing [143,145,146]. These surface layers exhibit higher hardness and compressive residual stresses. Cryogenic burnishing is a nontoxic, environmentally friendly, and sustainable means to enhance the quality of both manufactured products and the performance of manufacturing processes. The cryogenic burnishing process offers improved surface finish, increased tool life, elimination of conventional coolants, and improved machining performance [151].

Cryogenic burnishing continues to attract the attention of industrial and research communities [152]. In a recent study of cryogenic burnishing of Ti-6Al-4V alloy, Rotella et al. showed increased microhardness in the burnished subsurface. Kaynak et al. [153] made similar observations in terms of the hardness profiles of the surface and subsurface cryogenic burnishing of NiTi alloys. In a study by Pu et al. [154], an 87% increase in hardness was found in the surface layers after cryogenic burnishing. The compressive residual stresses were extended to a depth of around 200 μm beneath the magnesium alloy AZ31B's surface. The corrosion resistance of the selected magnesium alloy was also enhanced as a result of the basal plane's high intensity on the machined surface [153]. Compared to that of dry and minimum quantity lubrication burnishing, the hardness of the surface layer of the cryogenically burnished Ti-6Al-4V alloy increased. The lowest surface roughness was produced by cryogenic burnishing [155].

Huang and Kaynak [144] demonstrate the change in the surface layer during cryogenic burnishing of the Al 7050-T7451 alloy. The effects of cryogenic burnishing on an Al 7050-T7451 alloy were evaluated with a roller burnishing tool using liquid nitrogen as the coolant. In comparison to dry burnishing at corresponding burnishing rates of 25, 50, and 100 m/min, cryogenic burnishing produced average microhardness improvements of 9.5%, 17.5%, and 24.8% within the 200-μm depth. These harder layers with compressive residual stresses were induced after the dry and cryogenic burnishing processes.

Yang and Dillon [141] investigated the effect of cryogenic burnishing on the surface integrity modifications of Co-Cr-Mo biomedical alloys. Burnishing experiments involving the application of liquid nitrogen were conducted on an annealed Co-Cr-Mo alloy using a custom-made burnishing tool. Liquid nitrogen was applied to the workpiece on the flank side of the tool-workpiece contact. The application of liquid nitrogen during the burnishing process significantly curbed the temperature increase within and outside the nitrogen application zone. A better surface finish, a thick burnishing-influenced surface layer, a high hardness, and significant grain refinement were simultaneously achieved with the application of cryogenic cooling. The results showed that cryogenic burnishing can be effectively utilized for modifying the surface properties of Co-Cr-Mo biomedical alloys.

4.9 CONCLUSION

SPE of various components of thermal power plants cannot be totally avoided under extreme conditions, but surface coatings can be used to extend the working life of materials. An HVOF thermally sprayed cermet (WC-Co-Cr) coating displayed better performance against the deleterious effects of erosion. However, thermally sprayed coatings exhibit higher surface roughness along with porosity, voids, and cracks on the surface, which initiate erosive attacks. It can be concluded from the available studies that SPD processes can efficiently modify the surface irregularities of thermally sprayed coatings. This review paper presents the benefits of the novel technique of burnishing with a rotating tool instead of traditional roller burnishing. The rotating burnishing tool generates greater burnishing forces with increasing temperature on the coating surface. After burnishing with the rotating tool, the splat boundaries were broken and fused together, which resulted in an increase in the microhardness of the coating surface. Cryogenic burnishing with the application of liquid nitrogen is also reported to be beneficial for generating ultrafine grains on the surface of coatings because the heat generated while burnishing quickly dissipates. The physical and mechanical properties of the coating increase after surface treatment with cryogenic burnishing.

REFERENCES

1. Prozhega, M.V., et al., Experimental study of erosion-corrosion wear of materials: a review. *Journal of Friction and Wear*, 2014. **35**(2): 155–160.
2. Brownlie, F., et al., A study on the erosion-corrosion behavior of engineering materials used in the geothermal industry. *Wear*, 2021. **477**: 203821.
3. Anand, K. and H. Conard, Local impact damage and erosion mechanisms in WC-6wt.% Co alloys. *Materials Science and Engineering*, 1988. **A 105/106**: 411–421.
4. Barber, J., B.G. Mellor, and R.J.K. Wood, The development of subsurface damage during high energy solid particle erosion of a thermally sprayed WC–Co–Cr coating. *Wear*, 2005. **259**(1–6): 125–134.
5. Bala, N., H. Singh, and S. Prakash, High-temperature oxidation studies of cold-sprayed Ni–20Cr and Ni–50Cr coatings on SAE 213-T22 boiler steel. *Applied Surface Science*, 2009. **255**(15): 6862–6869.
6. Kumar, M. and R. Upadhyaya, Corrosion and wear analysis of HVOF sprayed WC-10CO-4Cr coating on geothermal turbine blade. *World Journal of Engineering*, 2019. **16**(6): 768–774.
7. Nitesh Vashishtha, R.K.K. and S.G. Sapate, Tribological behavior of HVOF sprayed WC12Co, WC-10Co-4Cr–25NiCr coating. *Tribology International*, 2016. **23**(2): 15–21.
8. Kumar, P. and B.S. Sidhu, Characterization and high-temperature erosion behavior of HVOF thermal spray cermet coatings. *Journal of Materials Engineering and Performance*, 2016. **25**(1): 250–258.

9. Chatha, S.S., H.S. Sidhu, and B.S. Sidhu, High temperature hot corrosion behavior of NiCr and Cr_3C_2–NiCr coatings on T91 boiler steel in an aggressive environment at 750°C. *Surface and Coatings Technology*, 2012. **206**(19): 3839–3850.
10. Gat, N. and W. Tabakoff, Some effects of temperature on the erosion of metals. *Wear*, 1978. **50**(1): 85–94.
11. Roy, M., K.K. Ray, and G. Sundararajan, An analysis of the transition from metal erosion to oxide erosion. *Wear*, 1998. **217**(2): 312–320.
12. Bugge, J., S. Kjær, and R. Blum, High-efficiency coal-fired power plants development and perspectives. *Energy*, 2006. **31**(10–11): 1437–1445.
13. Kumar, P. and B.S. Sidhu, Characterization and high-temperature erosion behavior of HVOF thermal spray cermet coatings. *Journal of Materials Engineering and Performance*, 2015. **25**(1): 250–258.
14. Feng, C., et al., Study of solid particle erosion wear resistance of WC–Co cemented carbide. *Journal of Failure Analysis and Prevention*, 2020. **20**(2): 543–554.
15. Patel, M., et al., Study of solid particle erosion behavior of SS 304 at room temperature. *Procedia Technology*, 2016. **23**: 288–295.
16. Li, G., et al., Numerical simulation of predicting and reducing solid particle erosion of solid–liquid two-phase flow in a choke. *Petroleum Science*, 2009. **6**(1): 91–97.
17. Kevin, J., B.S.S., Stein, and A.R. Marder, Erosion of thermal spray MCr–Cr_3C_2 cermet coatings. *Wear*, 1999. **224**: 153–159.
18. Hutchings, I.M., Ductile–Brittle transitions and wear maps for the erosion and abrasion of brittle materials. *Journal of Physics D: Applied Physics*, 1992. **25**: A212–A22.
19. Islam, M.A. and Z.N. Farhat, Effect of impact angle and velocity on erosion of API X42 pipeline steel under high abrasive feed rate. *Wear*, 2014. **311**(1): 180–190.
20. Finnie, I., G.R. Stevick, and J.R. Ridgely, The influence of impingement angle on the erosion of ductile metals by angular abrasive particles. *Wear*, 1992. **152**(1): 91–98.
21. Zhou, J. and B. Bahadur, Erosion-corrosion of Ti-6Al-4 V in elevated temperature air environment. *Wear*, 1995. **186–187**: 332–339.
22. Zhang, L., V. Sazonov, J. Kent, T. Dixon, and V. Novozhilov, Analysis of boiler-tube erosion by the technique of acoustic emission: part I. Mechanical erosion. *Wear*, 2001. **250**(1–12): 762–769.
23. Hutchings, I.M., Friction and wear of ceramics. *Tribology International*, 1995. **28**(6): 421–422.
24. Sundararajan, G., M. Roy, and B. Venkataraman, Erosion efficiency-a new parameter to characterize the dominant erosion micromechanism. *Wear*, 1990. **140**(2): 369–381.
25. Singh, J., S. Kumar, and S.K. Mohapatra, Tribological analysis of WC–10Co–4Cr and Ni–$20Cr_2O_3$ coating on stainless steel 304. *Wear*, 2017. **376–377**: 1105–1111.
26. Kumar, R.K., et al., Effect of spray particle velocity on cavitation erosion resistance characteristics of HVOF and HVAF processed 86WC-10Co4Cr hydro turbine coatings. *Journal of Thermal Spray Technology*, 2016. **25**(6): 1217–1230.
27. Nagaraj, T.A., et al., Solid particle erosion behavior of cast CY5SnBiM at room temperature. *Journal of the Institution of Engineers (India): Series D*, 2019. **100**(1): 91–96.
28. Wellman, R.G. and J.R. Nicholls, High temperature erosion–oxidation mechanisms, maps and models. *Wear*, 2004. **256**(9–10): 907–917.
29. Hidalgo, V.H., et al., A comparative study of high-temperature erosion wear of plasma-sprayed NiCrBSiFe and WC–NiCrBSiFe coatings under simulated coal-fired boiler conditions. *Tribology International*, 2001. **34**(3): 161–169.
30. Tillmann, W., et al., Adhesion of HVOF-sprayed WC-Co coatings on 316L substrates processed by SLM. *Journal of Thermal Spray Technology*, 2020. **29**(6): 1396–1409.
31. Sidhu, B.S. and S. Prakash, Erosion-corrosion of plasma as sprayed and laser remelted Stellite-6 coatings in a coal fired boiler. *Wear*, 2006. **260**(9–10): 1035–1044.

32. Chawla, V., et al., Hot corrosion & erosion problems in coal based power plants in India and possible solutions – a review. *Journal of Minerals and Materials Characterization and Engineering*, 2011. **10**: 367–386.
33. Taherkhani, B., et al., Erosion due to solid particle impact on the turbine blade: experiment and simulation. *Journal of Failure Analysis and Prevention*, 2019. **19**(6): 1739–1744.
34. Kumar, P. and B.S. Sidhu, Degradation of pulverized coal burner nozzles: a review. *Indian Journal of Engineering, Science and Technology*, 2010. **4**(1): 63–66.
35. Krishnamoorthy, P.R., S. Seetharamu, and P. Sampathkumaran, Influence of the mass flux and impact angle of the abrasive on the erosion resistance of materials used in pulverized fuel bends and other components in thermal power stations. *Wear*, 1993. **165**(2): 151–157.
36. Kim, J.J., et al., Solid particle erosion of Inconel 625 at elevated temperatures. *Journal of Materials Science Letters*, 2000. **19**(9): 759–761.
37. Krishnamoorthy, P.R., S. Seetharamu, and P. Sampathkumaran, Influence of the mass flux and impact angle of the abrasive on the erosion resistance of materials used in pulverized fuel bends and other components in thermal power stations. *Wear*, 1993. **165**: 151–157.
38. Levin, B.F., J.N. DuPont, and A.R. Marder, Solid particle erosion resistance of ductile wrought superalloys and their weld overlay coatings. *Journal of Materials Science*, 1998. **33**(8): 2153–2163.
39. Alidokht, S.A., et al., Effect of microstructure and properties of Ni-WC composite coatings on their solid particle erosion behavior. *Journal of Materials Engineering and Performance*, 2019. **28**(3): 1532–1543.
40. Chinnadurai, S. and S. Bahadur, High-temperature erosion of Haynes and Waspaloy: effect of temperature and erosion mechanisms. *Wear*, 1995. **186–187**: 299–305.
41. Cabral, T.S., et al., Influence of a cobalt-based wire injection in austenitic coating deposited via CW-GMAW. *Journal of the Brazilian Society of Mechanical Sciences and Engineering*, 2018. **40**(9): 461.
42. Cvijović, I., M.T. Jovanović, and D. Peruško, Cyclic oxidation behavior of Ti3Al-based alloy with Ni–Cr protective layer. *Corrosion Science*, 2008. **50**(7): 1919–1925.
43. Fernández-Lima, F.A., et al., Structural characterization of chemically deposited PbS thin films. *Materials Science and Engineering: B*, 2007. **136**(2–3): 187–192.
44. Panziera, R.C., et al., Study of the effects of the laser remelting process on the microstructure and properties of the WC–10Co–4Cr coating sprayed by HVOF. *Journal of the Brazilian Society of Mechanical Sciences and Engineering*, 2020. **42**(3): 119.
45. Tillmann, W., et al., Thermally sprayed finestructured WC-12Co coatings finished by ball burnishing and grinding as an innovative approach to protect forming tools against wear. *Surface and Coatings Technology*, 2015. **268**: 134–141.
46. Jalali Azizpour, M., Evaluation of through thickness residual stresses in thermal sprayed WC–Co coatings. *Journal of the Brazilian Society of Mechanical Sciences and Engineering*, 2017. **39**(2): 613–620.
47. Ahmad Alidokht, S., et al., Erosive wear behavior of cold-sprayed Ni-WC composite coating. *Wear*, 2017. **376–377**: 566–577.
48. Thakur, L. and N. Arora, A comparative study on slurry and dry erosion behavior of HVOF sprayed WC–CoCr coatings. *Wear*, 2013. **303**(1–2): 405–411.
49. Zhang, S., et al., Microstructure and high temperature erosion behavior of laser cladded CoCrFeNiSi high entropy alloy coating. *Surface and Coatings Technology*, 2021. **417**: 127218.
50. Praveen, A.S., et al., Erosion wear behavior of plasma sprayed NiCrSiB/Al_2O_3 composite coating. *International Journal of Refractory Metals and Hard Materials*, 2015. **52**: 209–218.

51. Hamilton, A., A. Sharma, and U. Pandel, Solid particle erosion resistance of HVAF-sprayed WC-10Co-4Cr coating on CA6NM steel. *Surface Review and Letters*, 2017. **24**(Supp01): 1850011.
52. Jegadeeswaran, N., M.R. Ramesh, and K.U. Bhat, Oxidation resistance HVOF sprayed coating 25% (Cr_3C_2-25(Ni20Cr)) + 75%NiCrAlY on titanium alloy. *Procedia Materials Science*, 2014. **5**: 11–20.
53. Bhosale, D.G., et al., High temperature solid particle erosion behavior of SS 316L and thermal sprayed WC-Cr3C2–Ni coatings. *Wear*, 2020. **462–463**: 203520.
54. Wang, Y., et al., APS prepared NiCrBSi-YSZ composite coatings for protection against cavitation erosion. *Journal of Alloys and Compounds*, 2017. **699**: 1095–1103.
55. Zhang, X., et al., Comparison on multiangle erosion behavior and mechanism of Cr3C2-NiCr coatings sprayed by SPS and HVOF. *Surface and Coatings Technology*, 2020. **403**: 126366.
56. Yang, Q. and R. McKellar, Nanolayered CrAlTiN and multilayered CrAlTiN–AlTiN coatings for solid particle erosion protection. *Tribology International*, 2015. **83**: 12–20.
57. Rao, K.V.S., et al., Erosion behavior of HVOF sprayed SiC-WC-Cr3C2 multilayer coating on 304 stainless steel. *Materials Today: Proceedings*, 2018. **5**(11): 24685–24690.
58. Hong, S., et al., Cavitation erosion characteristics at various flow velocities in NaCl medium of carbide-based cermet coatings prepared by HVOF spraying. *Ceramics International*, 2020. **47**: 1929–1939.
59. Vijaya Lakshmi, D., et al., Corrosion and erosion behavior of iron aluminide (FeAl(Cr)) coating deposited by detonation spray technique. *Advanced Powder Technology*, 2021. **32**: 2192–2201.
60. Shin, D. and A. Hamed, Influence of micro–structure on erosion resistance of plasma sprayed 7YSZ thermal barrier coating under gas turbine operating conditions. *Wear*, 2018. **396–397**: 34–47.
61. Wen, Z.H., et al., Effect of vacuum remelting on the solid particles erosion behavior of Ni60-NiCrMoY composite coatings prepared by plasma spraying. *Vacuum*, 2016. **134**: 73–82.
62. Singh, R., et al., Laser cladding of Stellite 6 on stainless steel to enhance solid particle erosion and cavitation resistance. *Surface and Coatings Technology*, 2014. **251**: 87–97.
63. Gu, J., et al., Improvement of solid particle erosion and corrosion resistance using TiAlSiN/Cr multilayer coatings. *Surface and Coatings Technology*, 2020. **402**: 126270.
64. Hong, S., et al., Microstructure and cavitation–silt erosion behavior of high-velocity oxygen–fuel (HVOF) sprayed Cr_3C_2–NiCr coating. *Surface and Coatings Technology*, 2013. **225**: 85–91.
65. Qu, Z., et al., An investigation on erosion behavior of nanostructured 7YSZ coatings at elevated temperature. *Surface and Coatings Technology*, 2016. **299**: 129–134.
66. Naveed, M., et al., Study of high velocity solid particle erosion behavior of Ti2AlC MAX phase coatings. *Wear*, 2015. **342–343**: 391–397.
67. Mathapati, M., M.R. Ramesh, and M. Doddamani, High temperature erosion behavior of plasma sprayed NiCrAlY/WC-Co/cenosphere coating. *Surface and Coatings Technology*, 2017. **325**: 98–106.
68. Wang, D., et al., Effects of laser remelting on microstructure and solid particle erosion characteristics of ZrO_2–7wt%Y_2O_3 thermal barrier coating prepared by plasma spraying. *Ceramics International*, 2014. **40**(6): 8791–8799.
69. Kumar, A., A. Sharma, and S.K. Goel, Erosion behavior of WC–10Co–4Cr coating on 23-8-N nitronic steel by HVOF thermal spraying. *Applied Surface Science*, 2016. **370**: 418–426.
70. Mao, Z., et al., Dry abrasion property of TiN-matrix coating deposited by reactive high velocity oxygen fuel (HVOF) spraying. *Journal of Coatings Technology and Research*, 2010. **7**(2): 253–259.

71. Chen, J., et al., Erratum to: Microstructural characterization and tribological behavior of HVOF sprayed NiMoAl coating from 20 to 800°C. *Journal of Thermal Spray Technology*, 2015. **24**(3): 281–281.
72. Matthews, S., B. James, and M. Hyland, High temperature erosion of Cr_3C_2-NiCr thermal spray coatings — the role of phase microstructure. *Surface and Coatings Technology*, 2009. **203**(9): 1144–1153.
73. Guilemany, J.M., et al., Microstructure characterization of WC-Ni coatings obtained by HVOF thermal spraying. *Scripta Metallurgica et Materialia*, 1995. **33**(1): 55–61.
74. Espallargas, N., et al., Cr_3C_2–NiCr and WC–Ni thermal spray coatings as alternatives to hard chromium for erosion–corrosion resistance. *Surface and Coatings Technology*, 2008. **202**(8): 1405–1417.
75. Stewart, S., R. Ahmed, and T. Itsukaichi, Rolling contact fatigue of posttreated WC–NiCrBSi thermal spray coatings. *Surface and Coatings Technology*, 2005. **190**(2–3): 171–189.
76. Dragos, U., et al., Improvement of the oxidation behavior of electron beam remelted MCrAlY coatings. *Solid State Sciences*, 2005. **7**(4): 459–464.
77. Rani, M., et al., Postprocessing of Ni-Cr-Al_2O_3 thermal spray coatings through friction stir processing for enhanced erosion–corrosion performance. *Journal of Thermal Spray Technology*, 2019. **28**(7): 1466–1477.
78. Uusitalo, M.A., P.M.J. Vuoristo, and T.A. Mäntylä, High temperature corrosion of coatings and boiler steels below chlorine-containing salt deposits. *Corrosion Science*, 2004. **46**(6): 1311–1331.
79. Venkatesh, L., et al., Microstructural response of various chromium carbide based coatings to erosion and nano impact testing. *Wear*, 2017. **386–387**: 72–79.
80. Robertson, A.L. and K.W. White, Microscale fracture mechanisms of a Cr_3C_2-NiCr HVOF coating. *Materials Science and Engineering: A*, 2017. **688**: 62–69.
81. Revankar, G.D., et al., Wear resistance enhancement of titanium alloy (Ti–6Al–4V) by ball burnishing process. *Journal of Materials Research and Technology*, 2017. **6**(1): 13–32.
82. Majumdar, J.D., Thermal and cold spraying technology in manufacturing, in *Handbook of Manufacturing Engineering and Technology*, A.Y.C. Nee, Editor. 2015, Springer London: London. pp. 2805–2850.
83. Kamal, S., R. Jayaganthan, and S. Prakash, High temperature oxidation studies of detonation-gun-sprayed Cr3C2–NiCr coating on Fe- and Ni-based superalloys in air under cyclic condition at 900°C. *Journal of Alloys and Compounds*, 2009. **472**(1–2): 378–389.
84. Padmavathi, G., et al., Effects of high velocity oxy fuel thermal spray coating on mechanical and tribological properties of materials–a review. *Materials Today: Proceedings*, 2020. **27**: 2152–2157.
85. Gorlach, I.A., A new method for thermal spraying of Zn–Al coatings. *Thin Solid Films*, 2009. **517**(17): 5270–5273.
86. Katsich, C., et al., Erosive wear of hardfaced Fe–Cr–C alloys at elevated temperature. *Wear*, 2009. **267**(11): 1856–1864.
87. Masoumi, H., S.M. Safavi, and M. Salehi, Grinding force, specific energy and material removal mechanism in grinding of HVOF-sprayed WC–Co–Cr coating. *Materials and Manufacturing Processes*, 2014. **29**(3): 321–330.
88. Masoumi, H., et al., Effect of grinding on the residual stress and adhesion strength of HVOF thermally sprayed WC–10Co–4Cr coating. *Materials and Manufacturing Processes*, 2014. **29**(9): 1139–1151.
89. Pu, Z., et al., Ultrafine-grained surface layer on Mg–Al–Zn alloy produced by cryogenic burnishing for enhanced corrosion resistance. *Scripta Materialia*, 2011. **65**(6): 520–523.
90. Maximov, J.T., et al., Spherical motion burnishing implemented on lathes. *International Journal of Machine Tools and Manufacture*, 2009. **49**(11): 824–831.

91. Kumar Goyal, D., et al., Slurry erosion behavior of HVOF sprayed WC–10Co–4Cr and Al_2O_3+13TiO_2 coatings on a turbine steel. *Wear*, 2012. **289**: 46–57.
92. Valiev, R.Z., et al., Correction to: Producing bulk ultrafine-grained materials by severe plastic deformation. *JOM*, 2020. **72**(9): 3304–3304.
93. Estrin, Y. and H.S. Kim, Modeling microstructure evolution toward ultrafine crystallinity produced by severe plastic deformation. *Journal of Materials Science*, 2008. **43**(3): 1174–1174.
94. H., S., The effect on fatigue life of residual stress and surface hardness resulting from different cutting conditions of 0.45% C steel. *International Journal of Machine Tools and Manufacture*, 2005. **45**(2): 131–136.
95. A. Javidi, U. Rieger, and W. Eichlseder, The effect of machining on the surface integrity and fatigue life. *International Journal of Fatigue*, 2008. **30**(10–11): 2050–2055.
96. Valiev, R.Z., et al., Producing bulk ultrafine-grained materials by severe plastic deformation. *Journal of Material*, 2006. **58**: 33–39.
97. Shi, Y.N. and Z. Han, Tribological behaviors of nanostructured surface layer processed by means of surface mechanical attrition treatment. *Key Engineering Materials*, 2008. **384**: 321–334.
98. Wang, Z.B., et al., Effect of surface nanocrystallization on friction and wear properties in low carbon steel. *Materials Science and Engineering: A*, 2003. **352**(1–2): 144–149.
99. Segal, V.M., Severe plastic deformation: simple shear versus pure shear. *Materials Science and Engineering: A*, 2002. **338**(1): 331–344.
100. Valiev, R.Z., N.A. Krasilnikov, and N.K. Tsenev, Plastic deformation of alloys with submicron-grained structure. *Materials Science and Engineering: A*, 1991. **137**: 35–40.
101. Huang, J.Y., et al., Microstructures and dislocation configurations in nanostructured Cu processed by repetitive corrugation and straightening. *Acta Materialia,* 2001. **49**(9): 1497–1505.
102. Korbel, A., F. Dobrzanski, and M. Richert, Strain hardening of aluminum at high strains. *Acta Metallurgica*, 1983. **31**(2): 293–298.
103. Neishi, K., et al., Low-temperature superplasticity in a Cu–Zn–Sn alloy processed by severe plastic deformation. *Materials Science and Engineering: A*, 2001. **307**(1): 23–28.
104. Xiong, Y., et al., Influence of compressive plastic zone at the crack tip upon fatigue crack propagation. *International Journal of Fatigue*, 2008. **30**(1): 67–73.
105. Morisada, Y., et al., Modification of thermally sprayed cemented carbide layer by friction stir processing. *Surface and Coatings Technology*, 2010. **204**(15): 2459–2464.
106. Hodder, K.J., et al., Fabrication of aluminum–alumina metal matrix composites via cold gas dynamic spraying at low pressure followed by friction stir processing. *Materials Science and Engineering: A*, 2012. **556**: 114–121.
107. Peat, T., et al., The erosion performance of cold spray deposited metal matrix composite coatings with subsequent friction stir processing. *Applied Surface Science*, 2017. **396**: 1635–1648.
108. Singh, P. and P. Kumar, An overview of biomedical materials and techniques for better functional performance, life, sustainability and biocompatibility of orthopedic implants. *Indian Journal of Engineering, Science and Technology*, 2018. **11**(28): 1–7.
109. Luo, H., et al., The effect of burnishing parameters on burnishing force and surface microhardness. *The International Journal of Advanced Manufacturing Technology*, 2006. **28**(7): 707–713.
110. Başak, H., S. Özkan, and A. Taşkesen, Application of burnishing process on friction stir welding and investigation of the effect of burnishing process on the surface roughness, hardness and strength. *Experimental Techniques*, 2011. **35**(1): 8–16.
111. Sánchez Egea, A.J., et al., Joining metrics enhancement when combining FSW and ball-burnishing in a 2050 aluminum alloy. *Surface and Coatings Technology*, 2019. **367**: 327–335.

112. Khalilpourazary, S. and J. Salehi, How alumina nanoparticles impact surface characteristics of Al7175 in roller burnishing process. *Journal of Manufacturing Processes*, 2019. **39**: 1–11.
113. Ulhe, P.N., U.D. Patil, and C.R. Patil, Optimization of roller burnishing process parameters on surface roughness using response surface methodology. *Materials Today: Proceedings*, 2019. **18**: 3632–3637.
114. Jagadeesh, G.V. and S. Gangi Setti, An experimental study on the surface finish of ball burnished magnesium (rare earth base) alloy. *Materials Today: Proceedings*, 2019. **18**: 4711–4716.
115. Prevey, P. and J. Cammett, Low cost corrosion damage mitigation and improved fatigue performance of low plasticity burnished 7075T6. *Journal of Materials Engineering and Performance*, 2001. **10**: 548–555.
116. Skalski, K., A. Morawski, and W. Przybylski, Analysis of contact elastic–plastic strains during the process of burnishing. *International Journal of Mechanical Sciences*, 1995. **37**(5): 461–472.
117. Tadic, B., et al., Using specially designed high-stiffness burnishing tool to achieve high-quality surface finish. *The International Journal of Advanced Manufacturing Technology*, 2013. **67**(1): 601–611.
118. Su, H., et al., Surface characteristics and corrosion behavior of TC11 titanium alloy strengthened by ultrasonic roller burnishing at room and medium temperature. *Journal of Materials Research and Technology*, 2020. **9**(4): 8172–8185.
119. Korzynski, M., A. Pacana, and J. Cwanek, Fatigue strength of chromium coated elements and possibility of its improvement with slide diamond burnishing. *Surface and Coatings Technology*, 2009. **203**(12): 1670–1676.
120. Sachin, B., S. Narendranath, and D. Chakradhar, Effect of working parameters on the surface integrity in cryogenic diamond burnishing of 17-4 PH stainless steel with a novel diamond burnishing tool. *Journal of Manufacturing Processes*, 2019. **38**: 564–571.
121. Lin, Y.C., S.W. Wang, and H.Y. Lai, The relationship between surface roughness and burnishing factor in the burnishing process. *The International Journal of Advanced Manufacturing Technology*, 2004. **23**(9): 666–671.
122. Thorat, S.R. and A.G. Thakur, Analysis of surface roughness and wear resistance in low plasticity burnishing process using multiobjective optimization technique. Materials Today: Proceedings, 2020. 41: 1082–1088.
123. R. Lee, Roller super finishing-a new method of imparting high quality surface finish. Engineering Designs, 1961. **4**.
124. Gladwell, G.M.L., Contact problems in classical theory of elasticity. *Sijthoff and Noordhoff*, 1980. **6**: 1–36.
125. Prevey, P. and J. Cammett, The influence of surface enhancement by low plasticity burnishing on the corrosion fatigue performance of AA7075-T6. *International Journal of Fatigue*, 2004. **26**(9): 975–982.
126. Hiegemann, L., et al., Analytical prediction of roughness after ball burnishing of thermally coated surfaces. *Procedia Engineering*, 2014. **81**: 1921–1926.
127. Nikitin, I., I. Altenberger, H.J. Maier, and B. Scholtes, Mechanical and thermal stability of mechanically induced near surface nanostructures. *Materials Science and Engineering: A*, 2005. **9**(403): 318–327.
128. El-Tayeb, N.S.M., G.M., The mechanical wear behavior of burnishing surfaces. *Materials Science and Engineering: A*, 1993. **9**: 19–209.
129. Hamadache, H., L. Laouar, N.E. Zeghib, and K. Chaoui, Characteristics of Rb40 steel superficial layer under ball and roller burnishing. *Journal of Materials Processing Technology*, 2006. **180**(1–3): 130–136.
130. Radziejewska, J., N.B., W. Kalita, Hybrid method for modification of surface layer. *Proceedings of the 13th International Scientific Conference CO-MAT-TECH*, Trnava, Slovakia, 2005, pp. 111–116.

131. Luca, L., S. Neagu-Ventzel, and I. Marinescu, Effects of working parameters on surface finish in ball-burnishing of hardened steels. *Precision Engineering*, 2005. **29**(2): 253–256.
132. Hassan, A.M. and A.S. Al-Bsharat, Influence of burnishing process on surface roughness, hardness, and microstructure of some nonferrous metals. *Wear*, 1996. **199**(2): 1–8.
133. Avilés, R., et al., Influence of low-plasticity ball burnishing on the high-cycle fatigue strength of medium carbon AISI 1045 steel. *International Journal of Fatigue*, 2013. **55**: 230–244.
134. Tugay, I.O., A. Hosseinzadeh, and G.G. Yapici, Hardness and wear resistance of roller burnished 316L stainless steel. Materials Today: Proceedings, 2021. **47**: 2405–2409.
135. Zhu, Y.-X., et al., A burnished and Al-alloyed magnesium surface with improved mechanical and corrosion properties. *Corrosion Science*, 2021. **184**: 109395.
136. Ramesh, S., et al., Surface modification of multidirectional forged biodegradable Mg-Zn alloy by ball burnishing process: modeling and analysis using deep neural network. *Journal of Manufacturing Processes*, 2021. **68**: 423–434.
137. Travieso-Rodríguez, J.A., et al., Hardening effect and fatigue behavior enhancement through ball burnishing on AISI 1038. *Journal of Materials Research and Technology*, 2019. **8**(6): 5639–5646.
138. Swirad, S. and R. Wdowik, Determining the effect of ball burnishing parameters on surface roughness using the Taguchi method. *Procedia Manufacturing*, 2019. **34**: 287–292.
139. Priyadarsini, C., et al., A review on ball, roller, low plasticity burnishing process. *Materials Today: Proceedings*, 2019. **18**: 5087–5099.
140. Świrad, S., et al., Influence of hydrostatic burnishing strategy on the surface topography of martensitic steel. *Measurement*, 2019. **138**: 590–601.
141. Yang, S., et al., Effect of cryogenic burnishing on surface integrity modifications of Co-Cr-Mo biomedical alloy. *Journal of Biomedical Materials Research Part B-Applied Biomaterials*, 2013. **101**(1): 139–52.
142. Singh, P. and P. Kumar, Improvement in surface integrity of thermally sprayed cermet coatings. *Materials Today: Proceedings*, 2021. **45**: 4431–4436.
143. Yang, S., et al., Surface layer modifications in Co-Cr-Mo biomedical alloy from cryogenic burnishing. *Procedia Engineering*, 2011. **19**: 383–388.
144 Huang, B., et al., Surface layer modification by cryogenic burnishing of Al 7050-T7451 alloy and validation with FEM-based burnishing model. *Procedia CIRP*, 2015. **31**: 1–6.
145. Caudill, J., et al., Enhancing the surface integrity of Ti-6Al-4 V alloy through cryogenic burnishing. *Procedia CIRP*, 2014. **13**: 243–248.
146. Sachin, B., S. Narendranath, and D. Chakradhar, Effect of cryogenic diamond burnishing on residual stress and microhardness of 17-4 PH stainless steel. *Materials Today: Proceedings*, 2018. **5**(9, Part 3): 18393–18399.
147. Tang, J., et al., The effect of cryogenic burnishing on the formation mechanism of corrosion product film of Ti-6Al-4V titanium alloy in 0.9% NaCl solution. *Surface and Coatings Technology*, 2018. **345**: 123–131.
148. Hiegemann, L., et al., Prediction of roughness after ball burnishing of thermally coated surfaces. *Journal of Materials Processing Technology*, 2015. **217**: 193–201.
149. Sachin, B., S. Narendranath, and D. Chakradhar, Enhancement of surface integrity by cryogenic diamond burnishing toward the improved functional performance of the components. *Journal of the Brazilian Society of Mechanical Sciences and Engineering*, 2019. **41**(10): 396.
150. Yang, S., et al., Cryogenic cooling effect on surface and subsurface microstructural modifications in burnishing of Co–Cr–Mo biomaterial. *Journal of Materials Processing Technology*, 2015. **217**: 211–221.

151. Saldaña-Robles, A., et al., Influence of ball-burnishing on roughness, hardness and corrosion resistance of AISI 1045 steel. *Surface and Coatings Technology*, 2018. **339**: 191–198.
152. Rotella, G., et al., The effects of cooling conditions on surface integrity in machining of Ti6al4v alloy. *Manufacturing Technology*, 2014. **71**(1): 47–55.
153. Kaynak, Y., H. Karaca, R.D. Noebe, and I.S. Jawahir, Tool wear analysis in cryogenic machining of NiTi shape memory alloys: a comparison of tool-wear performance with dry and MQL machining. *Wear*, 2013. **306**: 51–63.
154. Pu, Z., et al., Finite element simulation of residual stresses in cryogenic machining of AZ31B Mg alloy. *Procedia CIRP*, 2014. **13**: 282–287.
155. Sagbas, A., Analysis and optimization of surface roughness in the ball burnishing process using response surface methodology and desirabilty function. *Advances in Engineering Software*, 2011. **42**(11): 992–998.

5 Applications of Silicon-Based Coatings in Industrial High-Temperature Environments

Channabasavaiah Durga Prasad, Varun Kuppur Rajendra, Srinivasa Govindaiah, Aprameya Chokkadi Ramesh, Hulimane Bhaskara Shivaprasad, Puneet Mathur, Dandinashivara Rangadhamaiah Shankara, Jayatirtha Madhavarao Patil, and Suresh Gudala

5.1 INTRODUCTION

Silicon coatings are a new class of high-temperature coating materials with interesting properties that make them suitable for a broad range of applications in different sectors. Since World War II, silicon resin has been used commercially as a damping fluid in airplane engines to reduce corona discharge at high altitudes [1–4]. Other silicons used for insulation were adherent silicons obtained from trifunctional organosilicon intermediates, used in electric motors with glass films [5–8]. The characteristics of silicon resins are affected by crosslink density, phenyl-to-methyl ratio, molecular weight, and silanol content. Silicon resins are marketed in the form of emulsions, solid powders, and solutions in organic solvents [9–12]. New developments in silicon resin have seen many high-temperature-resistant coating compositions being produced. As additives, reactive intermediates, and carriers for aqueous systems, silicon resins have given rise to a new market. The development of the chemistry of silicon for high-temperature applications leads to increased corrosion and weathering, along with better color stability in both medium- and high-temperature applications [13–18].

During the early 1950s, silicon formulations were used on stove parts, boilers, engines, mufflers, heat exchangers, exhaust stacks, and aircraft sections where the protective feature operated at temperatures up to 816°C. While during this decade, prospective coloring of the pigments made the use of silicon coatings more versatile enough to cater the color requirement in colored items like olive-drab space heaters

 DOI: 10.1201/9781003542957-5

and military truck's exhaust pipes [19–25]. Some formulators started incorporating silicons into the organic polymers or the other way around, reducing costs but not on performance quality, with the alkyd resins being the most commonly used [26–30]. Like other silicon technologies, these innovations were launched by the space program; however, they were versatile and adaptable to serve in other areas. A 100% silicon coating, which is heat resistant up to 1,371°C, was applied on tiles used in the space shuttle orbiter to withstand extreme heat during the re-entry into Earth's atmosphere and was also used in the nuclear manufacturing industry. Later on for the chemical processing industry, this coating was patented for avoiding crystallization in fused glass structures at 982°C during the production of titanium oxide [31–36].

Silicon coatings are applied in aerospace industries and automotive industries, besides construction industries; because they are capable of withstanding high temperatures and can also resist corrosion, hence having excellent thermal stability for a long time. These coatings are used on substrate materials such as ceramics, polymers, and metals used at temperatures of 200°C–600°C used in applications such as corrosion and heat resistance and protection from abrasion without affecting the physical characteristics of the material [37–40]. Low coefficient of thermal conductivity reduces the rate of heat transfer from the substrate to the external environment, which makes silicon coatings insulators inside combustion chambers, exhaust systems, as well as furnace liners. They are also responsible for innovation and development and hence deemed significant in matters to do with technology [41–45]. One other advantage of using silicon coating in high-temperature areas is its high claw back to the different types of substrates, which can be improved by adding polymers to the products to facilitate higher bond strength. These coatings find more applications in severe climates where multiple reactions take place and are resistant to all forms of organic solvents, acids, and alkalis. Furthermore, they offer electrical protection, particularly from moisture, between the substrate and environment, thus finding use in electronics applications. When used, silicon coatings enjoy UV stability coupled with the ability to function in solar panels without degrading. The usage of silicon coatings remains applicable in numerous fields, with its operations set to benefit from the daily advancements in technology [46–53].

The potential application of silicon-based coatings across various industries has been depicted in Figure 5.1, including aerospace turbine blades, automotive exhaust systems, power generation boiler tubes, and oil and gas pipelines. Silicon coatings impart oxidation protection and thermal stability and improve the wear characteristics in these conditions. The graphic points out specific components in each industry, illustrating how silicon-based coatings add the function of lengthening the lifespan of materials in high-temperature conditions [54–58].

5.2 SILICON-BASED COATINGS AND THEIR EXCEPTIONAL PROPERTIES

Very often, materials are used in environments with high temperatures, and the demand is to have heat-resistant materials with stable mechanical properties. They have become an essential answer in many sectors due to their exceptional characteristics suitable for applications in high-temperature conditions [7,59–63]. This overview presents several types of silicon-based coatings that have been developed for high-temperature applications to distinguish their features and applications.

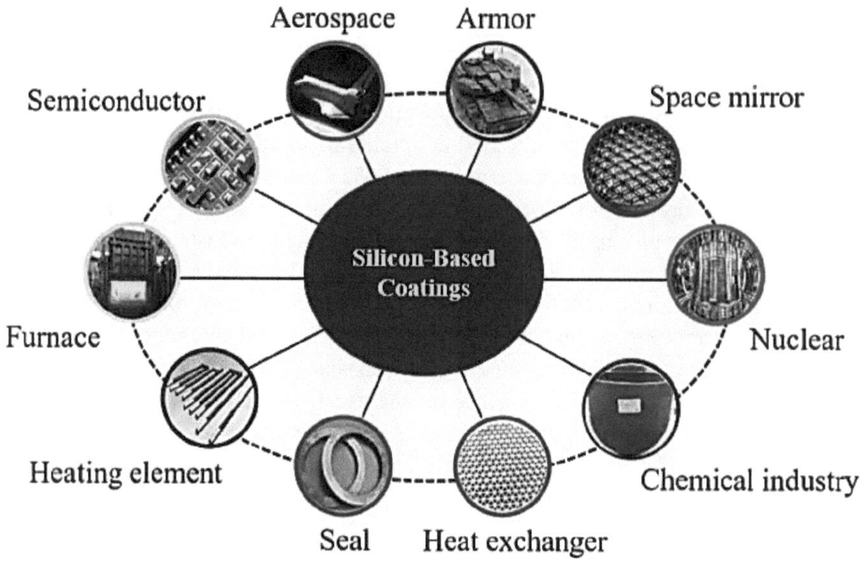

FIGURE 5.1 Application of silicon-based coatings.

5.2.1 Silicon Carbide Coatings

Silicon carbide (SiC) is very hard and thermally stable, making it suitable for high-temperature applications. SiC material can withstand temperatures over 1,600°C without registering much degradation. These coatings also display outstanding fatigue strength and corrosion protection as well as fairly high thermal stability, which are necessary for applications that include aerospace parts, gas turbines, and nuclear power plant equipment. Not only do they shield the base material from wear and thermal shock, but they also promote improvement in thermal performance in superior applications [8,64–66]. The resistance of these materials to chemical attack also strengthens this application where there is chemical aggression.

5.2.2 Silicon Nitride Coatings

The main material, silicon nitride (Si_3N_4), is memorable for its high mechanical characteristics and thermal stability that allow using a temperature of 1,400°C. The main features that can be highlighted in these coatings are their low thermal conductivity in order to isolate heat-sensitive components. This material is also immune to oxidation and therefore ideal for use in temperature applications, gas turbines, cutting tools, and automobile engines. These properties enable the material to perform with high tolerance during high thermal cycling, making it ideal for crucial stress elements. Moreover, Si_3N_4 coatings can also prolong the life of cutting tools and machine parts surfaces by providing a shielding effect against wear and heat [9,67–70].

5.2.3 Silicon Dioxide Coatings

Silicon dioxide (SiO_2)-based coatings possess high-temperature stability and minimal thermal expansion, making them suitable for high-temperature applications. They can withstand temperatures up to 1,200°C in the absence of any large-scale depreciation; they can be used as protective layers in the shielded areas of very high temperatures. SiO_2 coating films exhibit high dielectric performance, making them useful in electronics [10,71–74]. They also have fairly good anticorrosive properties, protecting the substrates against chemical attack. Applied often in aerospace and electronics industries, the SiO_2 coatings improve the reliability of the part when exposed to thermal and mechanical loads.

5.2.4 Silicon-Based Thermal Barrier Coatings

Thermal barrier coatings (TBCs) are developed to provide enhanced thermal barrier characteristics and are predominantly used in gas turbine engines and other high-temperature applications. Usually, these coatings can resist temperatures above 1,300°C and minimize heat conduction toward the substrates. TBCs are developed so as to shield important sections from heat, thus improving the performance and reliability of gas turbines. These coatings thus help to enhance the function of the engines, their efficiency, and rates of fuel utilization, making them very critical in aerospace and energy industries [11,75–78].

5.2.5 Silicon Phosphide Coatings

Silicon phosphide (SiP) coatings have remarkable properties, including oxidization resistance and thermal stability up to about 1,000°C. These coatings are required in equipment where parts operate in high-temperature and corrosive environments. In this way, due to the enhanced protective character of SiP coatings, the materials used in industrial applications can last longer and provide efficiency in harsh environments [12,79–81]. Due to their specialty, they are ideally used in industries such as chemical and aerospace, where performance under conditions of increased pressure is necessary.

Silicon-based coating materials refer to a comprehensive category of application methods on substrate materials, and each will properly operate with its own demands in applications. These techniques provide good applicability of silicon-based materials to get high durability and excellent adhesion or to produce materials with specific appearances for the identification of the materials in the aerospace, electronics, automobile, and medical sectors [13]. The selection of the technique frequently comes down to requirements like thickness, substrate type, and the conditions under which the coating will function, so specific solutions can be catered to many different industries.

5.3 THERMAL SPRAY APPROACHES TO APPLY SILICON-BASED COATINGS

Thermal spray coatings are mechanical surface coatings used with a view to improving the properties of materials as used in different sectors. These coatings are generated by applying molten or semi-molten materials on other surfaces, and the layer formed enhances characteristics such as wear, corrosion protection, and heatproofing. As described by the principle of thermal spray methods in Figure 5.2, feedstock materials are preheated to a molten or semi-molten state using a variety of energy sources, combustions, and electric arcs. The heated particles are therefore directed onto a substrate with high kinetic energies. In contact, the particles tend to spread out and fuse into a compact and bond well with a substrate. This process improves the substrate with some of its own properties, like wear resistance and corrosion protection, among others [14–16,82–85]. The behaviors of the final coating layer are a function of particle velocity, temperature, and properties of the material.

The widely used thermal spray processes are plasma spraying, high-velocity oxy-fuel (HVOF) spraying, flame spraying, cold spraying, and detonation gun spraying. Each technique provides desirable coating properties such as density, adhesion, and thermal properties, and is used in aerospace, automotive, or energy industries. Silicon-based coatings, to which the present invention particularly pertains, have the advantage in high-temperature services due to their thermal stability and oxidation resistance [17]. Decisions on the method depend on the desired properties, substrate materials, and applications. Altogether, thermal spray has a significant role in elongating the durability and efficiency of parts that are exposed to severe conditions.

5.3.1 Plasma Spraying

Plasma spraying is a thermal spray technique that operates at temperatures of about 10,000°C–15,000°C in the plasma arc. This technique generates dense films with low porosity and greatly improves their mechanical properties. The major benefit of plasma spraying is the first layer adhesion that results from spraying particles with high thermal energy. As a result, these coatings themselves demonstrate excellent oxidation characteristics, making them ideal for use in locations where the formation

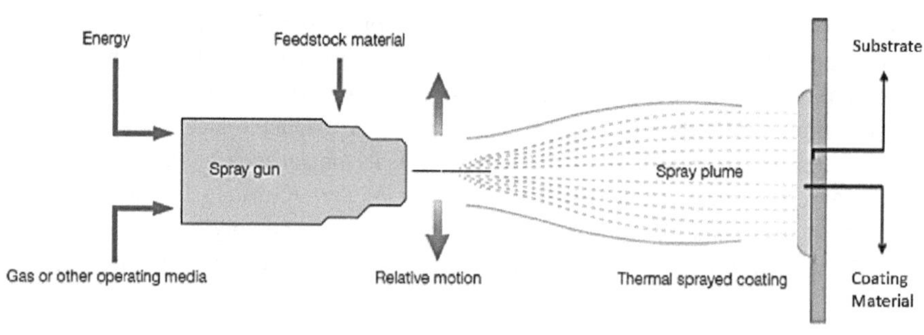

FIGURE 5.2 Principle of thermal spray process.

of rust might otherwise be an issue [18]. Nevertheless, plasma-sprayed coatings have high-temperature stability, making them attractive in some applications such as aerospace and automobiles. Another benefit of this method is its ability to apply thicker coatings compared to most thermal spray methods, increasing the benefits of using technology in terms of wear resistance and durability. In a broader context of employing various methods applicable in the deposition of coatings, plasma spraying has been considered one of the most efficient techniques, capable of delivering tough and durable coatings adequate for use in high-temperature applications [19].

5.3.2 High-Velocity Oxy-Fuel Spraying

A discrete feature of HVOF is recognized by its capacity to discharge high gaseous velocities that play an important role in depositing compact layers of materials with minimal chances of getting oxidized. A high impact velocity is achieved with HVOF coatings; this leads to high wear resistance, thus making it suitable for components that are used under harsh conditions where durability is key. Moreover, due to the high velocities, one gets good interfacial adhesion, which greatly boosts the performance of the coated component. In most cases, HVOF coatings have lower thermal conductivity compared to plasma spraying, which can be useful where a thermal barrier is needed. Moreover, the HVOF technology can use many silicon-based materials such as SiC and silicon alumina, allowing it to fulfill various industrial demands [20,33–38].

5.3.3 Flame Spraying

Flame spraying, or oxy-fuel spraying, is one of the oldest thermal spray techniques currently available in the market. It uses a fuel gas (like acetylene) and oxygen that results in formation of flame to soften the silicon-based powder. The molten droplets are then sprayed onto the substrate. The velocity of the particles depends on the liquid flow rate, lamp intensity, and many other factors. CVD is a highly flexible technique that applies to most of the silicon-containing materials. Nevertheless, flame spraying generally produces coatings with higher porosity than plasma spraying and HVOF, so they may not be suitable for some applications. However, it is often selected because of the scope and overall price of the solution [21,57–61].

5.3.4 Cold Spraying

Cold spraying is a relatively new thermal spray technique that works at comparatively low temperatures and thus reduces the problems of thermal degradation of silicon-based parts. In this process, high-velocity gas streams catapult solid powder particles toward the substrate, and the particles bond mechanically upon impact on the substrate. One of the particular advantages of cold spraying is that the material to be deposited may be in the form of a volatile oxide or just susceptible to degradation at high temperatures; thus, it may be used to apply some silicon-based coatings, maintaining all its properties. This results in low porosity and high bond strength of the coatings [22,74–78].

5.3.5 Detonation Gun Spraying

Detonation gun spraying utilizes a fuel oxidizer mixture that is ignited, which produces acoustic pressure within the combustion chamber that propels the silicon-based powder toward the substrate. This method produces high impact velocities, resulting in high-density, more adherent coatings. Detonation gun spraying is especially suitable for depositing high-performance coatings with optimal mechanical characteristics [23]. This technique can give a better surface having lesser porosity and higher density, making it applicable in strategic areas in aerospace as well as energy industries.

While each of the thermal spray techniques for applying silicon-based coatings has its own characteristic merits, each technique is applicable to particular uses. Choosing the right thermal spray method is thus determined by the properties of the deposited coating, the substrate material, and the application process.

5.4 HIGH-TEMPERATURE COATING APPLICATIONS OF SILICON MATERIALS

5.4.1 Aerospace and Aviation

Silicon coatings, particularly because of their high-temperature weathering capability, thermal stability, and high chemical resistance, are widely used in aerospace and aircraft applications. These coatings offer vital safeguards to aerospace parts, particularly those that are subjected to high environments, for instance, re-entry vehicles and rocket engines that are prone to high temperatures during high-speed flights. Silicon-based thermal systems are used here to the extent that they are made for operation up to 1,500°C while having a low coefficient of thermal conductivity [24]. It not only protects such parts but also assists in preserving the structure of construction under extremely high thermal conditions.

The deposition of silicon is useful in improving the performance of turbine engines. The protective layer offers a vast enhancement in anticorrosive/antiwear capability required for endorsing a high-temperature, abrasive atmosphere of turbines used in gas turbines. This protective function is not just for inner structures; important aircraft parts, including the fuselage and wings, are coated with silicon materials [25]. Here, silicon coatings shield the outer surfaces from environmental corrosion and aggressive factors, thus preserving the aircraft's shape, decreasing drag coefficients, and, consequently, fuel requirements.

Silicon coatings are characterized by good adhesion properties, particularly to aluminum alloy substrates that are common in many of the lightweight composite materials of modern aircraft. Furthermore, these coatings improve the durability in responding to weather conditions, a factor critical in planes flying in different climatic conditions and Mach numbers. The layer of silicon coatings adds to the flight protective layer, meaning that the shells of the aircraft receive minimum impact from UV rays, oxidation, and fluctuating temperatures in the future [26].

In particular, applications such as anti-slip coatings are used to prevent the formation of ice on aircraft surfaces to reduce safety and performance. The effectiveness of these coatings in providing long-term protection across a range of substrates is supported by their high bond strength. In addition, in simulated heat exchanger conditions, coatings with silicon have shown significant enhancement in thermal stability and resistance to corrosion; hence, their importance in strategic aerospace applications. In meeting all these diverse roles, silicon coatings play the role of safety, durability, and efficient operation to essentially affirm their place as one of the most crucial materials in aerospace and aviation industries [27–28].

5.4.2 Automotive Industry

Silicon heatproof coatings based on silicon have gained wide use in automobiles, offering protection and other performance attributes that are critical for today's efficient automobile engines and parts. These coatings are appreciated for their high heat resistance and protection from oxidation and wear, which are of prime importance in harsh automotive applications, for instance, in exhaust systems, cylinder heads, and pistons [29].

In parts of the engine, the silicon-based coatings work as a thermal barrier against heat transfer to the other components, which increases the thermal efficiency of the engine. It is especially useful in parts such as exhaust manifolds and turbochargers where high temperature is the normal working parameter. We have noticed that low coatings assist in enhancing fuel economy and emissions by controlling heat in the engine [30]. For instance, silicon-based materials can be used to coat exhaust manifolds, which stand to last the thermal cycles without degrading the performance of the engine.

Based on wear resistance properties, silicon-based coatings are applied in parts characterized by high friction, for example, piston rings and cylinder linings. These coatings minimize surface wear and extend the lifecycle of these components as a result of their hard, wearing surface that endures the mechanical load inherent in any engine. The silicon-based coatings improve corrosion protection in automobile exhaust systems wherein materials are known to corrode due to high oxidation propensity [31]. The silicon-based coatings can prevent corrosion, hence enabling the parts to maintain their undeniable mechanical properties as well as functionality after enduring the terrible effects of the exhaust products for the long term [32].

5.4.3 Oil and Gas Industries

Silicon-based thermal spray coatings have gained more demand in the oil and gas industry, where product components and equipment can undergo various kinds of rigors such as high temperatures, corrosive atmospheres, and abrasive products. These coatings provide superior characteristics, including abrasion resistance, hard-wearing properties, and anticorrosive products for needful uses like pipelines, drilling instruments, and valves.

In pipelines and other structural sections, which are exposed to certain harsh environments, the silicon-based coatings function as a barrier that protects against corrosion due to interaction with salts, high-pressure gases, and chemical substances. This is especially the case when it comes to an offshore oil extraction industry, as salt-water and high humidity will quickly ruin unprotected surfaces. The Si-containing coatings protect the internal walls of pipelines, increase their overall performance, and extend the corrosion and maintenance-free life of pipelines used to transport crude oil, gas, and other corrosive products [33].

Also, thermal spray coatings made of silicon are applied to valves and pumps that experience abrasive slurry and drilling fluids. These coatings are stiff, wear-resistant materials that prevent alterations to surface characteristics, including the loss of material by corrosion Analyses were carried out to check whether high-stress conditions such as high-speed, abrasive material retaining silicon-based coatings' adhesion and surface hardness will be excellent. This durability reduces costly standstill time and successive part replacements that are essential in oil drilling and extraction, where equipment must be highly reliable [34]. In high-temperature refineries, ambient or process temperatures, silicon-containing coatings offer high-temperature stability and protection from oxidation to components such as heat exchangers, separators, and other high-temperature equipment. These coatings, being thermally stable, do not develop micro-cracks and spalls during temperature variations, hence shielding the base material of the equipment. These coatings enhance the safety, reliability, and durability of equipment, leading to meeting the industry needs of performance-driven, low-maintenance products [35].

5.5 ELEMENTS INFLUENCING SILICON-BASED COATINGS' PERFORMANCE IN HIGH-TEMPERATURE SETTINGS

The overall effectiveness of silicon-based coatings in high-temperature conditions depends on factors such as ingredients, thermal stability, bonding, and mechanical features. These coatings are for use in sectors such as aerospace, automotive, and oil and gas due to the protection against heat, oxidation, and wear. For instance, SiC has dimensional stability coupled with superior hardness and wear resistance that recommends it for use where abrasion conditions exist; silicon oxide, however, forms a passive coat that shields the surface from further oxidation [36–38].

Thermal stability is the other factor considered by the system. However, prolonged exposure to high temperatures can erode this layer over time, and thus the benefit of this coating will be unrealized. The high-temperature oxidation resistance and high-temperature structural stability are important features of components such as turbine blades and exhaust systems, which undergo numerous cycles of expansion and contraction [39].

The coating's bond with the substrate also determines how long the silicon coatings can last within a high-temperature process. Failure in adhesion will result in a deformation of the coating by the formation of cracks or spalls as a result of thermal expansion differences or mechanical loads. Plasma spraying and HVOF spraying are useful as they are fast with a high energy impact on the surface they are bonding, thus promoting a sound coating-substrate interface [40].

Some properties, like hardness and abrasion or wear resistance, are important so that the coating could sustain in the most worn-out areas. Silicon-based coatings, especially with SiC, demonstrate high hardness and are useful in shielding from abrasive particles that are popular in high-temperature environments. A lower porosity, high-density microstructure adds to these characteristics and supports the prolonged use adhesion of the coating against mechanical wear as well as particle erosion [41].

The composition, thermal stability, adhesion, and mechanical properties are closely interacting with each other, and the fine-tuning of these factors is necessary to achieve high-performance silicon-based coating material in severe and high-temperature service conditions.

5.6 PERFORMANCE OF SILICON-BASED COATINGS IN HIGH-TEMPERATURE APPLICATIONS

5.6.1 Effects of Temperature, Stress, and Exposure Time on the Performance of Silicon-Based Coatings

Temperature plays a vital role in determining the longevity and robustness of silicon-based coatings. These coatings show outstanding thermal stability up to 400°C, enduring temperatures of up to 600°C for brief periods. Nevertheless, when exposed to high temperatures, the coatings' mechanical and physical attributes might deteriorate, causing problems like cracking, delamination, and oxidation. For instance, research on the heat resistance of silicon coatings on steel surfaces reveals that these coatings remain stable up to approximately 300°C. Degradation begins when this temperature is surpassed, leading to a notable reduction in their corrosion-prevention capabilities, suggesting a critical threshold for high-temperature applications [42–44]. The performance of silicon coatings is greatly impacted by the stress and strain that are encountered during heat exposure. Since the coating's thermal expansion coefficient is lower than that of some substrates, elevated temperatures may result in internal tensions. This disparity may cause the coating and substrate to adhere less strongly, which might eventually cause cracking and delamination. The coating undergoes several cycles of expansion and contraction in response to temperature changes. This can exacerbate tension in the coating over time, leading to problems like microscopic fractures that reduce the coating's durability and strength.

Another important factor that affects how effectively silicon coatings function at high temperatures is the length of exposure [45]. Prolonged exposure to high temperatures significantly weakens the coating's ability to withstand corrosion. When silicon coatings are applied on carbon steel substrates, prolonged exposure to high temperatures may eventually reduce their capacity to stop corrosion. Because of the continuous heat and oxidative stress that the coatings endure, their performance gradually deteriorates, highlighting the need to improve coating composition and application techniques for settings that require extended thermal stability [46]. Understanding the impacts of extended exposure, internal tensions brought on by variations in thermal expansion, and temperature thresholds can all help to optimize silicon-based coating designs. By addressing these performance-affecting elements, it is possible to improve the coatings' durability, adhesion, and general heat resistance, ensuring their reliable operation under demanding high-temperature circumstances.

5.6.2 Performance Comparison of Silicon-Based Coatings and Other High-Temperature Coating Substances

It is used in many different industries because of its unique properties, which include low surface energy, strong adhesion, high-temperature stability, and resistance to weather and chemicals. However, high-temperature coatings are also applied using organic, metallic, and ceramic coatings. Although zirconia and alumina ceramic coatings are generally less robust and more brittle than silicon coatings, they are known for their capacity to withstand high temperatures and offer thermal insulation [47]. Metallic coatings made of nickel and chrome are frequently utilized in hot conditions, although high temperatures have a tendency to oxidize and corrode them. Because of their poor heat durability, organic coatings such as epoxy and polyurethane are inappropriate for high-temperature applications. Numerous studies have assessed the performance of silicon-based coatings in comparison to other materials used for high-temperature coatings. One study focused on the thermal stability of silicon and ceramic coatings on carbon/carbon composites in response to high temperatures [48]. Although the study found that both coatings could withstand high temperatures, the silicon coating had better adherence to the substrate and less thermal conductivity. Generally speaking, silicon-based coatings have proven to be the best option for many sectors despite the existence of substitute materials for high-temperature coatings.

Silicon-based coatings have gained importance in industrial high-temperature environments due to their excellent thermal stability, oxidation resistance, and ability to form protective silica layers [86–94]. These coatings are widely used in power generation, aerospace, and petrochemical industries to combat degradation under extreme thermal and chemical stresses. Studies have shown that advanced processing techniques, such as thermal spraying and heat treatments, significantly improve their structural integrity and performance [95–99]. Additionally, optimizing coating parameters has been shown to enhance protection against oxidation and erosion, ensuring long-term reliability in these challenging applications [100–114].

5.7 CONCLUSIONS AND FUTURE DIRECTIONS

5.7.1 Conclusion in the Field of Silicon-Based Coatings for High-Temperature Applications

The silicon coatings are durable, corrosion- and oxidation-resistant, and have strong thermal stability; they show intriguing promise for usage at high temperatures. According to research, silicon-based coatings can resist temperatures ranging from 200°C to more than 1,000°C, making them suitable for a variety of high-temperature environments.

- Coatings containing silicon are increasingly essential for extending the life and effectiveness of industrial components in high-heat environments. They are ideal for demanding applications in sectors including manufacturing, energy, aircraft, and automobiles because of their remarkable properties, which include great resistance to oxidation, stability at high temperatures, and absence of reactivity to chemicals.
- Advanced thermal spray techniques, including HVOF, cold spraying, and plasma spraying, have enabled precise control of coating characteristics, resulting in tailored solutions that enhance heat control, corrosion prevention, and wear resistance.
- Despite significant advancements, silicon-based coatings may still be improved in terms of mechanical qualities, porosity reduction, and striking a balance between hardness and heat conductivity.
- The applications of these coatings will expand due to ongoing research and developments in silicon-based materials and deposition techniques, confirming their critical role in maintaining the dependability and extending the lifespan of critical industrial systems.

5.7.2 Future Study Directions

Research on silicon-based coatings at high temperatures should focus on creating novel silicon-based materials, enhancing thermal spray techniques, and looking into multipurpose hybrid coatings in the future. By addressing current challenges and future demands across a variety of sectors, these projects can help expand the usage of silicon-based coatings.

- The goal of developments in silicon-based materials, such as silicon oxycarbide and silicon borides, is to improve their performance, durability, and resistance to thermal shock in industrial settings.
- Increasing control over variables including particle velocity, temperature, and feed rate is the main focus of improving thermal spray technology. Suspension plasma spraying and high-velocity air-fuel are two techniques that allow for exact changes, resulting in coatings with better mechanical properties and reduced porosity.
- Researchers work to create hybrid and multifunctional coatings using a variety of materials in an effort to enhance performance in difficult situations. Better properties like wear resistance and thermal insulation can be achieved by mixing them with silicon-based coatings.

DECLARATIONS

Ethical Approval: The chapter accurately and thoroughly reflects the authors' research and analysis.

Conflict of Interest: There is no conflict of interest that the author has disclosed.

FUNDING

Author C. Durga Prasad would like to thank the Science and Engineering Research Board (SERB) for financial support to carry out this research work. Project File No.: CRG/2022/004140, under the Core Research Grant (CRG) scheme, Government of India.

Data and Material Accessibility: Since no new data were generated or examined in this study, data accessibility is not relevant to this publication.

REFERENCES

1. Andriot, M., Chao, S. H., Colas, A., Cray, S. E., de Buyl, F., DeGroot, J. V., Dupont, A., Easton, T., Garaud, J. L., Gerlach, E., & Gubbels, F. (2007). Silicones in industrial applications. *Inorganic Polymer*, 61–161.
2. Clope, R. W., Glaser, M., & Madson, W. H. (1970). *Silicone Resins for Organic Coatings*. Philadelphia: Federation of Societies for Coatings Technology.
3. Houstain, D. P. (1975). Silicone coating adds years to jet equipment service. *Modern Paint and Coatings*, 9, 42–43.
4. Wang, W., Zammarano, M., Shields, J. R., Knowlton, E. D., Kim, I., Gales, J. A., Hoehler, M. S., & Li, J. (2018). A novel application of silicone-based flame-retardant adhesive in plywood. *Construction and Building Materials*, 189, 448–459.
5. Eduok, U., Faye, O., & Szpunar, J. (2017). Recent developments and applications of protective silicone coatings: a review of PDMS functional materials. *Progress in Organic Coatings,* 111, 124–163.
6. Sati, P., Kumari, A., & Kumar, S. (2023). Silicon-based technologies for high-temperature coatings and their corrosion behaviours. In: Pakseresht, A. H., Mosas, K. K. A. (eds.), *Coatings for High-Temperature Environments: Anti-Corrosion and Anti-Wear Applications*, pp. 103–122. Cham: Springer Nature Switzerland.
7. Chruściel, J. J. (2022). *Silicon-Based Polymers and Materials*. Berlin: Walter de Gruyter GmbH & Co KG.
8. Xu, M., Girish, Y. R., Rakesh, K. P., Wu, P., Manukumar, H. M., Byrappa, S. M., & Byrappa, K. (2021). Recent advances and challenges in silicon carbide (SiC) ceramic nanoarchitectures and their applications. *Materials Today Communications,* 28, 102533.
9. Wang, Q., He, G., Deng, S., Liu, J., Li, J., Li, Y., Pan, L., Liu, G., Li, X., & Li, J. (2020). Wettability and infiltration of Si drop on silica substrate containing α-Si3N4 coating: influence of oxygen content in α-Si3N4 coating. *Ceramics International,* 46(8), 12695–12704.
10. Issa, E. (2022). *Novel Reactor Design and Method for Atmospheric Pressure Chemical Vapor Deposition of Micro and Nano SiO_2-x Films in Photovoltaic Applications*. Thuringia, German: BoD–Books on Demand.
11. Gildersleeve, E. J., Bakan, E., & Vaßen, R. (2024). Integrating state of the art zirconia thermal barriers with ytterbium silicate environmental barriers for silicon-based ceramic turbine components. *Journal of Thermal Spray Technology,* 33(1), 195–209.
12. Reinhold, R., Stoeck, U., Grafe, H.-J., Mikhailova, D., Jaumann, T., Oswald, S., Kaskel, S., & Giebeler, L. (2018). Surface and electrochemical studies on silicon diphosphide as easy-to-handle anode material for lithium-based batteries the phosphorus path. *ACS Applied Materials & Interfaces,* 10(8), 7096–7106.
13. Kwon, H.-T., Lee, C. K., Jeon, K.-J., & Park, C.-M. (2016). Silicon diphosphide: a Si-based three-dimensional crystalline framework as a high-performance Li-ion battery anode. *ACS Nano,* 10(6), 5701–5709.
14. Thakur, L., & Vasudev, H., (eds.). (2021). *Thermal Spray Coatings*. Boca Raton, FL: CRC Press.

15. Soboyejo, W. O., Obayemi, J. D., Annan, E., Ampaw, E. K., Daniels, L., & Rahbar, N. (2016). Review of high temperature ceramics for aerospace applications. *Advanced Materials Research,* 1132, 385–407.
16. Solntsev, S. S., Grashchenkov, D. V., & Evdokimov, S. A. (2014). Promising high-temperature composite materials and coatings for aerospace technology. *Glass and Ceramics,* 71, 23–27.
17. Al-Jothery, H. K. M., Albarody, T. M. B., Yusoff, P. S. M., Abdullah, M. A., & Hussein, A. R. (2020). A review of ultra-high temperature materials for thermal protection system. *IOP Conference Series: Materials Science and Engineering*, 863(1), 012003. IOP Publishing.
18. National Research Council, Division on Engineering, Physical Sciences, National Materials Advisory Board, Commission on Engineering, Technical Systems, and Committee on Coatings for High-Temperature Structural Materials. (1996). *Coatings for High-Temperature Structural Materials: Trends and Opportunities.* Washington DC: National Academies Press.
19. Sengupta, P., & Manna, I. (2019). Advanced high-temperature structural materials for aerospace and power sectors: a critical review. *Transactions of the Indian Institute of Metals,* 72, 2043–2059.
20. Mondal, K., Nuñez III, L., Downey, C. M., & Van Rooyen, I. J. (2021). Thermal barrier coatings overview: design, manufacturing, and applications in high-temperature industries. *Industrial & Engineering Chemistry Research,* 60(17), 6061–6077.
21. Okada, A. (2008). Automotive and industrial applications of structural ceramics in Japan. *Journal of the European Ceramic Society,* 28(5), 1097–1104.
22. Orłowicz, A. W., Mróz, M., Tupaj, M., & Trytek, A. (2015). Materials used in the automotive industry. *Archives of Foundry Engineering,* 15, 75–78.
23. Mishra, A. (2020). Automotive materials: an overview. *International Research Journal of Engineering and Technology (IRJET),* 7(08), 4852–4857.
24. Zarei, M., Eskandarzade, M., Babapoor, A., & Seyfaee, A. (2024). A review of recent advances and applications of inorganic coating for oil and gas pipe systems. *Surface and Coatings Technology*, 494, 131339.
25. Liskiewicz, T. W., & Al-Borno, A. (2014). DLC coatings in oil and gas production. *Journal of Coating Science and Technology,* 1(1), 59–68.
26. Obot, I. B., Onyeachu, I. B., Umoren, S. A., Quraishi, M. A., Sorour, A. A., Chen, T., Aljeaban, N., & Wang, Q. (2020). High temperature sweet corrosion and inhibition in the oil and gas industry: progress, challenges and future perspectives. *Journal of Petroleum Science and Engineering,* 185, 106469.
27. Sati, P., Kumari, A., & Kumar, S. (2023). Silicon-based technologies for high-temperature coatings and their corrosion behaviours. In: Pakseresht, A. H., Mosas, K. K. A. (eds.), *Coatings for High-Temperature Environments: Anti-Corrosion and Anti-Wear Applications*, pp. 103–122. Cham: Springer Nature.
28. Wang, Z., Wang, Y., Wang, S., Zou, Y., Chen, G., Wen, L., ... & Zhou, Y. (2023). Enhanced high-temperature oxidation resistance of HfSi2-modified silicon based multilayer ceramic coating on Nb alloy prepared by a novel strategy. *Journal of the European Ceramic Society,* 43(11), 4717–4730.
29. Barroso, G., Li, Q., Bordia, R. K., & Motz, G. (2019). Polymeric and ceramic silicon-based coatings–a review. *Journal of Materials Chemistry A,* 7(5), 1936–1963.
30. Alvi, S. A., & Akhtar, F. (2018). High temperature tribology of polymer derived ceramic composite coatings. *Scientific Reports,* 8(1), 15105.
31. Guo, L., He, W., Chen, W., Xue, Z., He, J., Guo, Y., ... & Gao, L. (2023). Progress on high-temperature protective coatings for aero-engines. *Surface Science and Technology,* 1(1), 6.

32. Stojanović, I., Škrlec, B., Kurtela, M., Alar, V., & Odeljan, M. (2023). Physical and chemical properties of high-temperature silicone-based polymer coatings applied on different surface roughnesses. *Coatings,* 13(12), 2100.
33. Sara, B. (2015). Si-SiC based materials obtained by infiltration of silicon: study and applications. PhD diss., Politecnico di Torino.
34. Sun, X., Zhang, J., Pan, W., Wang, W., & Tang, C. (2022). Research progress in surface strengthening technology of carbide-based coating. *Journal of Alloys and Compounds,* 905, 164062.
35. Bordia, R. (2009). *Nanoscale Reinforced, Polymer Derived Ceramic Matrix Coatings.* Seattle, WA: University of Washington.
36. Montero, W. W. H. (2017). Study of intrinsic and doped silicon-carbon films by PECVD of medium and high vacuum for solar cells. PhD diss., InstitutoNacional de Astrofísica, Óptica y Electrónica.
37. Buchwalder, S., Bourgeois, F., Leon, J. J. D., Hogg, A., & Burger, J. (2023). Parylene-AlOx stacks for improved 3D encapsulation solutions. *Coatings,* 13(11), 1942.
38. Asok, D. (2013). Study of Si (Al) CN functionalized carbon nanotube composite as a high temperature thermal absorber coating material.
39. Lee, T.-H. (2011). Silicon carbide high temperature logic. PhD diss., Case Western Reserve University.
40. Kollur, S., Shiva Prakash, S., Surendra, J., Kumar, G. S. P., Kumar, R. S., Prasad, C. D., Pavan Kumar, B. K., Kore, S., & Nusrathulla, M. (2024). Thermal cycle behavior of plasma sprayed thermal barrier coatings on cast iron substrate for the application of liner of internal combustion engine. *Results in Surfaces and Interfaces, Elsevier Science*, 17, 100300. https://doi.org/10.1016/j.rsurfi.2024.100300
41. Nayak, H., Varun, K. R., Venkatesh, M. K., Shamkuwar, S., Suresh Kumar, R., Sudarshan, T. A., Malge, A., Vijay, M., & Prasad, C. D. (2024). Thermal cycle behavior of plasma sprayed thermal barrier coatings on cast iron substrate for the application of liner of internal combustion engine. *Results in Surfaces and Interfaces, Elsevier Science*, 17, 100297. ISSN 2666–8459. https://doi.org/10.1016/j.rsurfi.2024.100297
42. Lakkannavar, V., Yogesha, K. B., Prasad, C. D., Srinivasa, G., & Mathapati, M. (2024). Assessment of the microstructure, adhesion and elevated temperature erosion resistance of plasma-sprayed NiCrAlY/Cr_3C_2/h-BN composite coating. *Results in Surfaces and Interfaces, Elsevier Science*, 17, 100289. ISSN 2666–8459. https://doi.org/10.1016/j.rsurfi.2024.100289
43. Kumar, G. S. P., Keshavamurthy, R., Harish Kumar, M., Yogaraju, R., Sachith, T. S., & Prasad, C. D. (2024). Studies on air jet erosion behavior of plasma-sprayed Cr_2C_3 coatings on Al6061 alloy. *Results in Surfaces and Interfaces, Elsevier Science*, 16, 100278. ISSN 2666–8459. https://doi.org/10.1016/j.rsurfi.2024.100278
44. Lakkannavar, V., Yogesha, K. B., Prasad, C. D., & Prasad, C. (2024). Investigation of cyclic oxidation and hot corrosion behaviour of plasma-sprayed NiCrAlY/Cr_3C_2/h-BN/Mo coatings on T22 boiler steel alloy. *Journal of Bio- and Tribo-Corrosion, Springer,* 10, 85. https://doi.org/10.1007/s40735-024-00890-w
45. Lakkannavar, V., Yogesha, K. B., Prasad, C. D., Suresh, R., Lakshmikanthan, A., & Hanumanthlal, S. (2024). Investigation of cyclic oxidation and corrosion behavior of plasma-sprayed NiCrAlY/Cr_3C_2/h-BN coatings on T22 boiler steel alloy. *Surface and Coatings Technology, Elsevier Science*, 492, 131205. ISSN 0257–8972. https://doi.org/10.1016/j.surfcoat.2024.131205
46. Lakkannavar, V., Yogesha, K. B., Prasad, C. D., Phanden, R. K., Srinivasa, G., & Prasad, C. (2024). Thermal spray coatings on high-temperature oxidation and corrosion applications – a comprehensive review. *Results in Surfaces and Interfaces, Elsevier Science*, 15, 100250. ISSN 2666–8459. https://doi.org/10.1016/j.rsurfi.2024.100250

47. Prasad, C. D., Soni, P. K., Nagaraja, K. C., Eswaran, A., Suresh Kumar, R., Deshmukh, K., Shrivastava, R., & Tiwari, A. (2024). Studies on wear and microstructure assessment of WC-Co reinforced iron based HVOF coating. *Results in Surfaces and Interfaces, Elsevier Science*, 15, 100237. ISSN 2666–8459. https://doi.org/10.1016/j.rsurfi.2024.100237
48. Kollur, S., Manjunath, K. V., Prasad, C. D., Reddy, B. B., Rudra, K. S., Kumar, P., & Aprameya, C. R. (2024). Evaluation of microstructural and mechanical properties of tribaloy based composite cladding by microwave heating. *Journal of the Institution of Engineers (India): Series D*, Springer. https://doi.org/10.1007/s40033-024-00670-3
49. Ramesh, M. R., Medabalimi, S., Kumar, R. S., Prasad, C. D., & Sollapur, S. B. (2024). Cyclic oxidation and hot corrosion behaviour of HVOF sprayed NiCrAl coating on industrial boiler tube steels. *JOM the Journal of the Minerals, Metals & Materials Society (TMS), Springer*, 76, 3172–3184. https://doi.org/10.1007/s11837-024-06526-1
50. Lakkannavar, V., Yogesha, K. B., Prasad, C. D., Mruthunjaya, M., & Suresh, R. (2024). A review on tribological and corrosion behavior of thermal spray coatings. *Journal of the Institution of Engineers (*India*):* Series D, Springer. https://doi.org/10.1007/s40033-024-00636-5
51. Nayak, H., Kollur, S., Prasad, K., Erannagari, S., Prasad, C. D., & Nagabhushana, N. (2024). Development of equal proportional YSZ+Al_2O_3 thermal barrier coating and effect of coating thickness on the corrosion behaviour on cast iron substrate. *High Temperature Corrosion of Materials, Springer*, 101, 291–308. https://doi.org/10.1007/s11085-024-10222-5
52. Prasad, C. D., Kollur, S., Aprameya, C. R., Chandramouli, T. V., Jagadeesha, T., & Prashanth, B. N. (2024). Investigations on tribological and microstructure characteristics of WC-12Co/FeNiCrMo composite coating by HVOF process. *JOM the Journal of the Minerals, Metals & Materials Society (TMS), Springer*, 76, 186–195. https://doi.org/10.1007/s11837-023-06242-2
53. Prasad. C. D., Kollur, S., Nusrathulla, M., Satheesh Babu, G., Hanamantraygouda, M. B., Prashanth, B. N., & Nagabhushana, N. (2024). Characterisation and wear behaviour of SiC reinforced FeNiCrMo composite coating by HVOF process. *Transactions of the IMF, Taylor and Francis*, 102(1), 22–28. https://doi.org/10.1080/00202967.2023.2246259
54. Sharanabasava, H., Raviprakash, M., Prasad, C. D., Ramesh, M. R., Phanibhushana, M. V., Vasudev, H., & Kumar, S. (2023). Microstructure, mechanical and wear properties of SiC and Mo reinforced NiCr microwave cladding. *Advances in Materials and Processing Technologies, Taylor and Francis.* https://doi.org/10.1080/2374068X.2023.2257937
55. MadhuSudana Reddy, G., Prasad, C. D., Kollur, S., Lakshmikanthan, A., Suresh, R., & Aprameya, C. R. (2023). Investigation of high temperature erosion behaviour of NiCrAlY/TiO_2 plasma coatings on titanium substrate. *JOM the Journal of the Minerals, Metals & Materials Society (TMS), Springer,* 75, 3317–3323. https://doi.org/10.1007/s11837-023-05894-4
56. Madhusudana Reddy, G., Prasad, C. D., Patil, P., Kakur, N., & Ramesh, M. R. (2023). High temperature erosion performance of NiCrAlY/Cr_2O_3/YSZ plasma spray coatings. *Transactions of the IMF, Taylor and Francis.* https://doi.org/10.1080/00202967.2023.2208899
57. Sharanabasva, H., Prasad, C. D., & Ramesh, M. R. (2024). Characterization and wear behavior of NiCrMoSi microwave cladding. *Journal of Materials Engineering and Performance,* 33, 763–775. https://doi.org/10.1007/s11665-023-07998-z
58. Madhusudana Reddy, G., Prasad, C. D., Patil, P., Kakur, N., & Ramesh, M. R. (2023). Investigation of plasma sprayed NiCrAlY/Cr_2O_3/YSZ coatings on erosion performance of MDN 420 steel substrate at elevated temperatures. *International Journal of Surface Science and Engineering*, 17(3), 180–194. https://dx.doi.org/10.1504/IJSURFSE.2023.10054266

59. Sharanabasva, H., Prasad, C. D., & Ramesh, M. R. (2023). Effect of Mo and SiC reinforced NiCr microwave cladding on microstructure, mechanical and wear properties. *Journal of the Institution of Engineers (India): Series D*, Springer. https://doi.org/10.1007/s40033-022-00445-8
60. Nithin, H. S., Nishchitha, K. M., Pradeep, D. G., Durga Prasad, C., & Mathapati, M. (2023). Comparative analysis of CoCrAlY coatings at high temperature oxidation behavior using different reinforcement composition profiles. *Welding in the World, Springer*, 67, 585–592. https://doi.org/10.1007/s40194-022-01405-2
61. Madhu Sudana Reddy, G., Prasad, C. D., Shetty, G., Ramesh, M. R., Nageswara Rao, T., & Patil, P. (2022). Investigation of thermally sprayed NiCrAlY/TiO_2 and NiCrAlY/Cr_2O_3/YSZ cermet composite coatings on titanium alloys. *Engineering Research Express, IOP*, 4, 025049. https://doi.org/10.1088/2631-8695/ac7946
62. Madhusudana Reddy, G., Prasad, C. D., Patil, P., Kakur, N., & Ramesh, M. R. (2022). Elevated Temperature erosion performance of plasma sprayed NiCrAlY/TiO_2 coating on MDN 420 steel substrate. *Surface Topography: Metrology and Properties, IOP*, 10, 025010. https://doi.org/10.1088/2051-672X/ac6a6e
63. Madhusudana Reddy, G., Prasad, C. D., Shetty, G., Ramesh, M. R., Nageswara Rao, T., & Patil, P. (2022). High temperature oxidation behavior of plasma sprayed NiCrAlY/TiO_2&NiCrAlY/Cr_2O_3/YSZ coatings on titanium alloy. *Welding in the World, Springer.* https://doi.org/10.1007/s40194-022-01268-7
64. Naik, T., Mathapathi, M., Prasad, C. D., Nithin, H. S., & Ramesh, M. R. (2022). Effect of laser post treatment on microstructural and sliding wear behavior of HVOF sprayed NiCrC and NiCrSi coatings. *Surface Review and Letters,* 29(1), 225000. https://doi.org/10.1142/S0218625X2250007X
65. Madhusudana Reddy, G., Prasad, C. D., Shetty, G., Ramesh, M. R., Nageswara Rao, T., & Patil, P. (2021). High temperature oxidation studies of plasma sprayed NiCrAlY/TiO_2&NiCrAlY/Cr_2O_3/YSZ cermet composite coatings on MDN-420 special steel alloy. *Metallography, Microstructure and Analysis,* Springer, 10, 642–651. https://doi.org/10.1007/s13632-021-00784-0
66. Dinesh, R., Rohan Raykar, S., Rakesh, T. L., Prajwal, M. G., Shashank Lingappa, M., & Durga Prasad, C. (2021). Feasibility study on MoCoCrSi/WC-Co cladding developed on austenitic stainless steel using microwave hybrid heating. *Journal of Mines, Metals & Fuels,* 69(12A). https://doi.org/10.18311/jmmf/2021/30113
67. Prasad, C. D., Lingappa, S., Joladarashi, S., Ramesh, M. R., & Sachin, B. (2021). Characterization and sliding wear behavior of CoMoCrSi+flyash composite cladding processed by microwave irradiation. *Materials Today Proceedings, Elsevier Science,* 46, 2387–2391. https://doi.org/10.1016/j.matpr.2021.01.156
68. Madhu, G., MrityunjayaSwamy, K. M., Kumar, D. K., Prasad, C. D., & Harish, U. (2021). Evaluation of hot corrosion behavior of HVOF thermally sprayed Cr_3C_2_35NiCr coating on SS 304 boiler tube steel. *American Institute of Physics*, 2316, 030014. https://doi.org/10.1063/5.0038279
69. Reddy, M. S., Prasad, C. D., Patil, P., Ramesh, M. R., & Rao, N. (2021). Hot corrosion behavior of plasma sprayed NiCrAlY/TiO_2 and NiCrAlY/Cr_2O_3/YSZ cermetscoatings on alloy steel. *Surfaces and Interfaces, Elsevier Science*, 22, 100810. https://doi.org/10.1016/j.surfin.2020.100810
70. Prasad, C. D., Joladarashi, S., Ramesh, M. R., & Srinath, M. S. (2020). Microstructure and tribological resistance of flame sprayed CoMoCrSi/WC-CrC-Ni and CoMoCrSi/WC-12Co composite coatings remelted by microwave hybrid heating. *Journal of Bio and Tribo-Corrosion, Springer*, 6, 124. https://doi.org/10.1007/s40735-020-00421-3
71. Prasad, C. D., Joladarashi, S., & Ramesh, M. R. (2020). Comparative investigation of HVOF and flame sprayed CoMoCrSi coating. *American Institute of Physics*, 2247, 050004. https://doi.org/10.1063/5.0003883

72. Prasad, C. D., Jerri, A., & Ramesh. M. R. (2020). Characterization and sliding wear behavior of iron based metallic coating deposited by HVOF process on low carbon steel substrate. *Journal of Bio and Tribo-Corrosion, Springer*, 6, 69. https://doi.org/10.1007/s40735-020-00366-7
73. Prasad, C. D., Joladarashi, S., Ramesh, M. R., Srinath, M. S., & Channabasappa, B. H. (2020). Comparison of high temperature wear behavior of microwave assisted HVOF sprayed CoMoCrSi-WC-CrC-Ni/WC-12Co composite coatings. *Silicon, Springer*, 12, 3027–3045, 1–19. https://doi.org/10.1007/s12633-020-00398-1
74. Prasad, C. D., Joladarashi, S., Ramesh, M. R., Srinath, M. S., & Channabasappa, B. H. (2019). Effect of microwave heating on microstructure and elevated temperature adhesive wear behavior of HVOF deposited CoMoCrSi-Cr_3C_2 composite coating. *Surface and Coatings Technology, Elsevier Science*, 374, 291–304. https://doi.org/10.1016/j.surfcoat.2019.05.056
75. Prasad, C. D., Joladarashi, S., Ramesh, M. R., Srinath, M. S., & Channabasappa, B. H. (2019). Development and sliding wear behavior of Co-Mo-Cr-Si cladding through microwave heating. *Silicon, Springer*, 11, 2975–2986. https://doi.org/10.1007/s12633-019-0084-5
76. Prasad, C. D., Joladarashi, S., Ramesh, M. R., Srinath, M. S., & Channabasappa, B. H. (2019). Microstructure and tribological behavior of flame sprayed and microwave fused CoMoCrSi/CoMoCrSi-Cr_3C_2 coatings. *Materials Research Express, IOP*, 6, 026512. https://doi.org/10.1088/2053-1591/aaebd9
77. Prasad, C. D., Joladarashi, S., Ramesh, M. R., Srinath, M. S., & Channabasappa, B. H. (2018). Influence of microwavehybrid heating on the sliding wear behaviour of HVOF sprayed CoMoCrSi coating. *Materials Research Express, IOP*, 5, 086519. https://doi.org/10.1088/2053-1591/aad44e
78. Prasad, C. D., Joladarashi, S., Ramesh, M. R., Sarkar, A. (2018). High temperature gradient cobalt based clad developed using microwave hybrid heating. *American Institute of Physics*, 1943, 020111. https://doi.org/10.1063/1.5029687
79. Girisha, K. G., Sreenivas Rao K. V., & Durga Prasad, C. (2018). Slurry erosion resistance of martenistic stainless steel with plasma sprayed Al_2O_3_40%TiO_2 coatings. *Materials Today Proceedings*, 5, 7388–7393. https://doi.org/10.1016/j.matpr.2017.11.409
80. Girisha, K. G., Durga Prasad, C., Anil, K. C., & Sreenivas Rao, K. V. (2015). Dry sliding wear behaviour of Al_2O_3 coatings for AISI 410 grade stainless steel. *Applied Mechanics and Materials*, 766–767, 585–589. https://doi.org/10.4028/www.scientific.net/AMM.766-767.585
81. Girisha, K. G, Rakesh, R., Durga Prasad, C., & Sreenivas Rao, K. V. (2015). Development of corrosion resistance coating for AISI 410 grade steel. *Applied Mechanics and Materials*, 813–814, 135–139. https://doi.org/10.4028/www.scientific.net/AMM.813-814.135
82. Prasad, C. D., Manjunath, K. V., Naik, P. K., Nagabhushana, N., & Kumar, P. (2024). Development of MoCoCrSi/Flyash composite cladding on stainless steel substrate through microwave irradiation. In: Thakur, L., Singh, J., Vasudev, H. (eds.), *Thermal Claddings for Engineering Applications*. Boca Raton, FL: CRC Press, Taylor & Francis Group, ISBN 9781032460543. https://doi.org/10.1201/9781032713830
83. Prasad, C. D., Mathapathi, M., Vasudev, H., & Thakur, L. (2022). Analysis of mechanical properties and microstructural characterisation of microwave cladding on stainless steel. In: Bansal, A., Vasudev, H. (eds.), *Advances in Microwave Processing for Engineering Materials*. Boca Raton, FL: CRC Press, Taylor & Francis Group, ISBN: 9781003248743. https://doi.org/10.1201/9781003248743
84. Prasad, C. D., Mathapathi, M., Joladarashi, S., & Ramesh, M. R. (2021). Investigation of microstructural and tribological behavior of metco 41C+WC-12Co composite coating sprayed by HVOF process. In: Thakur, L., Vasudev, H. (eds.), *Thermal Spray Coatings*. Boca Raton, FL: CRC Press, Taylor & Francis Group, ISBN: 9781032081489. https://doi.org/10.1201/9781003213185

85. Prasad, C. D., Jerri, A., & Ramesh, M. R. (2021). Evaluation of microstructural and dry sliding wear resistance of iron-based SiC-reinforced composite coating by HVOF process. In: Thakur, L., Vasudev, H. (eds.), *Thermal Spray Coatings*. Boca Raton, FL: CRC Press, Taylor & Francis Group. ISBN: 9781032081489. https://doi.org/10.1201/9781003213185
86. Prashar, G., Vasudev, H., & Thakur, L. (2022). High-temperature oxidation and erosion resistance of Ni-based thermally-sprayed coatings used in power generation machinery: a review. *Surface Review and Letters*, 29(03), 2230003.
87. Singh, G., Vasudev, H., Bansal, A., & Vardhan, S. (2021). Influence of heat treatment on the microstructure and corrosion properties of the Inconel-625 clad deposited by microwave heating. *Surface Topography: Metrology and Properties*, 9(2), 025019.
88. Prashar, G., & Vasudev, H. (2020). Hot corrosion behavior of super alloys. *Materials Today: Proceedings*, 26, 1131–1135.
89. Vasudev, H., Prashar, G., Thakur, L., & Bansal, A. (2022). Electrochemical corrosion behavior and microstructural characterization of HVOF sprayed inconel718-Al_2O_3 composite coatings. *Surface Review and Letters*, 29(02), 2250017.
90. Singh, P., Bansal, A., Vasudev, H., & Singh, P. (2021). In situ surface modification of stainless steel with hydroxyapatite using microwave heating. *Surface Topography: Metrology and Properties*, 9(3), 035053.
91. Vasudev, H., Singh, P., Thakur, L., & Bansal, A. (2020). Mechanical and microstructural characterization of microwave post processed Alloy-718 coating. *Materials Research Express*, 6(12), 1265f5.
92. Prashar, G., & Vasudev, H. (2022). Structure–property correlation of plasma-sprayed Inconel625-Al_2O_3 bimodal composite coatings for high-temperature oxidation protection. *Journal of Thermal Spray Technology*, 31(8), 2385–2408.
93. Singh, M., Vasudev, H., & Kumar, R. (2021). Corrosion and tribological behaviour of BN thin films deposited using magnetron sputtering. *International Journal of Surface Engineering and Interdisciplinary Materials Science (IJSEIMS)*, 9(2), 24–39.
94. Singh, M., Vasudev, H., & Singh, M. (2022). Surface protection of SS-316L with boron nitride based thin films using radio frequency magnetron sputtering technique. *Journal of Electrochemical Science and Engineering*, 12(5), 851–863.
95. Kulkarni, S. D., Manjunatha, Chandrasekhar, U., Manjunath, K. V., Prasad, C. D., & Vasudev, H. (2024). Design and optimization of polyvinyl-nitride rubber for tensile strength analysis. *International Journal on Interactive Design and Manufacturing (IJIDeM)*, 18(8), 5893–5908.
96. Prashar, G., & Vasudev, H. (2022). Surface topology analysis of plasma sprayed Inconel625-Al_2O_3 composite coating. *Materials Today: Proceedings*, 50, 607–611.
97. Sharanabasava, H., Raviprakash, M., Prasad, C. D., Ramesh, M. R., Phanibhushana, M. V., Vasudev, H., & Kumar, S. (2023). Microstructure, mechanical and wear properties of SiC and Mo reinforced NiCr microwave cladding. *Advances in Materials and Processing Technologies*, 10, 3620–3633.
98. Poojari, M., Hanumanthappa, H., Prasad, C. D., Jathanna, H. M., Ksheerasagar, A. R., Shetty, P., ... & Vasudev, H. (2024). Computational modelling for the manufacturing of solar-powered multifunctional agricultural robot. *International Journal on Interactive Design and Manufacturing (IJIDeM)*, 18(8), 5725–5736.
99. Prashar, G., & Vasudev, H. (2022). A review on the influence of process parameters and heat treatment on the corrosion performance of Ni-based thermal spray coatings. *Surface Review and Letters*, 29(01), 2230001.
100. Yedida, V. S., & Vasudev, H. (2022). A review on the development of thermal barrier coatings by using thermal spray techniques. *Materials Today: Proceedings*, 50, 1458–1464.

101. Kattimani, M. A., Venkatesh, P. R., Masum, H., Math, M. M., Bahadurdesai, V. N., Mustafkhadri, S., ... & Vasudev, H. (2024). Design and numerical analysis of tensile deformation and fracture properties of induction hardened inconel 718 superalloy for gas turbine applications. *International Journal on Interactive Design and Manufacturing (IJIDeM)*, 18(8), 5939–5949.
102. Dutta, V., Thakur, L., Singh, B., & Vasudev, H. (2022). A study of erosion–corrosion behaviour of friction stir-processed chromium-reinforced NiAl bronze composite. *Materials*, 15(15), 5401.
103. Majji, B. G. R., Vasudev, H., & Bansal, A. (2022). A review on the oxidation and wear behavior of the thermally sprayed high-entropy alloys. *Materials Today: Proceedings*, 50, 1447–1451.
104. Sunitha, K., & Vasudev, H. (2022). A short note on the various thermal spray coating processes and effect of post-treatment on Ni-based coatings. *Materials Today: Proceedings*, 50, 1452–1457.
105. Rao, G. S., Mukkamala, U., Hanumanthappa, H., Prasad, C. D., Vasudev, H., Shanmugam, B., & KishoreKumar, K. C. (2023). Evaluating and optimizing surface roughness using genetic algorithm and artificial neural networks during turning of AISI 52100 steel. *International Journal on Interactive Design and Manufacturing (IJIDeM)*, 18, 1–10.
106. Singh, J., Gill, H. S., & Vasudev, H. (2023). Computational fluid dynamics analysis on role of particulate shape and size in erosion of pipe bends. *International Journal on Interactive Design and Manufacturing (IJIDeM)*, 17(5), 2631–2646.
107. Anjaneya, G., Sunil, S., Kakkeri, S., Math, M. M., Vaibhav, M. N., Solaimuthu, C., ... & Vasudev, H. (2023). Numerical simulation of microchannel heat exchanger using CFD. *International Journal on Interactive Design and Manufacturing (IJIDeM)*, 18, 1–17.
108. Nagabhushana, P., Ramprasad, S., Prasad, C. D., Vasudev, H., & Prakash, C. (2024). Numerical investigation on heat transfer of a nano-fluid saturated vertical composite porous channel packed between two fluid layers. *International Journal on Interactive Design and Manufacturing (IJIDeM)*, 18(5), 2927–2944.
109. Singh, P., Vasudev, H., & Bansal, A. (2024). Effect of post-heat treatment on the microstructural, mechanical, and bioactivity behavior of the microwave-assisted alumina-reinforced hydroxyapatite cladding. *Proceedings of the Institution of Mechanical Engineers, Part E: Journal of Process Mechanical Engineering*, 238(3), 1060–1071.
110. Mahantesh, M. M., Rao, K. R., Chandra, A. P., Vijayakumar, M. N., Nandini, B., Prasad, C. D., & Vasudev, H. (2024). Design and modeling using finite element analysis for the sitting posture of computer users based on ergonomic perspective. *International Journal on Interactive Design and Manufacturing (IJIDeM)*, 18(8), 5875–5891.
111. Singh, J., Singh, R., Singh, S., Vasudev, H., & Kumar, S. (2023). Reducing scrap due to missed operations and machining defects in 90PS pistons. *International Journal on Interactive Design and Manufacturing (IJIDeM)*, 17(5), 2527–2539.
112. Arora, H., Kumar, V., Prakash, C., Pimenov, D., Singh, M., Vasudev, H., & Singh, V. (2021). Analysis of sensitization in austenitic stainless steel-welded joint. In: Prakash, C., Krolczyk, G., Singh, S., Pramanik, A. (eds.), *Advances in Metrology and Measurement of Engineering Surfaces: Select Proceedings of ICFMMP 2019* (pp. 13–23). Singapore: Springer.
113. Vasudev, H. (2020). Wear characteristics of Ni-WC powder deposited by using a microwave route on mild steel: microwave cladding of Ni-WC. *International Journal of Surface Engineering and Interdisciplinary Materials Science (IJSEIMS)*, 8(1), 44–54.
114. Prashar, G., Vasudev, H., Thakur, L., & Bansal, A. (2023). Performance of thermally sprayed hydroxyapatite coatings for biomedical implants: a comprehensive review. *Surface Review and Letters*, 30(01), 2241001.

6 Role of Silicon-Based Coatings in Combating High-Temperature Oxidation and Corrosion Material Degradation

Channabasavaiah Durga Prasad, N. Ramesh Babu, Mohd Hamid Hussain, Ranjeet Kumar Choudhary, S. Balaji, G. R. Gurunagendra, Saravana Bavan, G. Veeresha, and Suresh Gudala

6.1 INTRODUCTION

It has been known since the prewar era that the addition of silicon improved oxidation properties of heat-resistant steel and nichrome heater alloys [1–3]. In the case of silicon, the fact that it produces a passive oxide layer, silicon dioxide (SiO_2), on exposure to high temperature has made it indispensable in high-temperature-resistant alloys as well as high-temperature coatings [4–6]. Silicon is applied in heat-resistant steels and nichrome alloys to enhance their oxidation characteristics because it is a vital factor in the use of items that operate at high temperatures for long durations, as in furnaces, turbines, and heating components, among others [7–10]. In heat-resistant steels, silicon improves the oxide scale formation in such a way that it forms a stable, rigid scale that does not allow further metal to oxidize [11–13]. Due to existence of silicon in the framework of the alloy, formation of SiO_2 at the surface of the material hinders the penetration of oxygen and other corrosive components into the framework of the material [14–16]. In this case, the silicon oxide layer has extremely small and slow permeability to oxygen, and hence the process of diffusion, which leads to oxidation in most metals, is greatly impeded. This oxide layer is stable at high temperatures and remains protective even under conditions of high-temperature fluctuations, making it extremely important in extending the service life of components exposed to thermal cycling and high temperatures [17–21]. Silicon enhances the heat-resistant durability and ability of steels to withstand scaling, particularly in oxidative conditions that are typical in industrial uses of steels [22–26].

Similarly, in nichrome heater alloys, nickel (Ni) and chromium (Cr), being basic metals, also incorporate silicon, which has proven effective for high-temperature

 DOI: 10.1201/9781003542957-6

applications in these alloys [27–30]. Nichrome is most commonly utilized for electrical heating elements because it is very resistant to heat and does not readily degrade in structure under high-temperature conditions [31–34]. However, higher silicon content still further improves these properties for the alloys, as mentioned below, and a SiO_2 layer forms on the surface of the alloy with higher silicon content. This silicon oxide layer also protects the underlying layer from oxidation and scaling, typical problems characteristic of highly oxidizing environments at high temperatures [35–38]. The silicon-enriched oxide layer also prevents the consumption of chromium within the alloy since the formation of protective chromia (Cr_2O_3) scales is achieved with less use of chromium [39–42]. When used in conjunction with chromium, carburization preserves chromium, significantly increasing the durability and service life of heating elements in severe operating conditions, such as kilns, furnaces, and electric heaters [43–45].

Silicon as a coating material, a mainly popular form of silicide coatings, was of interest from the initial research and development concerning automatic possibilities of refractory metals at high-temperature technologies. For instance, a study covered molybdenum disilicide ($MoSi_2$) coatings and applications on refractory metals in different high-temperature oxidizing surroundings [13,46–50]. By virtue of oxidation, when exposed to high temperatures, $MoSi_2$ coatings develop a barrier layer of SiO_2 that acts as a barrier against oxidation and insulates the substrate against heat. Such a coating is valuable in applications where refractory metals such as molybdenum, tungsten, or tantalum may be used because this allows such materials to operate at temperatures higher than 1,200°C.

The coating of high-temperature corrosion-resistant materials using silicon-based compounds has received more emphasis in recent years. There is an outstanding need for materials that can withstand higher industrial conditions. Scientists and technologists are concerned with the formulation, microstructure, and method of application of these coatings to achieve their full protection potential. Thermal spray coating, chemical vapor deposition (CVD), and physical vapor deposition (PVD) provide the opportunity to control the coating's dimension, homogeneity, and adhesion to a base material, making silicon-based coatings suitable for high-temperature applications [14,51–58]. However, sustained research on improving silicon-carburized composite coatings shows promise in enhancing corrosion protection with silicon added to elements like aluminum, chromium, and rare earth metals [15,59–62].

6.2 MECHANISM OF HIGH-TEMPERATURE CORROSION

6.2.1 Oxidation

Oxidation is the most common form of high-temperature corrosion; it results from a reaction between a metal and oxygen; at high temperatures, the metal forms oxide. This process can cause a very high reduction of material, and it is often evidenced in parts used in very hot areas such as gas turbines, boilers, and furnaces. At first, metal oxidation happens quickly because a thin oxide layer is produced on its surface [16]. This oxide layer can become a shield inasmuch as it bonds to the foundation material and is cohesive, plugging the oxygen and decreasing the rate of further oxidation.

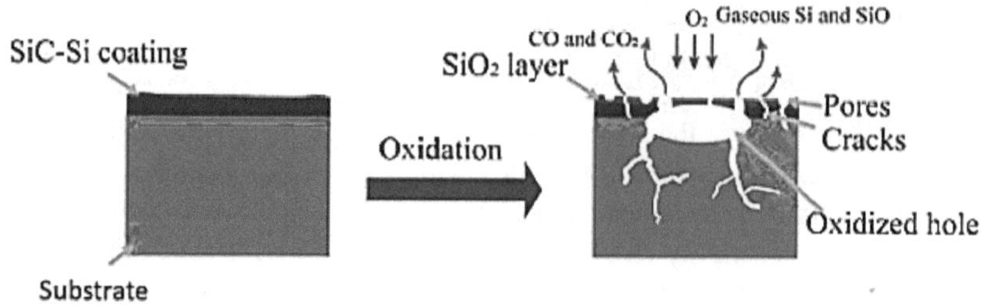

FIGURE 6.1 Oxidation mechanisms of Si-based coatings.

Nonetheless, in some cases, the oxide layer may disadhere or crack due to thermal cycling or mechanical loading; as a result, fresh metal is exposed to oxygen, which further enhances corrosion. The extent of the protective behavior of the oxide layer varies from one metal to the other and depends upon the stability of the oxide formed [17,63–65].

Figure 6.1 illustrates the oxidation behavior of a substrate coated with silicon carbide (SiC). The SiC coating forms a protective silica (SiO_2) film on the surface when exposed to high temperatures, which turns as a barrier against oxygen diffusion. A protective oxide layer significantly reduces the oxidation rate, enhancing the substrate's resistance to high-temperature degradation. As oxidation progress, the SiO_2 layer thickens, further limiting oxygen ingress and maintaining substrate integrity [18,66–70].

6.2.2 Sulfidation

Sulfidation is a specialized type of hot corrosion that occurs because of reaction of a metal with sulfur-containing derivatives like H_2S, SO_2, and metallic sulfates at high temperatures. This process results in the formation of metal sulfides rather than oxides as formed in the usual electrolytes, the latter offering less protection against subsequent corrosion. Sulfidation is severe in high-sulfur environments like coal combustion systems, petrochemical refining, and gas turbine applications. Unlike oxidation, the consequences of sulfide scales that form in sulfidation environments tend to be porous, non-adherent, and liable to cracking or spalling, and as such, the underlying metal exposed to the corrosive environment is subjected to severe corrosion, which tends to cause rapid material loss [19,71–74]. The degree of sulfidation also varies with the reaction temperature, the concentration of sulfur species, and the kind of metal or alloy.

6.2.3 Carburization

Carburization is high-temperature corrosion where the atoms of carbon penetrate and settle on the surface of the metal or an alloy and create carbides. This process is normally observed in facilities involving the processing of hydrocarbons, carbon monoxide, or other carbonaceous gases at high temperatures. Carbides, such as Cr_3C_2, iron carbide (Fe_3C), and others, change the mechanical properties of a material at an extreme level. The existence of carbides in the metal decreases the toughness of the metal, and when a component is exposed to mechanical pressure or thermal shocks, it becomes prone to cracking [20,75,76]. Carburization is widely experienced in petrochemical industries where the components are in contact with a carbon-containing environment in processes like catalytic reforming or hydrocarbon cracking.

6.3 TYPES OF SILICON-BASED COATINGS

Silicon-based coatings (Figure 6.2) stay widely used in various industrial sectors due to excellent assets, such as, oxidation resistance, high-temperature stability, and wear resistance. These coatings have been tailored for different environmental

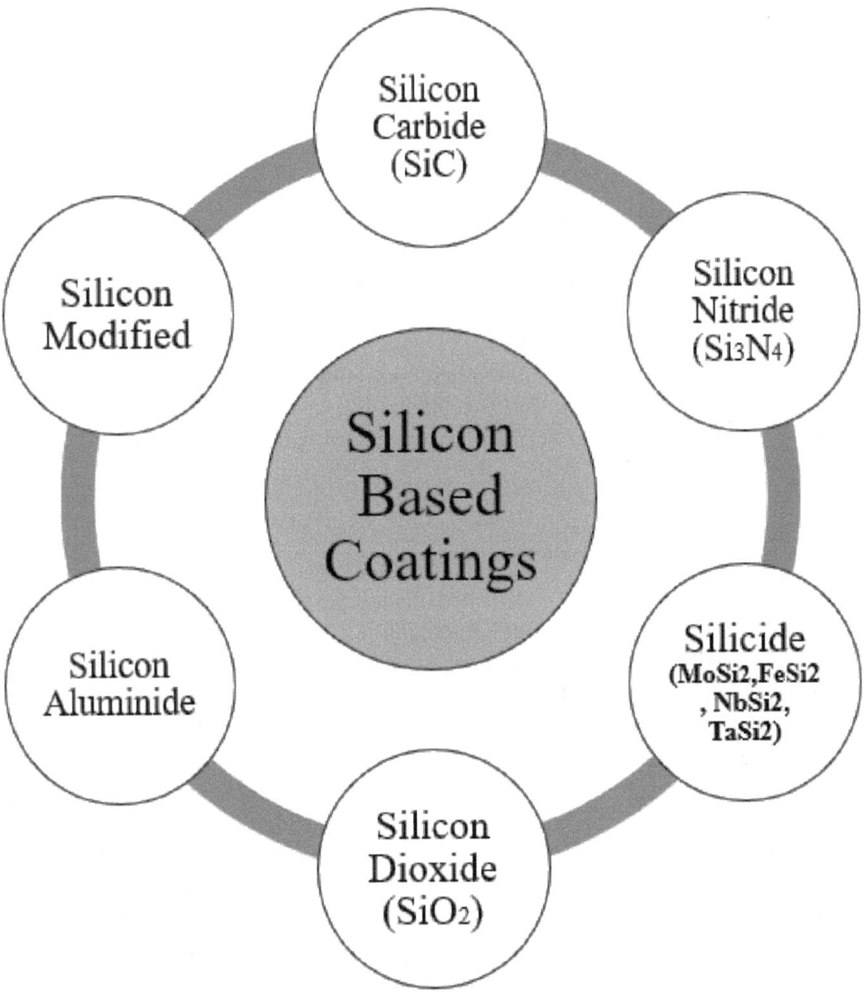

FIGURE 6.2 Silicon-based coatings at high temperatures.

and operational conditions, enhancing the durability and performance of materials exposed to extreme temperatures, corrosive environments, and mechanical wear [21].

The types of silicon-based coatings and all of these coactions have different attributes and uses. SiC coatings represent stability and maximum hardness at high temperatures second to none and are perfect for turbine blades and cutting tools. Si_3N_4 or silicon nitride is another layer coating identified by high hardness and wear resistance used in automobile engine parts and bearings. Hermetic silicide coatings show very high tolerance toward oxidation and hot corrosion at elevated temperatures; $MoSi_2$ face coatings have been found to be suitable for temperatures up to 1,700°C in power generation systems and gas turbines [22]. These types of coatings are therefore very appropriate for such areas of applications as high-temperature electronics and optical uses due to outstanding oxidation and general corrosion resistance properties of SiO_2. Silicon-aluminide coatings provide an alumina layer for oxidation and hot corrosion resistance in gas turbine and petrochemical areas [23]. M-Cr-Al-Y silicon coatings, known as silicon-modified coatings, can improve oxidation characteristics necessary for aerospace and power generation. For example, numerous works prove the efficiency of these types of coatings based on silicon in various applications, indicating their relevance for application in modern industry [24].

6.4 METHODS TO APPLY SILICON-BASED COATINGS

Silicon-based coatings, known for their exceptional resistance to high temperatures, oxidation, and corrosion, are applied using various techniques. The most protruding methods are thermal spray coatings, CVD, and PVD. These methods differ in relations of application processes, material compatibility, and the desired performance characteristics of the coatings. Each technique has its own advantages and specific industrial applications, with thermal spray coatings widely used for high-temperature environments [77–81].

6.4.1 Thermal Spray Coatings

Thermal spray coatings are a widely used technique that is applied to enhance the properties of substrate material by depositing a coat on its surface. This method involves converting a material into a powder or wire form and then heating the material, after which it is sprouted in the form of a high-velocity process, which is illustrated in Figure 6.3 [25]. Also, the method widely adopted for applying silicon-based coatings, particularly SiC and Si_3N_4, is used to protect components subjected to extreme conditions such as high temperatures and abrasive environments. The process involves heating the coating material, usually in powder or wire form, to a molten or semi-molten state and then propelling it onto the substrate's surface. The coating material cools and solidifies upon impact, forming a dense and protective layer [26].

This technique is particularly effective in enhancing wear resistance, corrosion protection, and thermal insulation of industrial components using thermal spray coatings. Plasma spraying, which is most common form, uses an electric arc to create a plasma jet that melts the silicon-based powder. The method was suitable for high-melting-point materials such as SiC, providing a dense, high-quality coating

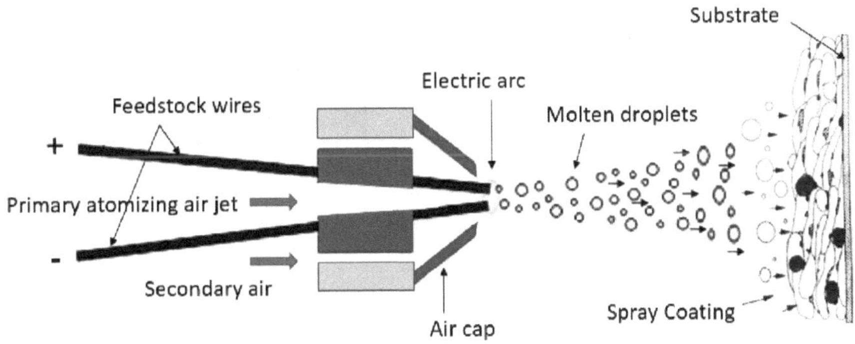

FIGURE 6.3 Thermal spray process.

that offers excellent protection against oxidation and corrosion at higher temperatures. It is extensively used in aerospace, automotive, and power generation industries for coating turbine blades, engine components, and wear parts [27]. Additionally, high-velocity oxy-fuel (HVOF) spraying is another thermal spray method in which fuel and oxygen are combusted at high pressure on the way to melt the coating material. The molten particles are then accelerated toward the substrate at supersonic speeds, resulting in a very dense and well-bonded coating. HVOF is particularly effective for applying silicon-based coatings in environments that require resistance to wear and oxidation, such as in industrial gas turbines and petrochemical applications. The high velocity of the sprayed particles ensures a compact coating with minimal porosity, making it suitable for high-temperature and corrosive conditions [28]. Moreover, in flame spraying, the silicon-based material is melted by an oxy-fuel flame before being sprayed onto the surface. While not as precise or controlled as plasma or HVOF spraying, flame spraying is a cost-effective method for applying silicon-based coatings in less demanding environments. It is commonly used for repairing worn parts and providing a basic layer of corrosion and wear protection on components like valves, shafts, and bearings [29,82–86].

Thermal spray coatings are favored for their ability to provide thick, durable coatings that can withstand harsh conditions. The process allows for flexibility in coating large surface areas and, unlike the use of silicon-based materials such as SiC and Si_3N_4, allows for tailored solutions depending on the specific needs of the application. They also provide significant advantages in terms of flexibility, thickness control, and the ability to apply a wide range of materials.

6.4.2 Chemical Vapor Deposition

CVD is another method for applying coatings, particularly for high-precision silicon-based applications. In CVD, gaseous reactants are presented into a reaction chamber containing the substrate shown in Figure 6.4. A chemical reaction occurs on the substrate surface, depositing a thin layer of silicon-based material, such as SiC or silicon nitride. The reaction is often initiated by heat, plasma, or light, depending on the specific CVD process used [30,87–90].

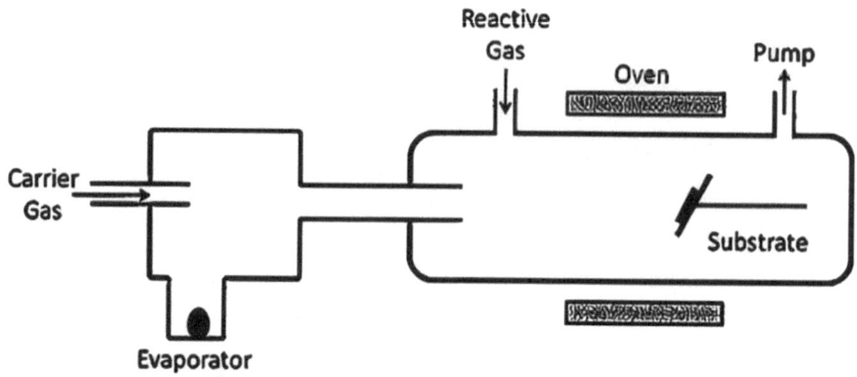

FIGURE 6.4 Chemical vapor deposition method.

CVD offers several advantages, including the ability to produce highly uniform coatings with excellent adhesion and low porosity. This method is suitable for complex geometries and components that require coatings with precise thickness control. CVD is widely used in the semiconductor industry, where high-purity silicon-based coatings are essential for protecting electronic components from oxidation and corrosion [31].

6.4.3 Physical Vapor Deposition

PVD remains a technique where the coating material is vaporized in a vacuum chamber and then condensed onto the substrate to form a thin film. PVD methods, including sputtering and evaporation, are commonly used for applying silicon-based coatings, particularly in high-precision industries like electronics, aerospace, and medical devices (Figure 6.5).

One of the main advantages of PVD is its ability to produce a very thin, dense coating with exceptional adhesion and smooth surfaces. This makes it superlative for applications where surface finish and coating thickness must be tightly controlled. The application of SiO_2 coatings using PVD for high-temperature electronics highlights the protective role these coatings play against oxidation and their optical transparency, which is crucial for optical devices exposed to extreme conditions [32]. While PVD typically produces thinner coatings than thermal spray methods, it is a highly effective application that has control over coating properties. Additionally, PVD processes can be conducted at relatively lower temperatures, making them suitable for substrates that may be sensitive to heat.

6.5 SILICON-BASED COATINGS FOR HIGH-TEMPERATURE APPLICATIONS

Silicon is a commonly beneficent element, and the functionality for the formation of stoichiometric oxide layers, mainly SiO_2 in a creeping temperature and oxidizing atmosphere, boosts the oxidation endurance. Silicon-based coatings are resistant

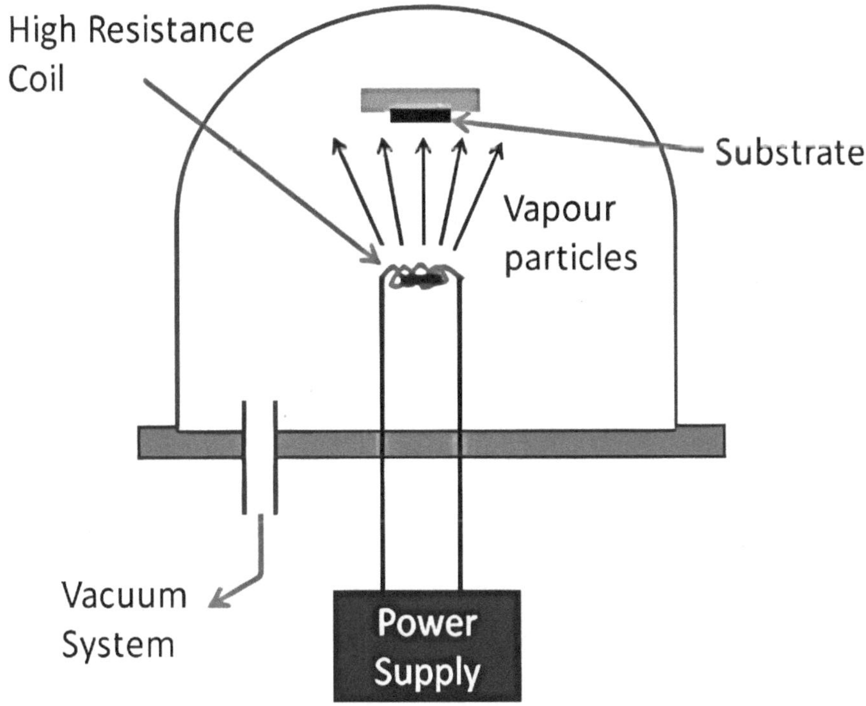

FIGURE 6.5 Physical vapor deposition technique.

to heat and do not degrade their mechanical properties and protective features at high-functioning temperatures, which makes them suitable for the elements in gas turbines, combustion chambers, and heat exchangers [33]. For instance, $MoSi_2$ coatings have been extensively investigated and used for the oxidation protection of refractory metals at high temperatures. Besides, $MoSi_2$ coatings also protect the surface of a solid material from oxidation and offer the added capability of insulating the heat from the substrate at high temperatures [34]. This sort of paint is a predominant causative in applications where such refractory metals as molybdenum, tungsten, or tantalum are used, making it possible for these to work at temperatures over 1,200°C. The instrument illustrations of $MoSi_2$ layer before and after oxidation at 1,300°C are shown in Figure 6.6 [35]. The coating generated SiO_2 oxide and Mo_5Si_3 segments on the surface with increasing oxidation time, and thickness of Mo_5Si_3 diffusion layer gradually increased.

Silicide diffusion coatings on nickel-based superalloys have been recognized for their superior mechanical strength and resistance to creep at high temperatures, making them perfect for turbine blades and other important parts in aerospace and power generation applications. However, these superalloys are prone to oxidation and hot corrosion in the harsh environments typical of gas turbines, where fuel combustion produces corrosive gases. Silicide diffusion coatings provide a protective barrier that enhances the oxidation and hot corrosion resistance of these

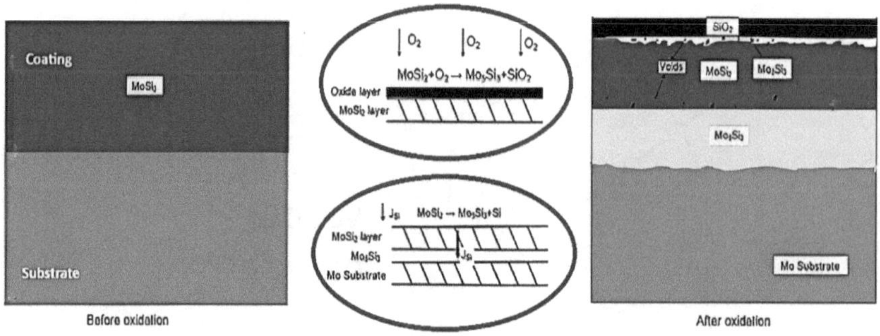

FIGURE 6.6 Representation design of oxidation process of $MoSi_2$ layer at 1,300°C.

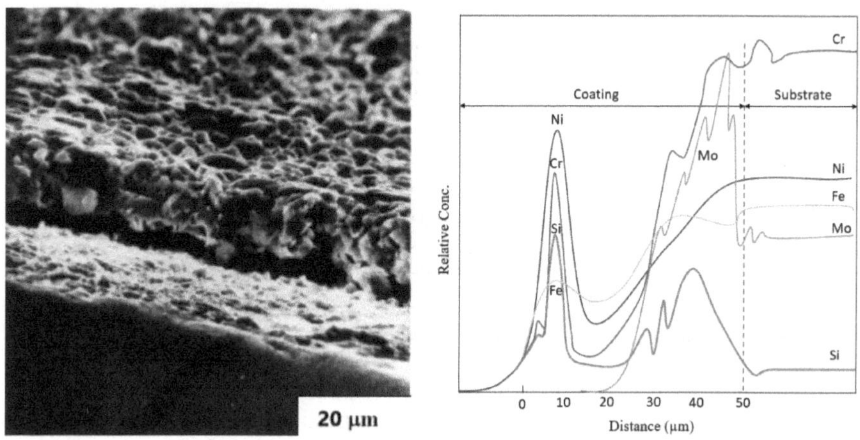

FIGURE 6.7 (a) SEM morphology and (b) electron microprobe analysis (EMPA) profile cross-section of silicon-comprising coating produced at 950°C as of $SiCl_4$-H_2 gas blend. SEM, Scanning electron microscopy.

superalloys, thus extending their service life [36]. In Figure 6.7a, the microstructures of the coatings produced by a gas mix of $SiCl_4$-H_2 at 900–1,000°C consist of two layers: a top layer mainly made up of nickel-rich mixed silicides and a lower layer predominantly composed of molybdenum-rich mixed silicides. It is thought that stable silicides are created through the siliconizing process and gather under the surface. Figure 6.7b illustrates the findings of the electron microprobe analysis of cross-sectional samples.

Evaluations of chromium and aluminum-based coatings, silicide, and low-silicon content coatings have demonstrated superior hot corrosion resistance. Chromium-based coatings, while effective in providing oxidation resistance through the formation of Cr_2O_3 scales, tend to suffer from spallation and depletion over time, especially in environments containing sulfur or chlorine.

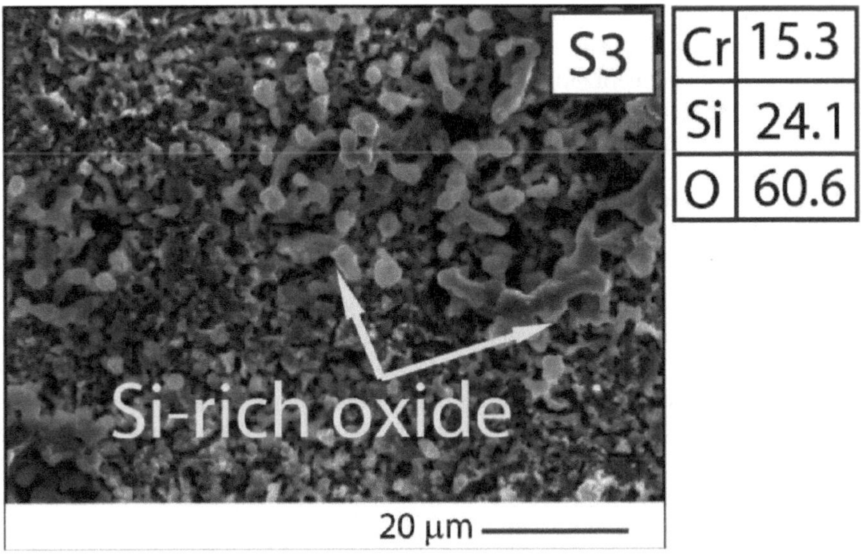

FIGURE 6.8 The internal structure of the scales that peeled off from alloys S3 after being exposed at 1,200°C for 100 hours. The tables provide the EDS composition of the specified region in atomic percent.

The inner surface appearance of scales that flaked off alloy S3 (Cr 97.0% ± 0.2%, Si 3.0% ± 0.2%) after 100 hours of exposure at 1,200°C is depicted in Figure 6.8. In the following figures, the EDS spectrum of marked areas shows the presence of elements in atomic percentages. The morphology reveals significant variances in the oxide coating adhesion and chemical concentration of two alloys. Alloy S16 (Cr 84.2% ± 0.3%, Si 15.8% ± 0.3%) has a continuous and protective oxide layer, while Alloy S3 has regions of spallation and oxide depletion [37].

Similarly, aluminum-based coatings form protective Al_2O_3 scales, but they can be vulnerable to aggressive hot corrosion at high temperatures. Silicide coatings, with their SiO_2 protective layers, offer a more robust solution, particularly in environments where hot corrosion is a primary concern. Silicon-based coatings also exhibit better thermal stability, ensuring that the protective oxide layer remains intact even after prolonged exposure to extreme temperatures [38,91–95].

6.6 PERFORMANCE OF SILICON-BASED COATINGS IN HARSH ENVIRONMENTS

6.6.1 Potential Behavior of Silicides in the Gaseous/Molten Salt Matrices

Silicides encounter two main challenges: oxidation and corrosion in gas-molten salt environments, particularly in areas such as gas turbines, combustion chambers, and other energy-related systems. The presence of molten salts, such as sodium sulfate (Na_2SO_4) or sodium chloride (NaCl), in combination with oxidative gases, can result in various corrosion mechanisms. The behavior of silicides under these conditions depends largely on the stability of their oxide layers, the solubility of their components in molten salts, and the formation of protective or harmful intermediate phases [39].

In an oxidative atmosphere, silicides tend to form a protective silica (SiO_2) layer on the surface. This silica layer acts as a barrier against further oxidation by preventing oxygen from penetrating the material. For example, $MoSi_2$ forms a dense and adherent SiO_2 scale that protects the underlying metal from oxidation at temperatures as high as 1,600°C. The efficiency of this oxide layer depends on the environmental temperature and the stability of SiO_2 under the given conditions. However, in the presence of molten salts, the silica layer can be disrupted. Molten salts may dissolve or penetrate the protective SiO_2 layer, leading to accelerated corrosion. The dissolution of SiO_2 into molten salts can result in the formation of volatile species, such as silicon oxychlorides, that degrade the protective layer [40]. The continuous exposure to molten salts and oxidative gases can thus erode the silicide surface, leading to rapid material degradation.

Molten salts also facilitate sulfidation processes, where sulfur-containing species react with the silicide surface. Sulfidation is particularly damaging in environments where the molten salts are derived from sodium sulfate (Na_2SO_4) or other sulfur-based compounds [41]. These salts can interact with the protective SiO_2 layer, causing its breakdown and enabling sulfur to penetrate the silicide. This results in the formation of non-protective compounds, such as sulfides, which degrade the material's integrity. Another mode of behavior in gas-molten salt environments involves the formation of volatile silicate compounds. In environments containing chlorine or sulfur, silicides can react to form volatile silicon oxychlorides ($SiCl_4$) or silicon oxysulfides. These compounds can easily evaporate, causing material loss and weakening the silicide surface [42]. This volatilization process can be accelerated by high temperatures and the presence of molten salts, leading to rapid material failure. To mitigate the degradation of silicides in gas-molten salt environments, advanced coating techniques are often employed. For instance, applying protective coatings, such as aluminides or chromium-based layers, can prevent direct exposure to corrosive species. Additionally, modifying the composition of silicides to include elements like titanium (Ti) or aluminum (Al) can improve the solidity of the protective SiO_2 coat and reduce the risk of sulfidation or oxidation [43].

The common phase constitution and structure for diffusion coatings are listed in Table 6.1.

6.6.2 Refractory Metals and Graphite Silicide Coatings

Refractory alloys and graphite are typically only used in high-temperature applications in environments that do not contain oxygen. Substances like molybdenum (Mo), niobium (Nb), and tantalum (Ta) cannot create protective oxide layers because of problems like oxide evaporation or the development of porous oxide structures. Efforts to improve oxidation resistance by alloying these materials with elements that form protective scales have been unsuccessful [50]. Significant research has been dedicated to applying coatings that allow these materials to be used in oxidizing

TABLE 6.1
Diffusion Coating

Element	Phase Structures	Base Material	Coating Methods	Appearance Tests
Si [44]		Ni and Fe based	Plasma spray coating	Cyclic oxidation, V_2O_5 corrosion
Si [45]	Ni_5Si_2	AKN 21	Plasma spray, CVD	Cyclic oxidation, V_2O_5 corrosion
Si [46]	MSi, M_2Si, M_5Si_3, indication of M_3Si_2 and Laves phase	Nimonic 90,105; IN 713 LC, 738 X; MAR M 246,421,432	CVD	Hot corrosion rig at 850°C. thermal stability, creep rupture
Si [47]	NiSi, Ni_3Si_2	IN 738 LC	Plasma spray, CVD	Oxidation at 900°C electrochemical deterioration, deterioration rig at 850°C
Si + Ti [48]	$Ni_{16}Ti_{16}$ Si_7(G), NiTiSi(E), $Ni_{49}Ti_{14}Si_{37}(T_3)$, $NiTiSi_2(T_2)$, Ni-Si phases	IN 738 LC	Plasma spray, CVD, molten salt	Thermal stability, hot bend, thermal cycling
Si, B [49]	γ-(Ni-5Si), Ni_3Si, Ni_5Si_2, NiB,CrB	IN 738 LC, 939; Nimonic 105	CVD	Oxidation at 900°C, thermal cycling, creep rupture

CVD, chemical vapor deposition.

conditions. Among various coatings, $MoSi_2$ has emerged as the most promising material due to its exceptional oxidation resistance and properties of its oxide scale at temperatures up to approximately 1,700°C.

Figure 6.9a and b shows the appearance of the coating's surface prior to oxidation. The coating's surface is evenly applied and unbroken, with minimal variation except for occasional microcracks and holes. After heating to 1,650°C for 4 hours, the coating's visual appearance post-oxidation can be seen in Figure 6.9c and d. The reason for these phenomena is that the coating density is higher on the surface due to the SiO_2 formed during the high-temperature oxidation process being in a molten state, resulting in the original coating. The self-repairing properties of the microcracks and holes it creates are referred to as self-healing [52].

6.6.3 Silicon-Bearing Coatings for Superalloys

Silicon has been extensively researched and utilized in coatings to safeguard materials from high-temperature oxidation, hot corrosion, and carburization. It has been used in two main ways: as a key ingredient in silicide diffusion and overlay coatings and as a component in standard coating systems like aluminum-silicon,

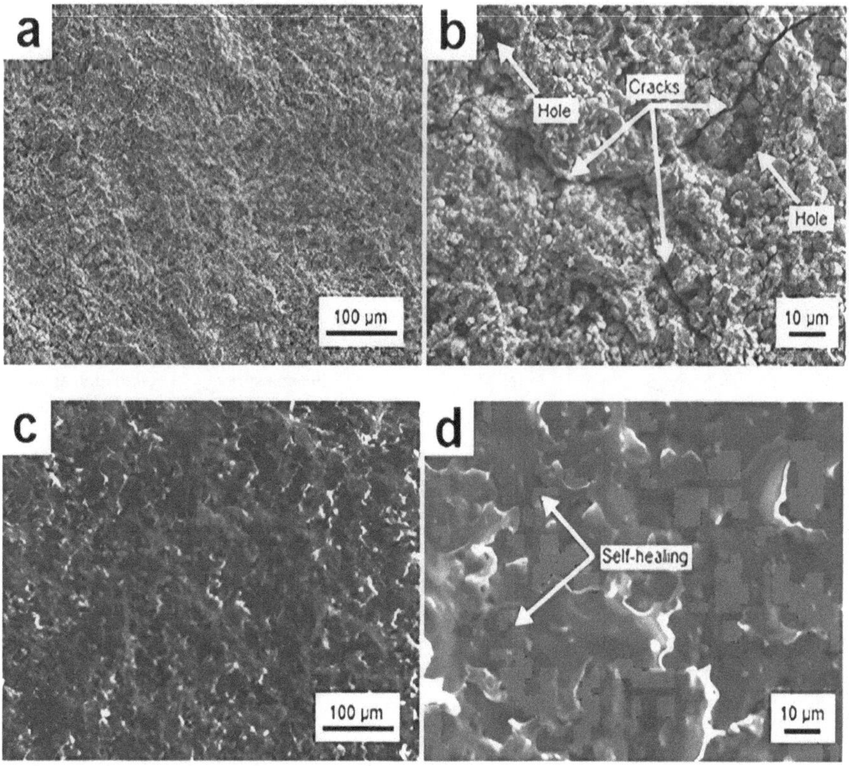

FIGURE 6.9 Shows surface pictures of coatings pre- and post-oxidation: Images (a) and (b) depict the specimen before oxidation, while images (c) and (d) show it after 4 hours of oxidation at 1,650°C.

nickel-chromium-silicon, and M-Cr-Al-Y-silicon. Silicide diffusion coatings involve introducing brittle silicide phases into the substrate surfaces using techniques such as CVD or pack cementation [53]. These coatings are created to provide customized phase structures and shapes to meet particular application needs, improving the material's functionality in challenging conditions [54].

Oxidation rate at elevated temperature of materials under conditions demonstrated for superalloy application, as depicted in Figure 6.10. Cyclic oxidation kinetic curves of two specimens were tested at 1,150°C for 100 hours. The bare superalloy gained approximately 6.99 mg/cm^2 in mass, whereas the superalloy with lower Mo content and additions of Cr, Y, and Ta showed an oxidation rate of around 60 mg/cm^2 under identical conditions. By incorporating chromium, yttrium, and tantalum into the alloy, the maximum amount of molybdenum content in the new design has been reduced. The addition of more Mo content during transformation facilitates its migration to the surface of the alloy at elevated temperatures, thereby impacting the durability of the oxide layer that provides protection. Therefore, the protective layers are compromised, leading to significant spalling observed after 20 hours of oxidation with decreased Mo and the inclusion of Cr, Y, and Ta in the alloy [55].

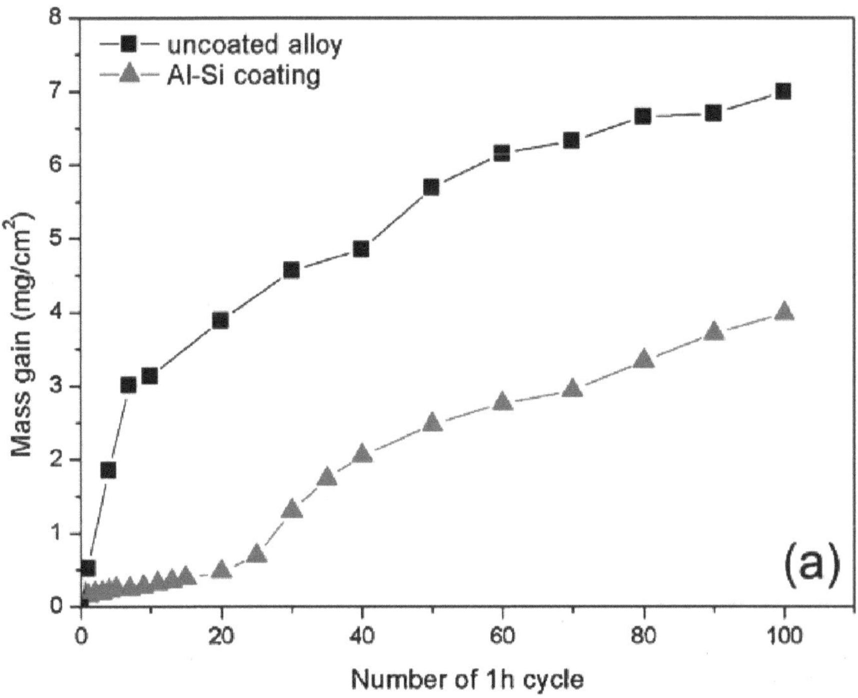

FIGURE 6.10 Mass gain for specimen after cyclic oxidation at 1,150°C used for 100 hours.

Silicon-based coatings are essential in reducing material degradation caused by high-temperature oxidation and corrosion [96–105]. Their ability to form a stable silica layer provides effective protection in aggressive environments, particularly in power generation and petrochemical industries [106–112]. Research has demonstrated that deposition techniques such as thermal spraying and microwave heating enhance the microstructural integrity and performance of these coatings [113–118]. Additionally, optimizing processing parameters and incorporating heat treatments have proven to significantly improve their resistance to oxidation and corrosion, making them a reliable choice for high-temperature applications [119,120].

6.7 SUMMARY

Silicon-based coatings were used in high-temperature applications due to their excellent oxidation resistance, thermal solidity, and protective properties. Silicon forms stable oxide layers, primarily SiO_2, which act as barriers against oxidation, hot corrosion, and carburization. These coatings are essential for components in heat exchangers, combustion chambers, gas turbines, and other high-temperature environments. Silicon-based coatings, such as silicides, SiC, Si_3N_4, and silicon-aluminides, have been extensively researched and applied in various industries. These coatings are applied using methods like thermal spray coatings, CVD, and PVD, each with its advantages and specific applications. In harsh environments containing molten salts and oxidative gases, silicides face challenges such as oxidation, hot corrosion, and sulfidation.

The behavior of silicides in these conditions depends on the stability of their oxide layers and the formation of protective or harmful intermediate phases. Strategies to mitigate degradation include applying protective coatings and modifying silicide compositions. Silicon-bearing coatings, such as silicide diffusion and overlay coatings, have been widely studied and applied for the protection of superalloys exposed to oxidation, hot corrosion, and carburization.

DECLARATIONS

Ethical Approval: The chapter accurately and thoroughly reflects the authors' research and analysis.

Conflict of Interest: There is no conflict of interest that the author has disclosed.

FUNDING

Author C. Durga Prasad would like to thank the Science and Engineering Research Board (SERB) for financial support to carry out this research work. Project File No.: CRG/2022/004140, under the Core Research Grant (CRG) scheme, Government of India.

Data and Material Accessibility: Since no new data were generated or examined in this study, data accessibility is not relevant to this publication.

REFERENCES

1. George, William F. *Antique Electric Waffle Irons 1900–1960: A History of the Appliance Industry in 20th Century America.* Bloomington, IN: Trafford Publishing (2003).
2. Sequeira, César AC. *High Temperature Corrosion: Fundamentals and Engineering.* Hoboken, NJ: John Wiley & Sons (2019).
3. Biscay, Nicolas, Lucile Henry, Tadafumi Adschiri, Masahiro Yoshimura, and Cyril Aymonier. Behavior of silicon carbide materials under dry to hydrothermal conditions. *Nanomaterials* 11, no. 5 (2021): 1351.
4. Kahl, Bruno Alexander. Ultra-High Temperature Ceramic Coatings for Hypersonic and Space Applications. PhD diss., Swinburne (2023).
5. Fanicchia, Francesco, and Sigrún Nanna Karlsdottir. Research and development on coatings and paints for geothermal environments: a review. *Advanced Materials Technologies* 8, no. 18 (2023): 2202031.
6. Yu, Xi. High-Temperature Steam and Pressurised Water Corrosion Behaviours of Nitride Ceramic Coatings for Accident Tolerant Fuel Claddings. Master's thesis, The University of Manchester (United Kingdom) (2019).
7. Bošnjaković, Mladen. Advance of sustainable energy materials: technology trends for silicon-based photovoltaic cells. *Sustainability* 16, no. 18 (2024): 7962.
8. Grilli, Maria Luisa, Daniele Valerini, Anca Elena Slobozeanu, Bogdan O. Postolnyi, Sebastian Balos, Antonella Rizzo, and Radu Robert Piticescu. Critical raw materials saving by protective coatings under extreme conditions: a review of last trends in alloys and coatings for aerospace engine applications. *Materials* 14, no. 7 (2021): 1656.
9. Morris, Tom. Resistors for high temperature applications. *Additional Conferences (Device Packaging, HiTEC, HiTEN, & CICMT) 2019*, no. HiTen (2019): 000099–000106. International Microelectronics Assembly and Packaging Society.

10. Duminica, Florin D., Muthu Karuppasamy, Florian Dawance, Jens Baber, Holger Friedrich, and Philippe Guaino. Printed electronics by plasma spraying: case study for high temperature sensors. *Journal of Thermal Spray Technology* 33 (2024): 1–14.
11. Owston, C. N. An Investigation of the Degradation of Ceramic Coatings on Metals Using a Microwave Technique (Report No. 157). (2022).
12. Majumdar, Sanjib, J. Kishor, Ananta Narayan Behera, Bathula Vishwanadh, Ananta Borgohain, R. Kapoor, R. Tewari, et al. Preparation and characterization of Nb-1Zr-0.1 C alloy suitable for liquid metal coolant channels of high temperature reactors. *Journal of Nuclear Engineering and Radiation Science* 7, no. 1 (2021): 011603.
13. Yu, Daliang, Xiaofei Ma, Huanyu Li, Shangzhou Zhang, Jianxun Qiu, Haihua Huang, Weiwei Wang, Xin Zhou, Mao Wen, and Ping Ren. Hard-yet-tough Mo-Si-C protective films for simultaneous friction and wear performance enhancement. *Applied Surface Science* 665 (2024): 160302.
14. Guo, Lei, Wenting He, Wenbo Chen, Zhaolu Xue, Jian He, Yiqian Guo, Yang Wu, et al. Progress on high-temperature protective coatings for aero-engines. *Surface Science and Technology* 1, no. 1 (2023): 6.
15. Dharini, T., Anand Krishnamoorthy, and P. Kuppusami. Silicone-based coatings for high-temperature applications. In: Amir Hossein Pakseresht and Kamalan Kirubaharan Amirtharaj Mosas (eds.), *Coatings for High-Temperature Environments: Anti-Corrosion and Anti-Wear Applications*, pp. 385–401. Cham: Springer Nature Switzerland (2023).
16. Wenga, Terrence, Digby D. Macdonald, and Wenchao Ma. Multi-scale computational study of high-temperature corrosion and the design of corrosion-resistant alloys. *Progress in Materials Science* 148 (2024): 101359.
17. Xu, Donghai, and Shuwei Guo. *Corrosion Characteristics, Mechanisms and Control Methods of Candidate Alloys in Sub-and Supercritical Water.* Springer Nature (2021).
18. Wang, Peipei, Menglin Zhang, Wanchang Sun, and Xuanru Ren. Oxidation protection of B4C modified HfB2-SiC coating for C/C composites at 1073–1473 K. *Ceramics International* 48, no. 3 (2022): 3206–3215.
19. Rey, J. R. C., A. Longo, B. Rijo, C. M. Pedrero, L. A. C. Tarelho, P. S. D. Brito, and C. Nobre. A review of cleaning technologies for biomass-derived syngas. *Fuel* 377 (2024): 132776.
20. Bawane, Kaustubh Krishna. Silicon Carbide-Nanostructured Ferritic Alloy Composites for Nuclear Applications. (2020).
21. Xu, Yongzhao, Xiduo Hu, Sudip Kundu, Anindya Nag, Nasrin Afsarimanesh, Samta Sapra, Subhas Chandra Mukhopadhyay, and Tao Han. Silicon-based sensors for biomedical applications: a review. *Sensors* 19, no. 13 (2019): 2908.
22. Sha, Bo, Anatolii N. Lukianov, Mykhailo G. Dusheiko, Volodymyr B. Lozinskii, Andriy N. Klyui, Dmytro V. Korbutyak, Sergey E. Pritchin, and Nickolai Klyui. Carbon-rich amorphous silicon carbide and silicon carbonitride films for silicon-based photoelectric devices and optical elements: application from UV to mid-IR spectral range. *Optical Materials* 106 (2020): 109959.
23. Yang, Yang, Wei Yuan, Wenquan Kang, Yintong Ye, Qiqi Pan, Xiaoqing Zhang, Yuzhi Ke, Chun Wang, Zhiqiang Qiu, and Yong Tang. A review on silicon nanowire-based anodes for next-generation high-performance lithium-ion batteries from a material-based perspective. *Sustainable Energy & Fuels* 4, no. 4 (2020): 1577–1594.
24. Anton, Ronja, Vito Leisner, Philipp Watermeyer, Michael Engstler, and Uwe Schulz. Hafnia-doped silicon bond coats manufactured by PVD for SiC/SiC CMCs. *Acta Materialia* 183 (2020): 471–483.
25. Thakur, Lalit, and Hitesh Vasudev, (eds.) *Thermal Spray Coatings.* CRC Press (2021).
26. Prashar, Gaurav, Hitesh Vasudev, and Lalit Thakur. Thermal spraying fundamentals: process applications, challenges, and future market. In: Lalit Thakur and Hitesh Vasudev (eds.), *Thermal Spray Coatings*, pp. 1–36. Boca Raton, FL: CRC Press (2021).

27. Sati, Priyanka, Ankita Kumari, and Sudesh Kumar. Silicon-based technologies for high-temperature coatings and their corrosion behaviours. In: Amir Hossein Pakseresht and Kamalan Kirubaharan Amirtharaj Mosas (eds.), *Coatings for High-Temperature Environments: Anti-Corrosion and Anti-Wear Applications,* pp. 103–122. Cham: Springer Nature Switzerland (2023).
28. Kovaleva, M. G., V. Yu Novikov, O. N. Vagina, and V. V. Sirota. Microstructure evolution and phase transformation of ZrB_2-45$MoSi_2$-10Al coating at high temperature. *Journal of Physics: Conference Series* 2021, no. 1 (2021): 012010.
29. Han, Jia-Yu, Li Wang, Ping Hu, Boliang Hu, Sheng-jie Ma, Lili Gao, Run Bai, Qiang Wang, Rui Feng, Bo Jin, and Kuai-she Wang. Research progress in modification of $MoSi_2$ coatings on surface of refractory metals and their alloys: a review. *Rare Metals* (2024).
30. Itzhak, D., F. Tuler, and M. Schieber. Silicon-containing coatings produced by a chemical vapour deposition method on nickel-based superalloys. *Thin Solid Films* 73, no. 2 (1980): 379–384.
31. Solimani, A., T. Nguyen, J. Zhang, D. J. Young, M. Schütze, and M. C. Galetz. Morphology of oxide scales formed on chromium-silicon alloys at high temperatures. *Corrosion Science* 176 (2020): 109023.
32. Rajendran, Susai, Saeid Kakooei, Mahdi Yeganeh, Yongxin Li, and Tuan Anh Nguyen, (eds.) *Corrosion Protection at the Nanoscale.* Elsevier (2020).
33. Pan, Yong. Role of hydrogen on the structural stability, mechanical and thermodynamic properties of the cubic TM3Si silicides. *International Journal of Hydrogen Energy* 46, no. 55 (2021): 28338–28345.
34. Dong, Tianhui, Xudong Zhang, Linmei Yang, and Feng Wang. Investigation of the structural stability, mechanical, and thermodynamic properties of Hf_5BSi_3 silicide with vacancies defects. *Vacuum* 191 (2021): 110349.
35. de Padoue Shyikira, Antoine, Naureen Akhtar, Gunstein Skomedal, and Peter Hugh Middleton. The effect of Mo and Ge reactive elements on high-temperature oxidation of higher manganese silicide. *Corrosion Science* 194 (2022): 109920.
36. Jacobsen, George M., Hangbok Choi, James A. Turso, Amanda M. Johnsen, Andrew J. Bascom, Xialu Wei, and Eugene A. Olevsky. Fabrication and characterization of zirconium silicide for application to gas-cooled fast reactors. *Nuclear Technology* 208, no. 1 (2022): 27–36.
37. Chen, Guanghui, Wei Shao, Aohan Zhou, Yan Zhang, Yang Wu, Zheng Zhou, Xingye Guo, Dingyong He, and Chungen Zhou. Effect of B on hot corrosion behavior of silicide coatings under different corrosive environments at 900°C. *Surface and Coatings Technology* 489 (2024): 131160.
38. Wang, Qiang, Xing Rui, Qing-Juan Wang, Yu Bai, Zhong-Ze Du, Wen-Juan Niu, Wei Wang, Kuai-She Wang, and Yuan Gao. Bonding and wear behaviors of supersonic plasma sprayed Fe-based coatings on Al-Si alloy substrate. *Surface and Coatings Technology* 367 (2019): 288–301.
39. Fushimi, Koji, Nozomi Ishii, Ayako Nakajima, Yuichi Kitagawa, Yasuchika Hasegawa, and Yasutaka Matsuo. Photolithographic fabrication of a micro-electrode surface on a carbon steel sheet for local hydrogen permeation measurements. *ISIJ International* 61, no. 4 (2021): 1112–1119.
40. Šulák, Ivo, Karel Obrtlík, Simona Hutařová, Martin Juliš, Tomáš Podrábský, and Ladislav Čelko. Low cycle fatigue and dwell-fatigue of diffusion coated superalloy Inconel 713LC at 800°C. *Materials Characterization* 169 (2020): 110599.
41. Najafizadeh, Mojtaba, Mehran Ghasempour-Mouziraji, Sajad Shahrani, Mansoor Bozorg, Constantinos Goulas, Morteza Hosseinzadeh, Pasquale Cavaliere, Angelo Perrone, and Serena Perrone. Optimization of the Si_{3N4} coating formation through plasma spraying on Inconel 738. *Tribology-Materials, Surfaces & Interfaces* 17, no. 4 (2023): 338–351.

42. Targhi, Vahid Tavakoli, Hamid Omidvar, Seyed Mohammad Mehdi Hadavi, and Fariborz Sharifianjazi. Microstructure and hot corrosion behavior of hot dip siliconized coating on Ni-base superalloy IN738LC. *Materials Research Express* 7, no. 5 (2020): 056527.
43. Azizzadeh, Hesamaldin, Seyed Abdolkarim Sajjadi, and Arezoo Sezavar. Mullite/YSZ hybrid thermal barrier coatings plasma sprayed on IN-738LC: microstructure, phase analysis and oxidation resistance. *Surface and Coatings Technology* 482 (2024): 130701.
44. Han, Jia-Yu, Li Wang, Ping Hu, Bo-Liang Hu, Sheng-Jie Ma, Li-Li Gao, Run Bai, Qiang Wang, Rui Feng, Bo Jin, and Kuai-she Wang. Research progress in modification of MoSi2 coatings on surface of refractory metals and their alloys: a review. *Rare Metals* (2024): 1–29.
45. Reszka, Kazimierz, Jerzy Morgiel, Tomasz Rydzkowski, and Vijay Kumar Thakur. On the morphological investigation of Pt dispersion and structure of alumina-platinum composites obtained by thermal oxidation of Al-Pt nano thin layers. *Nano-Structures & Nano-Objects* 17 (2019): 229–238.
46. Chen, Chuantong, Dongjin Kim, Zheng Zhang, Naoki Wakasugi, Yang Liu, Ming-Chun Hsieh, Shuaijie Zhao, Aiji Suetake, and Katsuaki Suganuma. Interface-mechanical and thermal characteristics of ag sinter joining on bare DBA substrate during aging, thermal shock and 1200 W/cm^2 power cycling tests. *IEEE Transactions on Power Electronics* 37, no. 6 (2022): 6647–6659.
47. Kochmańska, Agnieszka Elżbieta, Aneta Jarlaczyńska, and Jolanta Baranowska. Formation of silicide and silicide-aluminide coatings on molybdenum alloy during slurry cementation process: influence of slurry volume. *Materials* 14, no. 22 (2021): 6940.
48. Harris, Chad. Mo-Si-B Based Coatings for the Suppression of High Temperature Water Vapor Attack. PhD diss., The University of Wisconsin-Madison (2024).
49. Li, Shuang, Min Min Xu, Chun Yang Zhang, Yong Sheng Niu, Zhi Bin Bao, Shao Long Zhu, and Fu Heng Wang. Co-doping effect of Hf and Y on improving cyclic oxidation behavior of (Ni, Pt) Al coating at 1150°C. *Corrosion Science* 178 (2021): 109093.
50. Shanthala Kollur, S. Shiva Prakash, Jasti Surendra, G. S. Pradeep Kumar, R. Suresh Kumar, C. Durga Prasad, B. K. Pavan Kumar, Sandeep Kore, and M. Nusrathulla. Thermal cycle behavior of plasma sprayed thermal barrier coatings on cast iron substrate for the application of liner of internal combustion engine. *Results in Surfaces and Interfaces, Elsevier Science* 17 (2024): 100300. https://doi.org/10.1016/j.rsurfi.2024.100300
51. Haridasa Nayak, K. R. Varun, M. K. Venkatesh, Sonal Shamkuwar, R. Suresh Kumar, T. A. Sudarshan, Abhijeet Malge, M. Vijay, and C. Durga Prasad. Thermal cycle behavior of plasma sprayed thermal barrier coatings on cast iron substrate for the application of liner of internal combustion engine. *Results in Surfaces and Interfaces, Elsevier Science* 17 (2024): 100297. ISSN 2666–8459. https://doi.org/10.1016/j.rsurfi.2024.100297
52. Virupakshappa Lakkannavar, K. B. Yogesha, C. Durga Prasad, G. Srinivasa, and Mahantayya Mathapati. Assessment of the microstructure, adhesion and elevated temperature erosion resistance of plasma-sprayed NiCrAlY/Cr_3C_2/h-BN composite coating. *Results in Surfaces and Interfaces, Elsevier Science* 17 (2024): 100289. ISSN 2666–8459. https://doi.org/10.1016/j.rsurfi.2024.100289
53. Pradeep Kumar, G. S., R. Keshavamurthy, M. Harish Kumar, Raghu Yogaraju, T. S. Sachith, and C. Durga Prasad. Studies on air jet erosion behavior of plasma-sprayed Cr_2C_3 coatings on Al6061 alloy. *Results in Surfaces and Interfaces, Elsevier Science* 16 (2024): 100278. ISSN 2666–8459. https://doi.org/10.1016/j.rsurfi.2024.100278
54. Virupakshappa Lakkannavar, K. B. Yogesha, C. Durga Prasad, and Chandrashekar Prasad. Investigation of cyclic oxidation and hot corrosion behaviour of plasma-sprayed NiCrAlY/Cr3C2/h-BN/Mo coatings on T22 boiler steel alloy. *Journal of Bio- and Tribo-Corrosion, Springer* 10 (2024): 85. https://doi.org/10.1007/s40735-024-00890-w

55. Virupakshappa Lakkannavar, K. B. Yogesha, C. Durga Prasad, R. Suresh, Avinash Lakshmikanthan, and S. Hanumanthlal. Investigation of cyclic oxidation and corrosion behavior of plasma-sprayed NiCrAlY/Cr3C2/h-BN coatings on T22 boiler steel alloy. *Surface and Coatings Technology, Elsevier Science* 492 (2024): 131205. ISSN 0257–8972. https://doi.org/10.1016/j.surfcoat.2024.131205
56. Virupakshappa Lakkannavar, K. B. Yogesha, C. Durga Prasad, Rakesh Kumar Phanden, G. Srinivasa, and Chandrashekar Prasad. Thermal spray coatings on high-temperature oxidation and corrosion applications – a comprehensive review. *Results in Surfaces and Interfaces, Elsevier Science* 15 (2024): 100250. ISSN 2666–8459. https://doi.org/10.1016/j.rsurfi.2024.100250
57. Prasad, C. D., Piyush Kumar Soni, K. C. Nagaraja, A. Eswaran, R. Suresh Kumar, Khemraj Deshmukh, Rajeev Shrivastava, and Amit Tiwari. Studies on wear and microstructure assessment of WC-Co reinforced iron based HVOF coating. *Results in Surfaces and Interfaces, Elsevier Science* 15 (2024): 100237. ISSN 2666–8459. https://doi.org/10.1016/j.rsurfi.2024.100237
58. Shanthala Kollur, K. V. Manjunath, C. Durga Prasad, B. Byra Reddy, K. S. Rudra, Prakash Kumar, and C. R. Aprameya. Evaluation of microstructural and mechanical properties of tribaloy based composite cladding by microwave heating. *Journal of the Institution of Engineers (India): Series D, Springer* (2024). https://doi.org/10.1007/s40033-024-00670-3
59. Ramesh, M. R., Subbarao Medabalimi, R. Suresh Kumar, C. Durga Prasad, and Shrishail B. Sollapur. Cyclic oxidation and hot corrosion behaviour of HVOF sprayed NiCrAl coating on industrial boiler tube steels. *JOM the Journal of the Minerals, Metals & Materials Society (TMS), Springer* 76 (2024): 3172–3184. https://doi.org/10.1007/s11837-024-06526-1
60. Virupakshappa Lakkannavar, K. B. Yogesha, C. Durga Prasad, M. Mruthunjaya, and R. Suresh. A review on tribological and corrosion behavior of thermal spray coatings. *Journal of the Institution of Engineers (India): Series D, Springer* (2024). https://doi.org/10.1007/s40033-024-00636-5
61. Haridasa Nayak, Shanthala Kollur, K. Prasad, Suresh Erannagari, C. Durga Prasad, and N. Nagabhushana. Development of equal proportional YSZ+Al_2O_3 thermal barrier coating and effect of coating thickness on the corrosion behaviour on cast iron substrate. *High Temperature Corrosion of Materials, Springer* 101 (2024): 291–308. https://doi.org/10.1007/s11085-024-10222-5
62. Prasad, C. Durga, Shanthala Kollur, C. R. Aprameya, T. V. Chandramouli, T. Jagadeesha, and B. N. Prashanth. Investigations on tribological and microstructure characteristics of WC-12Co/FeNiCrMo composite coating by HVOF process. *JOM the Journal of the Minerals, Metals & Materials Society (TMS), Springer* 76 (2024): 186–195. https://doi.org/10.1007/s11837-023-06242-2
63. Prasad, C. Durga, S. Kollur, M. Nusrathulla, G. Satheesh Babu, M. B. Hanamantraygouda, B. N. Prashanth, and N. Nagabhushana. Characterisation and wear behaviour of SiC reinforced FeNiCrMo composite coating by HVOF process. *Transactions of the IMF, Taylor & Francis* 102, no. 1 (2024): 22–28. https://doi.org/10.1080/00202967.2023.2246259
64. Sharanabasava, H., M. Raviprakash, C. Durga Prasad, M. R. Ramesh, M. V. Phanibhushana, Hitesh Vasudev, and Sandeep Kumar. Microstructure, mechanical and wear properties of SiC and Mo reinforced NiCr microwave cladding. *Advances in Materials and Processing Technologies, Taylor & Francis* (2023). https://doi.org/10.1080/2374068X.2023.2257937
65. Madhu Sudana Reddy, G., C. Durga Prasad, Shanthala Kollur, Avinash Lakshmikanthan, R. Suresh, and C. R. Aprameya. Investigation of high temperature erosion behaviour of NiCrAlY/TiO_2 plasma coatings on titanium substrate. *JOM the Journal of the Minerals, Metals & Materials Society (TMS), Springer* 75 (2023): 3317–3323. https://doi.org/10.1007/s11837-023-05894-4

66. Madhu Sudana Reddy, G., C. Durga Prasad, Pradeep Patil, Naresh Kakur, and M. R. Ramesh. High temperature erosion performance of NiCrAlY/Cr_2O_3/YSZ plasma spray coatings. *Transactions of the IMF, Taylor & Francis* (2023). https://doi.org/10.1080/00202967.2023.2208899
67. Sharanabasva, H., C. Durga Prasad, and M. R. Ramesh. Characterization and wear behavior of NiCrMoSi microwave cladding. *Journal of Materials Engineering and Performance* 33 (2024): 763–775. https://doi.org/10.1007/s11665-023-07998-z
68. Madhu Sudana Reddy, G., C. Durga Prasad, Pradeep Patil, Naresh Kakur, and M. R. Ramesh. Investigation of plasma sprayed NiCrAlY/Cr_2O_3/YSZ coatings on erosion performance of MDN 420 steel substrate at elevated temperatures. *International Journal of Surface Science and Engineering* 17, no. 3 (2023): 180–194. https://dx.doi.org/10.1504/IJSURFSE.2023.10054266
69. Sharanabasva, H., C. Durga Prasad, and M. R. Ramesh. Effect of Mo and SiC reinforced NiCr microwave cladding on microstructure, mechanical and wear properties. *Journal of the Institution of Engineers (India): Series D, Springer* (2023). https://doi.org/10.1007/s40033-022-00445-8
70. Nithin, H. S., K. M. Nishchitha, D. G. Pradeep, C. Durga Prasad, and Mahantayya Mathapati. Comparative analysis of CoCrAlY coatings at high temperature oxidation behavior using different reinforcement composition profiles. *Welding in the World, Springer* 67 (2023): 585–592. https://doi.org/10.1007/s40194-022-01405-2
71. Madhu Sudana Reddy, G., C. Durga Prasad, Gagan Shetty, M. R. Ramesh, T. Nageswara Rao, and Pradeep Patil. Investigation of thermally sprayed NiCrAlY/TiO_2 and NiCrAlY/Cr_2O_3/YSZ cermet composite coatings on titanium alloys. *Engineering Research Express, IOP* 4 (2022): 025049. https://doi.org/10.1088/2631-8695/ac7946
72. Madhu Sudana Reddy, G., C. Durga Prasad, Pradeep Patil, Naresh Kakur, and M. R. Ramesh. Elevated temperature erosion performance of plasma sprayed NiCrAlY/TiO_2 coating on MDN 420 steel substrate. *Surface Topography: Metrology and Properties, IOP* 10 (2022): 025010. https://doi.org/10.1088/2051-672X/ac6a6e
73. Madhu Sudana Reddy, G., C. Durga Prasad, Gagan Shetty, M. R. Ramesh, T. Nageswara Rao, and Pradeep Patil. High temperature oxidation behavior of plasma sprayed NiCrAlY/TiO_2 & NiCrAlY/Cr_2O_3/YSZ coatings on titanium alloy. *Welding in the World, Springer* (2022). https://doi.org/10.1007/s40194-022-01268-7
74. Thavaraya Naik, Mahantayya Mathapathi, C. Durga Prasad, H. S. Nithin, and M. R. Ramesh. Effect of laser post treatment on microstructural and sliding wear behavior of HVOF sprayed NiCrC and NiCrSi coatings. *Surface Review and Letters* 29, no. 1 (2022): 225000. https://doi.org/10.1142/S0218625X2250007X
75. Madhu Sudana Reddy, G., C. Durga Prasad, Gagan Shetty, M. R. Ramesh, T. Nageswara Rao, and Pradeep Patil. High temperature oxidation studies of plasma sprayed NiCrAlY/TiO_2 & NiCrAlY/Cr_2O_3/YSZ cermet composite coatings on MDN-420 special steel alloy. *Metallography, Microstructure and Analysis, Springer* 10 (2021): 642–651. https://doi.org/10.1007/s13632-021-00784-0
76. Dinesh, Rayirth, S. Rohan Raykar, T. L. Rakesh, M. G. Prajwal, M. Shashank Lingappa, and C. Durga Prasad. Feasibility study on MoCoCrSi/WC-Co cladding developed on austenitic stainless steel using microwave hybrid heating. *Journal of Mines, Metals & Fuels* 69, no. 12A (2021). https://doi.org/10.18311/jmmf/2021/30113
77. Prasad, C. Durga, Shashank Lingappa, Sharnappa Joladarashi, M. R. Ramesh, and B. Sachin. Characterization and sliding wear behavior of CoMoCrSi + Flyash composite cladding processed by microwave irradiation. *Materials Today Proceedings, Elsevier Science* 46 (2021): 2387–2391. https://doi.org/10.1016/j.matpr.2021.01.156
78. Madhu, G., K. M. Mrityunjaya Swamy, Durga Ajay Kumar, C. Durga Prasad, and U. Harish. Evaluation of hot corrosion behavior of HVOF thermally sprayed Cr_3C_2_35NiCr coating on SS 304 boiler tube steel. *American Institute of Physics* 2316 (2021): 030014. https://doi.org/10.1063/5.0038279

79. Madhu Sudana Reddy, G., C. Durga Prasad, Pradeep Patil, M. R. Ramesh, and T. Nageswara Rao. Hot corrosion behavior of plasma sprayed NiCrAlY/TiO_2 and NiCrAlY/Cr_2O_3/YSZ cermets coatings on alloy steel. *Surfaces and Interfaces, Elsevier Science* 22 (2021): 100810. https://doi.org/10.1016/j.surfin.2020.100810
80. Prasad, C. Durga, Sharnappa Joladarashi, M. R. Ramesh, and M. S. Srinath. Microstructure and tribological resistance of flame sprayed CoMoCrSi/WC-CrC-Ni and CoMoCrSi/WC-12Co composite coatings remelted by microwave hybrid heating. *Journal of Bio and Tribo-Corrosion, Springer* 6 (2020): 124. https://doi.org/10.1007/s40735-020-00421-3
81. Prasad, C. Durga, Sharnappa Joladarashi, and M. R. Ramesh. Comparative investigation of HVOF and flame sprayed CoMoCrSi coating. *American Institute of Physics* 2247 (2020): 050004. https://doi.org/10.1063/5.0003883
82. Prasad, C. Durga, Akhil Jerri, and M. R. Ramesh. Characterization and sliding wear behavior of iron based metallic coating deposited by HVOF process on low carbon steel substrate. *Journal of Bio and Tribo-Corrosion, Springer* 6 (2020): 69. https://doi.org/10.1007/s40735-020-00366-7
83. Prasad, C. Durga, Sharnappa Joladarashi, M. R. Ramesh, M. S. Srinath, and B. H. Channabasappa. Comparison of high temperature wear behavior of microwave assisted HVOF sprayed CoMoCrSi-WC-CrC-Ni/WC-12Co composite coatings. *Silicon, Springer* 12 (2020): 3027–3045, 1–19. https://doi.org/10.1007/s12633-020-00398-1
84. Prasad, C. Durga, Sharnappa Joladarashi, M. R. Ramesh, M. S. Srinath, and B. H. Channabasappa. Effect of microwave heating on microstructure and elevated temperature adhesive wear behavior of HVOF deposited CoMoCrSi-Cr_3C_2 composite coating. *Surface and Coatings Technology, Elsevier Science* 374 (2019): 291–304. https://doi.org/10.1016/j.surfcoat.2019.05.056
85. Prasad, C. Durga, Sharnappa Joladarashi, M. R. Ramesh, M. S. Srinath, and B. H. Channabasappa. Development and sliding wear behavior of Co-Mo-Cr-Si cladding through microwave heating. *Silicon, Springer* 11 (2019): 2975–2986. https://doi.org/10.1007/s12633-019-0084-5
86. Prasad, C. Durga, Sharnappa Joladarashi, M. R. Ramesh, M. S. Srinath, and B. H. Channabasappa. Microstructure and tribologicalbehavior of flame sprayed and microwave fused CoMoCrSi/CoMoCrSi-Cr_3C_2 coatings. *Materials Research Express, IOP* 6 (2019): 026512. https://doi.org/10.1088/2053-1591/aaebd9
87. Prasad, C. Durga, Sharnappa Joladarashi, M. R. Ramesh, M. S. Srinath, and B. H. Channabasappa. Influence of microwavehybrid heating on the sliding wear behaviour of HVOF sprayed CoMoCrSi coating. *Materials Research Express, IOP* 5 (2018): 086519. https://doi.org/10.1088/2053-1591/aad44e
88. Prasad, C. Durga, Sharnappa Joladarashi, M. R Ramesh, and Anunoy Sarkar. High temperature gradient cobalt based clad developed using microwave hybrid heating. *American Institute of Physics* 1943 (2018): 020111. https://doi.org/10.1063/1.5029687
89. Girisha, K. G., K. V. Sreenivas Rao, and C. Durga Prasad. Slurry erosion resistance of martenistic stainless steel with plasma sprayed Al_2O_3-40%TiO_2 coatings. *Materials Today Proceedings* 5 (2018): 7388–7393. https://doi.org/10.1016/j.matpr.2017.11.409
90. Girisha, K. G., C. Durga Prasad, K. C. Anil, and K. V. Sreenivas Rao. Dry sliding wear behaviour of Al_2O_3 coatings for AISI 410 grade stainless steel. *Applied Mechanics and Materials* 766–767 (2015): 585–589. https://doi.org/10.4028/www.scientific.net/AMM.766-767.585
91. Girisha, K. G., R. Rakesh, C. Durga Prasad, and K. V. Sreenivas Rao. Development of corrosion resistance coating for AISI 410 grade steel. *Applied Mechanics and Materials* 813–814 (2015): 135–139. https://doi.org/10.4028/www.scientific.net/AMM.813-814.135
92. Prasad, C. Durga, K. V. Manjunath, Prem Kumar Naik, N. Nagabhushana, and Prakash Kumar. Development of MoCoCrSi/Flyash composite cladding on stainless steel substrate through microwave irradiation. In: Lalit Thakur, Jasbir Singh, and Hitesh Vasudev

(eds.), *Thermal Claddings for Engineering Applications.* Milton: CRC Press, Taylor & Francis Group (2024). ISBN 9781032460543. https://doi.org/10.1201/9781032713830
93. Prasad, C. Durga, Mahantayya Mathapathi, Hitesh Vasudev, and Lalit Thakur. Analysis of mechanical properties and microstructural characterisation of microwave cladding on stainless steel. In: Amit Bansal and Hitesh Vasudev (eds.), *Advances in Microwave Processing for Engineering Materials.* Boca Raton, FL: CRC Press, Taylor & Francis Group (2022). ISBN: 9781003248743. https://doi.org/10.1201/9781003248743
94. Prasad, C. Durga, Mahantayya Mathapathi, Sharanappa Joladarashi, and M. R. Ramesh. Investigation of microstructural and tribological behavior of metco 41C+WC-12Co composite coating sprayed by HVOF process. In: Lalit Thakur and Hitesh Vasudev (eds.), *Thermal Spray Coatings.* Boca Raton, FL: CRC Press, Taylor & Francis Group (2021). ISBN: 9781032081489. https://doi.org/10.1201/9781003213185
95. Prasad, C. Durga, Akhil Jerri, and M. R. Ramesh. Evaluation of microstructural and dry sliding wear resistance of iron-based SiC-reinforced composite coating by HVOF process. In: Lalit Thakur and Hitesh Vasudev (eds.), *Thermal Spray Coatings.* Boca Raton, FL: CRC Press, Taylor & Francis Group (2021). ISBN: 9781032081489. https://doi.org/10.1201/9781003213185
96. Dutta, Vineet, Lalit Thakur, Baljinder Singh, and Hitesh Vasudev. A study of erosion–corrosion behaviour of friction stir-processed chromium-reinforced NiAl bronze composite. *Materials* 15, no. 15 (2022): 5401.
97. Vasudev, Hitesh. Wear characteristics of Ni-WC powder deposited by using a microwave route on mild steel: microwave cladding of Ni-WC. *International Journal of Surface Engineering and Interdisciplinary Materials Science (IJSEIMS)* 8, no. 1 (2020): 44–54.
98. Singh, Prateek, Hitesh Vasudev, and Amit Bansal. Effect of post-heat treatment on the microstructural, mechanical, and bioactivity behavior of the microwave-assisted alumina-reinforced hydroxyapatite cladding. *Proceedings of the Institution of Mechanical Engineers, Part E: Journal of Process Mechanical Engineering* 238, no. 3 (2024): 1060–1071.
99. Prashar, Gaurav, and Hitesh Vasudev. Surface topology analysis of plasma sprayed Inconel625-Al_2O_3 composite coating. *Materials Today: Proceedings* 50 (2022): 607–611.
100. Singh, Gurpreet, Hitesh Vasudev, Amit Bansal, and Sandeep Vardhan. Influence of heat treatment on the microstructure and corrosion properties of the Inconel-625 clad deposited by microwave heating. *Surface Topography: Metrology and Properties* 9, no. 2 (2021): 025019.
101. Singh, Maninder, Hitesh Vasudev, and Manjeet Singh. Surface protection of SS-316L with boron nitride based thin films using radio frequency magnetron sputtering technique. *Journal of Electrochemical Science and Engineering* 12, no. 5 (2022): 851–863.
102. Singh, Prateek, Amit Bansal, Hitesh Vasudev, and Prateek Singh. In situ surface modification of stainless steel with hydroxyapatite using microwave heating. *Surface Topography: Metrology and Properties* 9, no. 3 (2021): 035053.
103. Yedida, V. S., and Hitesh Vasudev. A review on the development of thermal barrier coatings by using thermal spray techniques. *Materials Today: Proceedings* 50 (2022): 1458–1464.
104. Mahantesh, M. M., K. R. Rao, A. P. Chandra, M. N. Vijayakumar, B. Nandini, C. D. Prasad, and Hitesh Vasudev. Design and modeling using finite element analysis for the sitting posture of computer users based on ergonomic perspective. *International Journal on Interactive Design and Manufacturing (IJIDeM)* 18, no. 8 (2024): 5875–5891.
105. Kulkarni, S. D., Manjunatha, U. Chandrasekhar, K. V. Manjunath, C. D. Prasad, and Hitesh Vasudev. Design and optimization of polyvinyl-nitride rubber for tensile strength analysis. *International Journal on Interactive Design and Manufacturing (IJIDeM)* 18, no. 8 (2024): 5893–5908.

106. Singh, Jatinder, Rupinder Singh, Sukhwinder Singh, Hitesh Vasudev, and Sandeep Kumar. Reducing scrap due to missed operations and machining defects in 90PS pistons. *International Journal on Interactive Design and Manufacturing (IJIDeM)* 17, no. 5 (2023): 2527–2539.
107. Rao, G. S., U. Mukkamala, H. Hanumanthappa, C. D. Prasad, Hitesh Vasudev, B. Shanmugam, and K. C. Kishore Kumar. Evaluating and optimizing surface roughness using genetic algorithm and artificial neural networks during turning of AISI 52100 steel. *International Journal on Interactive Design and Manufacturing (IJIDeM)* 18 (2023): 1–10.
108. Jatinder Singh, Hardeep Singh Gill, and Hitesh Vasudev. Computational fluid dynamics analysis on role of particulate shape and size in erosion of pipe bends. *International Journal on Interactive Design and Manufacturing (IJIDeM)* 17, no. 5 (2023): 2631–2646.
109. Sharanabasava, H., M. Raviprakash, C. D. Prasad, M. R. Ramesh, M. V. Phanibhushana, Hitesh Vasudev, and S. Kumar. Microstructure, mechanical and wear properties of SiC and Mo reinforced NiCr microwave cladding. *Advances in Materials and Processing Technologies* 10 (2023): 3620–3633.
110. Prashar, Gaurav, Hitesh Vasudev, Lalit Thakur, and Amit Bansal. Performance of thermally sprayed hydroxyapatite coatings for biomedical implants: a comprehensive review. *Surface Review and Letters* 30, no. 01 (2023): 2241001.
111. Prashar, Gaurav, and Hitesh Vasudev. Hot corrosion behavior of super alloys. *Materials Today: Proceedings* 26 (2020): 1131–1135.
112. Majji, B. G. R., Hitesh Vasudev, and Amit Bansal. A review on the oxidation and wear behavior of the thermally sprayed high-entropy alloys. *Materials Today: Proceedings* 50 (2022): 1447–1451.
113. Anjaneya, G., S. Sunil, Shashikant Kakkeri, M. M. Math, M. N. Vaibhav, C. Solaimuthu, C. Durga Prasad, and Hitesh Vasudev. Numerical simulation of microchannel heat exchanger using CFD. *International Journal on Interactive Design and Manufacturing (IJIDeM)* 18 (2023): 1–17.
114. Sunitha, K., and Hitesh Vasudev. A short note on the various thermal spray coating processes and effect of post-treatment on Ni-based coatings. *Materials Today: Proceedings* 50 (2022): 1452–1457.
115. Prashar, Gaurav, Hitesh Vasudev, and Lalit Thakur. High-temperature oxidation and erosion resistance of Ni-based thermally-sprayed coatings used in power generation machinery: a review. *Surface Review and Letters* 29, no. 03 (2022): 2230003.
116. Prashar, Gaurav, and Hitesh Vasudev. Structure–property correlation of plasma-sprayed Inconel625-Al_2O_3 bimodal composite coatings for high-temperature oxidation protection. *Journal of Thermal Spray Technology* 31, no. 8 (2022): 2385–2408.
117. Singh, Maninder, Hitesh Vasudev, and Raj Kumar. Corrosion and tribological behaviour of BN thin films deposited using magnetron sputtering. *International Journal of Surface Engineering and Interdisciplinary Materials Science (IJSEIMS)* 9, no. 2 (2021): 24–39.
118. Arora, Harinder, Vikram Kumar, Chander Prakash, Dmitry Pimenov, Maninder Singh, Hitesh Vasudev, and Vikram Singh. Analysis of sensitization in austenitic stainless steel-welded joint. In: Chander Prakash, Grzegorz Krolczyk, Sunpreet Singh, and Alokesh Pramanik (eds.), *Advances in Metrology and Measurement of Engineering Surfaces: Select Proceedings of ICFMMP 2019,* pp. 13–23. Singapore: Springer (2021).
119. Nagabhushana, P., S. Ramprasad, C. D. Prasad, Hitesh Vasudev, and C. Prakash. Numerical investigation on heat transfer of a nano-fluid saturated vertical composite porous channel packed between two fluid layers. *International Journal on Interactive Design and Manufacturing (IJIDeM)* 18, no. 5 (2024): 2927–2944.
120. Parkash, Jatinder, Hardeep Singh Saggu, and Hitesh Vasudev. A short review on the performance of high velocity oxy-fuel coatings in boiler steel applications. *Materials Today: Proceedings* 50 (2022): 1442–1446.

7 Oxide-Based Composite Coatings for Gas Turbines to Combat Corrosion

Amrinder Mehta, Hitesh Vasudev, and Natarajan Jeyaprakash

7.1 INTRODUCTION

The conditions to which individual components of a gas turbine are subjected include high temperature, corrosive gases, and cyclic load. To overcome this problem, enhanced oxide-based composite coatings have been promoted and integrated, usually for sensitive parts such as turbine blades. These coatings provide an armor layer to the components, increasing their performance and longevity [1]. The multilayered characteristic of these coatings is that material characteristics can be optimized for the application while offering the advantages of other materials to cope with the conditions they are exposed to. These advanced protection coatings enable the original equipment manufacturers of gas turbines to increase part durability and efficiency and decrease maintenance expenses, thus increasing the dependability and effectiveness of the gas turbine systems [2].

Figure 7.1 illustrates a jet engine with three main sections: the compressor, combustor, and turbine. Air enters the compressor, is compressed, and mixed with fuel in the combustor, where it ignites. The hot gases expand through the turbine, generating thrust.

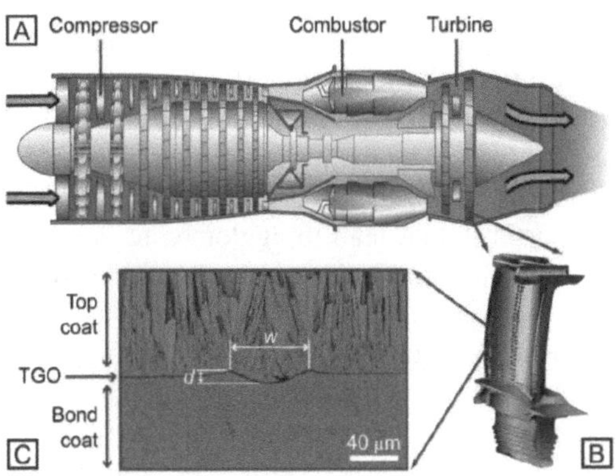

FIGURE 7.1 (a) Schematic diagram of a gas turbine, (b) blade of the turbine, and (c) cross-section TBC system [3]. TBC, thermal barrier coating.

DOI: 10.1201/9781003542957-7

A close-up of a turbine blade shows a thermal barrier coating (TBC) system with three layers: the top coat, thermally grown oxide (TGO), and bond coat. These layers protect the blade from high temperatures. The diagram highlights the microstructure of these coatings, showing dimensions "*w*" and "*d*" and a scale of 40 μm for reference [3].

Oxide-based composite coatings provide protection to gas turbine parts and are especially effective against oxidation and hot corrosion. As a result, these coatings form a protective shield that greatly enhances the service life of essential components, e.g., turbine blades. This enhanced durability keeps the engine reliable for a longer period and less likely to be replaced or repaired as often as other engines [4]. High temperature and corrosion from chemical ensembles are critical in protecting the portions of the components being coated. As a result, it brings about enhanced total engine performance, lower expenses on engine maintenance, and increased reliability. These advanced coatings constitute a vanguard of technological innovation in gas turbine structure. [5].

For its application in gas turbines, oxide-based composite coatings are designed in multilayer systems where every layer possesses a unique function. This structure normally consists of a bond coat for adhesion, a TGO layer for oxidation protection, and a ceramic final coat for thermal barrier. Notably, this design minimizes exposure to unfavorable operational conditions [6]. The bond coat in gas turbine coatings, which is ordinarily MCrAlY (where M is Nickel and/or Cobalt), acts as an essential substrate layer. It also reinforces the substrate by increasing the coefficient of adhesion of subsequent layers, provides equal oxidation protection as titanium, and expands evenly within layers. Its composition offers marvelous high-temperature applications and corrosion resistance in severe turbine operating conditions [7]. Yttria-stabilized zirconia (YSZ) is commonly used as the ceramic top coat in the TBC for the gas turbine coatings. This layer serves an important thermally insulating function, as they temperate extremes affecting lower assemblies. Ceramic material YSZ possesses low thermal conductivity, high thermal expansion factor, and steady-phase behavior at high operating temperatures, which is suitable for turbine conditions [8].

Figure 7.2 shows field emission scanning electron microscope (FE-SEM) micrographs of four different powder materials: (Figure 7.2a) YSZ: Appears as spherical particles with a relatively uniform size distribution, likely in the micron range; (Figure 7.2b) Al_2O_3 (Alumina): Shows irregularly shaped particles with sharp edges and a more varied size distribution compared to YSZ; (Figure 7.2c) NiCrAlY: Exhibits spherical particles similar to YSZ but with a larger size range, some particles appearing significantly larger than others; and (Figure 7.2d) CeO_2 (Cerium Oxide): Displays very fine, nano-sized particles that tend to agglomerate into larger clusters. These micrographs provide valuable information about the morphology, size distribution, and surface characteristics of each powder, which are crucial factors in determining their behavior in various applications such as TBCs or catalysts [9].

Figure 7.3 illustrates a TBC system used in high-temperature environments, such as gas turbines. It consists of three layers: an YSZ topcoat (200–250 μm), a bond coat (100–200 μm), and a Ni-based superalloy substrate (1–2 mm). The YSZ layer can withstand temperatures up to 1,300°C, while the superalloy endures up to 1,000°C. Hot gases from combustion can reach 1,500°C. Internal and outer cooling air films help manage temperatures, maintaining structural integrity by preventing overheating. The temperature profile shows how heat is gradually reduced across the layers [10].

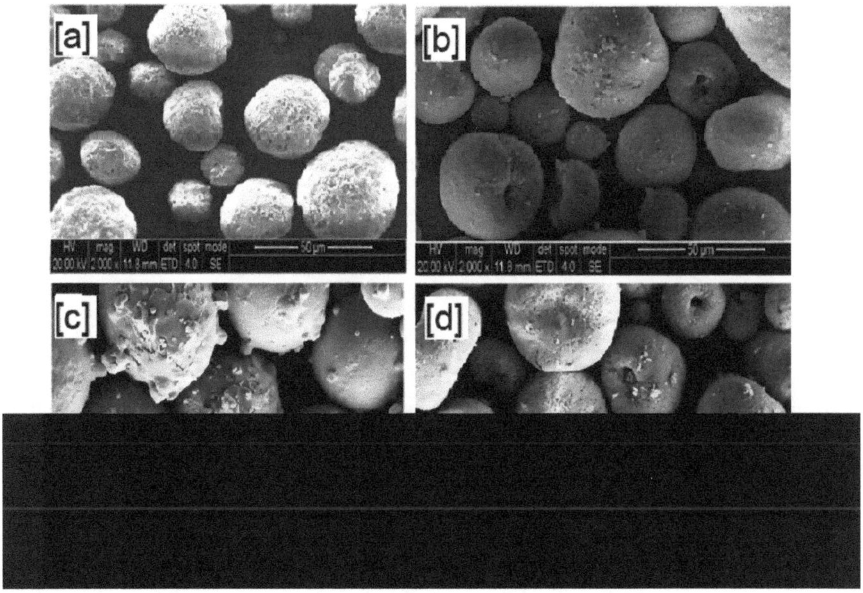

FIGURE 7.2 Typical morphologies of powders by FE-SEM: (a) YSZ, (b) Al_2O_3, (c) NiCrAlY, and (d) CeO_2 [9]. FE-SEM, field emission scanning electron microscope; YSZ, yttria-stabilized zirconia.

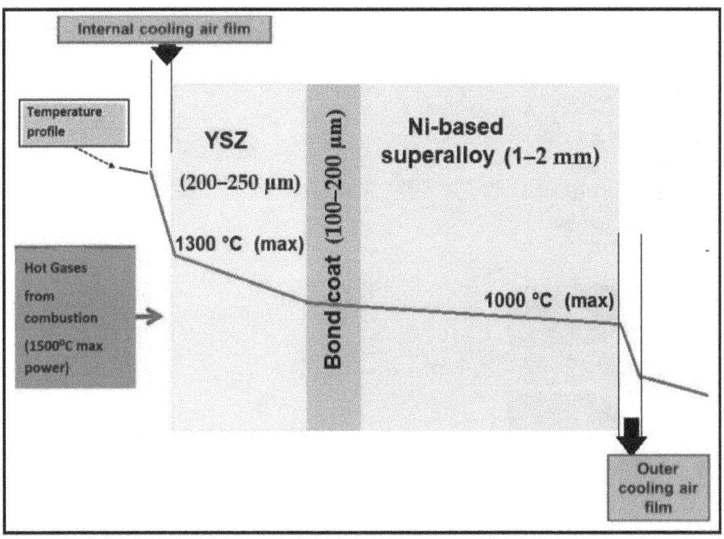

FIGURE 7.3 Thermal barrier coating system [10]. (*Under CC-4.0*).

The TBC system of gas turbines is a multilayered application with the goal of offering full protection of the surface. It provides thermal protection, protects parts from heat, and affords corrosion protection. Such a dual role enables the component to operate under extreme operating conditions, helping increase the efficiency of the turbine and the lifespan of the components as well. In TBC systems, the bond coat serves dual purposes, preventing penetration of corrosives to the underlying metal, improving adhesion between the substrate and the top coat, and providing a protective oxide scale. This scale offers the feature of oxidation resistance [11]. The ceramic top coat, normally YSZ, provides a thermal barrier by isolating the metal of the substrate from the high temperatures in the gas turbine setting.

7.2 DEPOSITION METHODS

7.2.1 Plasma Spraying

Thermal spraying methods are a group of thermochemical processes of applying various powdered materials to the surface of a substrate. The process starts with the creation of a plasma jet using electric arc discharge, where gases are heated to high temperatures [12]. Figure 7.4 shows the plasma spraying coating process. It deals with the injection of particles regulated by means of the powder size as well as the carrier gas flow into the plasma stream developed from the gases argon (Ar) together with hydrogen (H_2). Both voltage and current control the rate of the primary and the secondary gas. Thereby, the spray stream directs particles toward a substrate at different distances. These are position, velocity, temperature, and size of the particles. This technique is deployed utilizing a vacuum to apply various coatings for improved surface characteristics, including wear or heat shielding, by splashing molten or partially melted particles onto a substrate [13]. The fine powder coating material is then fed into this plasma jet, where the material is melted and accelerated toward the target substrate surface at very high velocities. The impact process is when molten particles flatten and solidify instantly through overlapping of small splats, which in sum constitute

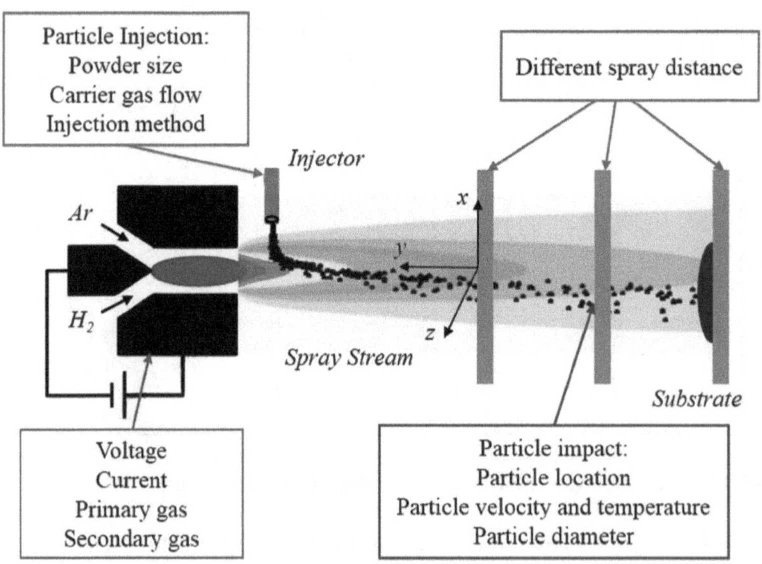

FIGURE 7.4 Process scheme of plasma spraying [13].

the entire coating. This versatile method can apply varying kinds of material, such as metals and cermet, for dense coatings with limited substrate tempering [14].

It finds a wide application in aerospace, power, and automobile industries for thermal and wear applications.

7.2.2 Suspension Plasma Spraying

Suspension plasma spraying (SPS) is a complex thermal spray—surface treatment and coating technique—to which the feedstock is a liquid suspension of fine ceramic particles as opposed to a dry powder. This method has the possibility to deposit the coatings with special microstructures and improved characteristics compared with plasma spraying [15]. In SPS, the ceramic is involved as submicron or nano-sized particles and is dispersed in a liquid phase, usually water or ethanol. This suspension is then sprayed into a plasma jet in which the liquid is evaporated, and the particles are sputtered and accelerated toward the substrate. The finer the particles, the greater the ability to produce coatings with high and detailed microstructures such as columnar structures, vertically cracked patterns, and high porosity [16]. Figure 7.5 illustrates the process of SPS. It starts with a dispersion of particles and agglomerates. This suspension is then introduced to a plasma jet in which fragmentation takes place. The process results in two outcomes: Axe and cadaverous droplets are harder characterized into (i) individual particles and (ii) re-agglomerated clusters. The SEM

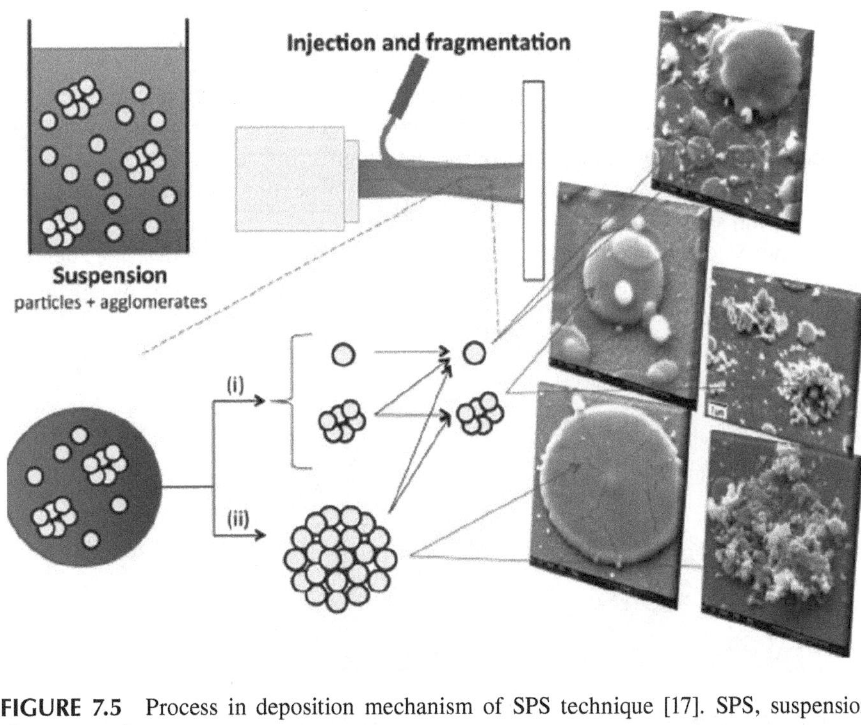

FIGURE 7.5 Process in deposition mechanism of SPS technique [17]. SPS, suspension plasma spraying.

images on the right illustrate the microstructural evolutions, where coatings of different appearances are formed. The use of nano- and submicron-sized particles is also permitted in SPS, thus obtaining coatings with special microstructures, enhanced heat and mechanical characteristics, and porosity [17].

Some of these characteristics make SPS coatings ideal for deployment in TBCs in gas turbines, wear-resistant coatings, and electrodes for solid oxide fuel cells. This technique helps exert a more precise control on microstructure and specific characteristics of a coating that make for enhanced performance and service life in diverse mainstream application areas.

7.3 CHALLENGES

There are still several important questions and challenges that composite coatings face for which researchers and businesses continue to seek answers. During structural design, a lot of challenges are faced for the determination of the composition and distribution of the diverse materials embedded in the covering to possess certain characteristics or qualities. Of all the nanomaterials, graphene and boron nitride nanosheets are particularly problematic when it comes to alignment and placement [18]. This remains questionable in some applications due to the inability to ensure that there is a good bond between the coated material and the substrate material and between the two or more layers incorporated in multilayered coatings [19].

Figure 7.6 illustrates degradation damages on (a) compressor blades and (b) turbine blades. In Figure 7.6a, the compressor blades show signs of surface erosion and corrosion, likely due to exposure to harsh operational environments, leading to material loss and rough surfaces. Figure 7.6b depicts turbine blades with more severe corrosion, characterized by pitting and scaling, which can significantly affect the efficiency and lifespan of the turbine. These damages are typically caused by high temperatures, corrosive gases, and particulate matter impacting the blades during operation, necessitating regular maintenance and inspection to ensure optimal performance and safety [20].

It is convenient to maintain the thickness, and to minimize less desirable defects, such as pores or gaps. Such structures are particularly important when the surface is large or complex.

FIGURE 7.6 Illustration of degradation damages on the surface of: (a) compressor blades and (b) turbine blades [20].

Cohesive techniques of coating often employ noxious reagents, or they release toxic by-products at some phase of the procedure, making ecological concerns arise [17]. One of them is the problem of elaboration of the substitutes that will be less dangerous for the environment. Performance in extreme conditions: The enhancement of the longevity and practical utility of coatings in unfavorable conditions, including high temperatures and the presence of corrosive agents, remains research in progress. New approaches to structure and composition analysis as well as to performance evaluation of coatings should be more effective, and their further elaboration is crucial for new materials [21]. Transferring existing research laboratory coatings to industrial-scale production while maintaining material quality and reasonable cost can be quite testing. The production of coatings that can afford multiple uses, for instance, corrosion and wear resistance and self-healing, is one of the most discussed topics in research on coatings today.

7.4 FUTURE DIRECTIONS

Future research activities may be directed toward advanced material synergy, new multilayer constructions, and higher performances of coatings that additionally provide self-healing abilities to further prolong the lifetimes of the turbine components in ever more severe service conditions. The oxide-formed composite coatings have an important role in preventing the corrosion and oxidation of the gas turbine parts [22]. Some important coatings have evolved in chemical composition, microstructure, and deposition methodology, rendering improved turbine characteristics. Better characteristics are being added to nanostructured composite coatings. With respect to fundamental research and practical application, recent increasing interest has been paid to the concept of synthesis for a surface protection layer with an unprecedented set of mechanical and tribological properties not frequently observed even in nanocrystalline materials, including high hardness and toughness, superior wear resistance, and low friction coefficient [23]. This is because nanocomposite coatings containing both crystalline and amorphous nanophases have recently received an increasing attention. The coating has to be mechanically robust to carry high-load, sliding or rolling contact without cohesive pull-apart or adhesive interface debonding. This is something that any material used for a wear-resistant coating cannot afford to lack [24]. As a rule, it is necessary for the most part of the time to have a low friction coefficient, which, in turn, would reduce the amount of friction losses as well as enhance the capacity to transport loads to a greater extent. It will be seen that the shearing stress, which is equivalent to the contact load multiplied by the coefficient of friction, is the main culprit responsible for the conventional types of coating failure, including cracking, chipping, and de-lamination [25]. This is evidenced by the fact that those failures are some of the most frequently reported coating failures. These requirements have only become possible in the recent past to be met in nanostructured or nanocomposite coatings.

7.4.1 Composite Coatings

Rare earth-doped coatings are modern coatings that have been developed to increase the hot corrosion of metal components predominantly at high temperatures. These often combine such compounds as lanthanum and cerium within the conventional

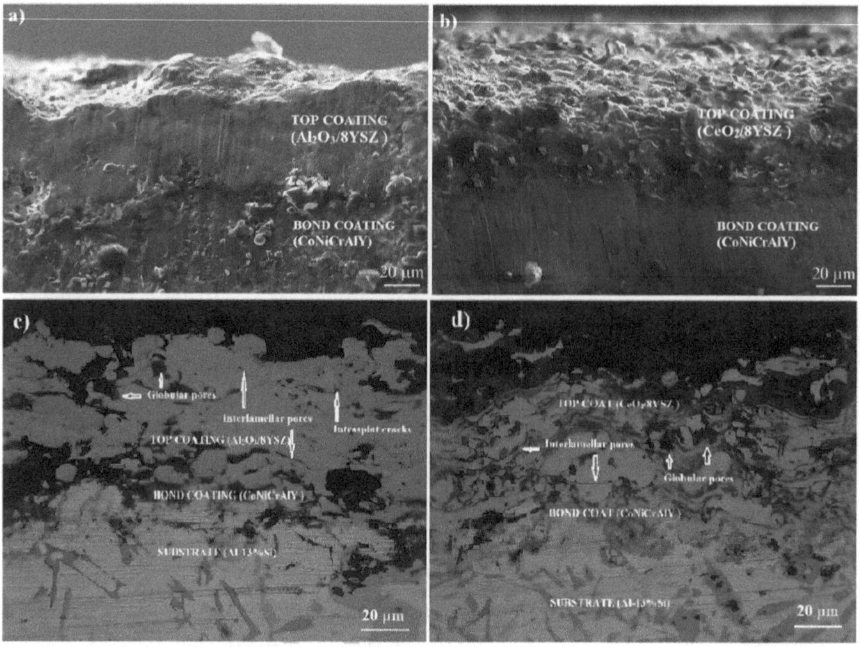

FIGURE 7.7 Cross-sectional morphologies of as-sprayed TBCs: (a) Cross-section of Al_2O_3/8YSZ TBCs and (b) cross-section of CeO_2/8YSZ TBCs. (c) Backscattered electron images of Al_2O_3/8YSZ TBCs and (d) backscattered electron images of CeO_2/8YSZ TBCs [26]. TBCs, thermal barrier coatings.

base of protective coatings, including zinc, aluminum, and magnesium. Figure 7.7 shows cross-sectional morphologies of TBCs.

Figure 7.7a depicts Al_2O_3/8YSZ TBCs with a top coating over a CoNiCrAlY bond coating. Figure 7.7b illustrates CeO_2/8YSZ TBCs with a similar structure. Figures 7.7c and d are backscattered electron images highlighting the microstructural features of Al_2O_3/8YSZ and CeO_2/8YSZ TBCs, respectively. Both images reveal globular and interlamellar pores, as well as intrasplat cracks, which are typical in TBCs for thermal insulation and stress accommodation [26].

The transition of the phases to microstructure through the incorporation of rare earths improves the stability of the coating and boosts corrosion resistance. For instance, Zn–Al–Mg coatings can be doped with rare earth elements such as La, causing the formation of $LaAl_3$ or $Al11Ce_3$, which enhances overall corrosion protection [27]. As for the mentioned industries, aerospace and power generation in particular, the usage of these coatings increases as components that are subject to high temperatures and aggressive working environments are used.

7.4.2 Multilayer Architectures TBCs

Figure 7.8 illustrates three types of TBCs used on a Hastelloy-X substrate: Single Layer TBC: Consists of a 300 µm thick YSZ layer over a 220 µm bond coat. Double-Layered TBC: Features a 60 µm YSZ layer topped with a 260 µm GZ

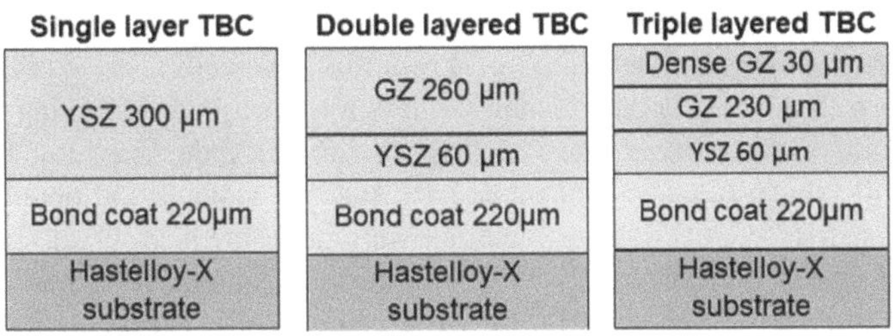

FIGURE 7.8 Coating structures of the samples under investigation. The spray parameters are the same for coatings with comparable shades [28].

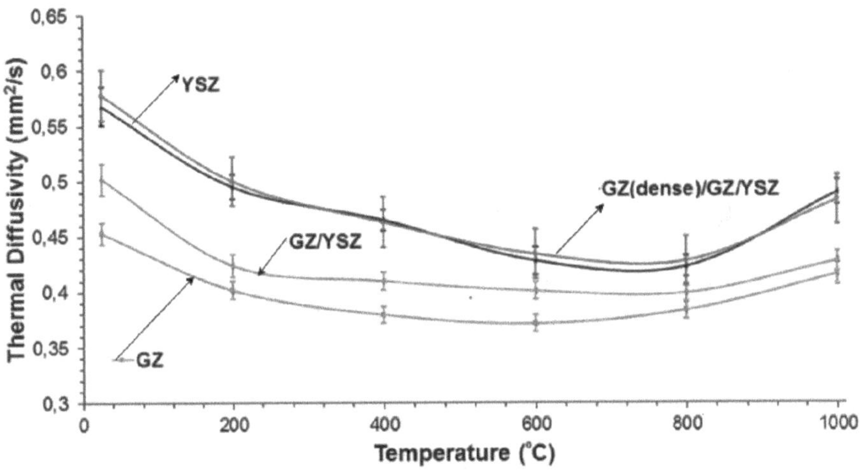

FIGURE 7.9 Plot of the TBCs' thermal diffusivity against temperature as sprayed [28]. TBCs, thermal barrier coatings.

(Gadolinium Zirconate) layer, both over the same bond coat. Triple Layered TBC: Comprises a 60 μm YSZ layer, a 230 μm GZ layer, and an additional dense 30 μm GZ layer, all over the bond coat [28].

Figure 7.9 compares the thermal diffusivity of four materials—YSZ, GZ, GZ/YSZ, and GZ(dense)/GZ/YSZ—across temperatures from 0°C to 1,000°C. In both graphs, YSZ shows the highest thermal diffusivity, decreasing slightly with temperature. GZ has the lowest values, also decreasing with temperature. GZ/YSZ and GZ(dense)/GZ/YSZ exhibit intermediate values, with GZ(dense)/GZ/YSZ showing slightly higher diffusivity than GZ/YSZ. The trends indicate a general decrease in thermal diffusivity as temperature increases for all materials. Error bars represent variability in measurements [28].

It should also be noted that the field of composite coatings provides several opportunities for research and development. Explorations of purity, particle size, and agglomerate distribution in the nanostructured feedstock powders are important. These results will mean that the next priority will be given to the understanding of nanozone's contribution to the improvements of the coating characteristics and revelation of performance features [29]. These studies will not only enhance our understanding of various outcomes but will also promote the development of further strides in the engineering of nanostructured powders and coatings. From this perspective, these points present researchers with avenues upon which to tap into novel ways of enhancing the performance, stability, and utility of these coatings in different industries [30]. This understanding will inform the development of new material systems and possibly change industries as diverse as aerospace, automotive, and biomedical. Oxide-based composite coatings are crucial for protecting gas turbine components from corrosion in high-temperature environments [31–36]. These coatings offer excellent thermal stability and corrosion resistance by forming dense oxide layers that act as barriers against aggressive gases and molten salts [37–43]. Research demonstrates that deposition techniques such as thermal spraying and microwave heating significantly enhance the structural integrity and performance of these coatings [44,45].

7.5 CONCLUSIONS

There is the use of composite coatings based on oxides whereby components like the turbine blades in the gas turbines are protected from heat, corrosive gases, and cycling stresses. These coatings are composed of a thermally generated oxide layer, a bond coat that can be MCrAlY, and a ceramic top coat of generally YSZ. In the bond coat, the adhesion and surface oxide resistance are enhanced, and during the top coat, thermal insulation is provided. Such coatings are typically applied by means of plasma spraying and SPS, which are typical cup methods. Thus, much work must be done to reach the maximum adhesion, enhance the deposition procedure, and create environmentally friendly techniques at the same time. Future work involves the study of creating coatings that are not only multifunctional but also possess better qualities. These coatings are, for example, rare-earth-doped coatings that are hot corrosion resistant and multilayered architecture applications of materials such as gadolinium zirconate, which possess better thermal properties. Also, further development of the study field of nanostructured coatings is stated as an opportunity.

REFERENCES

1. Almeida E. Surface treatments and coatings for metals. A general overview. 1. Surface treatments, surface preparation, and the nature of coatings. *Industrial Engineering Chemistry Research*. 2001;40:3–14.
2. Ataie A, Mostaghimi J, Pershin L, Xu P. Fabrication of nanostructured cobalt ferrite coatings using suspension plasma spraying (SPS) technique. *Surface Coatings Technology*. 2017;328:451–61.
3. Hille TS, Nijdam TJ, Suiker AS, Turteltaub S, Sloof WG. Damage growth triggered by interface irregularities in thermal barrier coatings. *Acta Materialia*. 2009;57:2624–30.
4. Bakan E, Vaßen R. Crack healing mechanisms in atmospheric plasma sprayed Yb-silicate coatings during post-process heat treatment. *Journal of the European Ceramic Society*. 2023;43:3684–93.
5. Bakhsheshi-Rad H, Hamzah E, Ismail A, Daroonparvar M, Kasiri-Asgarani M, Jabbarzare S, Medraj M. Microstructural, mechanical properties and corrosion behavior of plasma sprayed NiCrAlY/nano-YSZ duplex coating on Mg–1.2 Ca–3Zn alloy. *Ceramics International*. 2015;41:15272–7.

6. Bobzin K, Öte M, Knoch M. Surface pre-treatment for thermally sprayed ZnAl15 coatings. *Journal of Thermal Spray Technology*. 2017;26:464–72.
7. Bobzin K, Öte M, Linke T, Sommer J, Liao X. Influence of process parameter on grit blasting as a pretreatment process for thermal spraying. *Journal of Thermal Spray Technology*. 2016;25:3–11.
8. Brossard S, Munroe P, Tran A, Hyland M. Study of the splat formation for plasma sprayed NiCr on aluminum substrate as a function of substrate condition. *Surface Coatings Technology*. 2010;204:2647–56.
9. Mehta A, Vasudev H, Singh S. Sustainable manufacturing approach with novel thermal barrier coatings in lowering CO_2 emissions: performance analysis with probable solutions. *International Journal on Interactive Design Manufacturing*. 2023;18:1–13.
10. Mehta A, Vasudev H, Singh S, Prakash C, Saxena KK, Linul E, Buddhi D, Xu J. Processing and advancements in the development of thermal barrier coatings: a review. *Coatings*. 2022;12:1318.
11. Chen H, Fan M, Zhu W, Norton A. High temperature oxidation behaviour of combustion flame sprayed CoNiCrAlY coatings. *Surface Coatings Technology*. 2020;385:125431.
12. Cinca N, Lima CRC, Guilemany JM. An overview of intermetallics research and application: status of thermal spray coatings. *Journal of Materials Research Technology*. 2013;2:75–86.
13. Zhu J, Wang X, Kou L, Zheng L, Zhang H. Prediction of control parameters corresponding to in-flight particles in atmospheric plasma spray employing convolutional neural networks. *Surface Coatings Technology*. 2020;394:125862.
14. Daroonparvar M, Bakhsheshi-Rad HR, Saberi A, Razzaghi M, Kasar AK, Ramakrishna S, Menezes PL, Misra M, Ismail AF, Sharif S. Surface modification of magnesium alloys using thermal and solid-state cold spray processes: challenges and latest progresses. *Journal of Magnesium Alloys*. 2022;10:2025–61.
15. Di Girolamo G, Blasi C, Brentari A, Schioppa M. Microstructural, mechanical and thermal characteristics of zirconia-based thermal barrier coatings deposited by plasma spraying. *Ceramics International*. 2015;41:11776–85.
16. Di Girolamo G, Blasi C, Pagnotta L, Schioppa M. Phase evolution and thermophysical properties of plasma sprayed thick zirconia coatings after annealing. *Ceramics International*. 2010;36:2273–80.
17. Fan W, Bai Y. Review of suspension and solution precursor plasma sprayed thermal barrier coatings. *Ceramics International*. 2016;42:14299–312.
18. Di Girolamo G, Blasi C, Schioppa M, Tapfer L. Structure and thermal properties of heat treated plasma sprayed ceria–yttria co-stabilized zirconia coatings. *Ceramics International*. 2010;36:961–8.
19. Ding Y, Li H, Tian Y. Bristle blasting surface preparation in thermal spraying. *Journal of Thermal Spray Technology*. 2019;28:378–90.
20. Brun K, Kurz R, Simmons HR, Aerodynamic instability and life limiting effects of inlet and interstage water injection into gas turbines. *Turbo Expo: Power for Land, Sea, and Air*. 2005:47276:237–45.
21. Fan X, Zou B, Gu L, Wang C, Wang Y, Huang W, Zhu L, Cao X. Investigation of the bond coats for thermal barrier coatings on Mg alloy. *Applied Surface Science*. 2013;265:264–73.
22. Fanicchia F, Axinte D, Kell J, McIntyre R, Brewster G, Norton A. Combustion flame spray of CoNiCrAlY & YSZ coatings. *Surface Coatings Technology*. 2017;315:546–57.
23. Fanicchia F, Axinte DA. Transient three-dimensional geometrical/thermal modelling of thermal spray: normal-impinging jet and single straight deposits. *International Journal of Heat Mass Transfer*. 2018;122:1327–42.

24. Fanicchia F, Maeder X, Ast J, Taylor A, Guo Y, Polyakov M, Michler J, Axinte D. Residual stress and adhesion of thermal spray coatings: microscopic view by solidification and crystallisation analysis in the epitaxial CoNiCrAlY single splat. *Materials Design*. 2018;153:36–46.
25. Fauchais PL, Heberlein JV, Boulos MI, Fauchais PL, Heberlein JV, Boulos MI. Overview of thermal spray. In: MI Boulos, PL Fauchais, JVR Heberlein (eds.), *Thermal Spray Fundamentals: From Powder to Part*. Springer International Publishing AG, Cham, 2014:17–72.
26. Venkadesan G, Muthusamy J. Experimental investigation of Al_2O_3/8YSZ and CeO2/8YSZ plasma sprayed thermal barrier coating on diesel engine. *Ceramics International*. 2019;45:3166–76.
27. Gan JA, Berndt CC. Nanocomposite coatings: thermal spray processing, microstructure and performance. *International Materials Reviews*. 2015;60:195–244.
28. Mahade S, Curry N, Björklund S, Markocsan N, Nylén P. Thermal conductivity and thermal cyclic fatigue of multilayered $Gd_2Zr_2O_7$/YSZ thermal barrier coatings processed by suspension plasma spray. *Surface Coatings Technology*. 2015;283:329–36.
29. Garcia-Alonso D, Serres N, Demian C, Costil S, Langlade C, Coddet C. Pre-/during-/post-laser processes to enhance the adhesion and mechanical properties of thermal-sprayed coatings with a reduced environmental impact. *Journal of Thermal Spray Technology*. 2011;20:719–35.
30. Koivuluoto H. A review of thermally sprayed polymer coatings. *Journal of Thermal Spray Technology*. 2022;31:1750–64.
31. Prashar G, Vasudev H, Thakur L. High-temperature oxidation and erosion resistance of Ni-based thermally-sprayed coatings used in power generation machinery: a review. *Surface Review and Letters*. 2022;29(03):2230003.
32. Singh G, Vasudev H, Bansal A, Vardhan S. Influence of heat treatment on the microstructure and corrosion properties of the Inconel-625 clad deposited by microwave heating. *Surface Topography: Metrology and Properties*. 2021;9(2):025019.
33. Prashar G, Vasudev H. Hot corrosion behavior of super alloys. In Singh, S.K., Akinlabi, E.T., Kumar, K., Davim, J.P., Saxena, K.K. (Eds.), *Materials Today: Proceedings, 10th International Conference of Materials Processing and Characterization*. 2020;26:1131–5.
34. Vasudev H, Prashar G, Thakur L, Bansal A. Electrochemical corrosion behavior and microstructural characterization of HVOF sprayed inconel718-Al_2O_3 composite coatings. *Surface Review and Letters*. 2022;29(02):2250017.
35. Singh P, Bansal A, Vasudev H, Singh P. In situ surface modification of stainless steel with hydroxyapatite using microwave heating. *Surface Topography: Metrology and Properties*. 2021;9(3):035053.
36. Vasudev H, Singh P, Thakur L, Bansal A. Mechanical and microstructural characterization of microwave post processed Alloy-718 coating. *Materials Research Express*. 2020;6(12):1265f5.
37. Prashar G, Vasudev H. Structure–property correlation of plasma-sprayed Inconel625-Al_2O_3 bimodal composite coatings for high-temperature oxidation protection. *Journal of Thermal Spray Technology*. 2022;31(8):2385–2408.
38. Prashar G, Vasudev H. Surface topology analysis of plasma sprayed Inconel625-Al_2O_3 compositecoating.InPrakash,C.,Singh,S.,Rathi,R.,Krolczyk,G.(Eds.),*MaterialsToday: Proceedings, 2nd International Conference on Functional Material, Manufacturing and Performances (ICFMMP-2021)*. 2022;50:607–611.
39. Prashar G, Vasudev H. A review on the influence of process parameters and heat treatment on the corrosion performance of Ni-based thermal spray coatings. *Surface Review and Letters*. 2022;29(01):2230001.

40. Yedida VS, Vasudev H. A review on the development of thermal barrier coatings by using thermal spray techniques. *Materials Today: Proceedings.* 2022;50:1458–1464.
41. Majji BGR, Vasudev H, Bansal A. A review on the oxidation and wear behavior of the thermally sprayed high-entropy alloys. In Prakash, C., Singh, S., Rathi, R., Krolczyk, G. (Eds.), *Materials Today: Proceedings, 2nd International Conference on Functional Material, Manufacturing and Performances (ICFMMP-2021).* 2022;50:1447–1451.
42. Sunitha K, Vasudev H. A short note on the various thermal spray coating processes and effect of post-treatment on Ni-based coatings. In Prakash, C., Singh, S., Rathi, R., Krolczyk, G. (Eds.), *Materials Today: Proceedings, 2nd International Conference on Functional Material, Manufacturing and Performances (ICFMMP-2021).* 2022;50:1452–1457.
43. Vasudev H. Wear characteristics of Ni-WC powder deposited by using a microwave route on mild steel: microwave cladding of Ni-WC. *International Journal of Surface Engineering and Interdisciplinary Materials Science (IJSEIMS).* 2020;8(1):44–54.
44. Prashar G, Vasudev H, Thakur L, Bansal A. Performance of thermally sprayed hydroxy-apatite coatings for biomedical implants: a comprehensive review. *Surface Review and Letters.* 2023;30(01):2241001.
45. Parkash J, Saggu HS, Vasudev H. A short review on the performance of high velocity oxy-fuel coatings in boiler steel applications. In Prakash, C., Singh, S., Rathi, R., Krolczyk, G. (Eds.), *Materials Today: Proceedings, 2nd International Conference on Functional Material, Manufacturing and Performances* (ICFMMP-2021). 2022;50:1442–1446.

8 Development of TBCs for Enhanced Hot Corrosion Performance in Aero Engines

Amrinder Mehta and Hitesh Vasudev

8.1 INTRODUCTION

Thermal barrier coatings (TBCs) are essential protective layers applied to critical components in aero engines, such as turbine blades and combustion chambers. These coatings serve as thermal insulators, shielding the underlying metal parts from extreme temperatures during engine operation. By reducing heat transfer to the metal substrate, TBCs allow engines to run at higher temperatures, improving efficiency and performance [1]. Additionally, TBCs protect against oxidation and corrosion caused by the harsh combustion environment, significantly extending the lifespan of engine components. This protection is crucial for maintaining engine integrity and safety while also reducing maintenance costs and improving overall engine reliability [2].

Figure 8.1 illustrates a jet engine's operation through four main stages: Intake, Compression, Combustion, and Exhaust. This is because Air enters through the Air Inlet to the Intake, where the initial stage of the compression takes place. In the Compression stage, there is an enormous rise in the pressure of air. The combustion chambers mix this high-pressure air with fuel, igniting it to generate high-energy gases. The Turbine's operating gases are extracted from these gases and used to drive

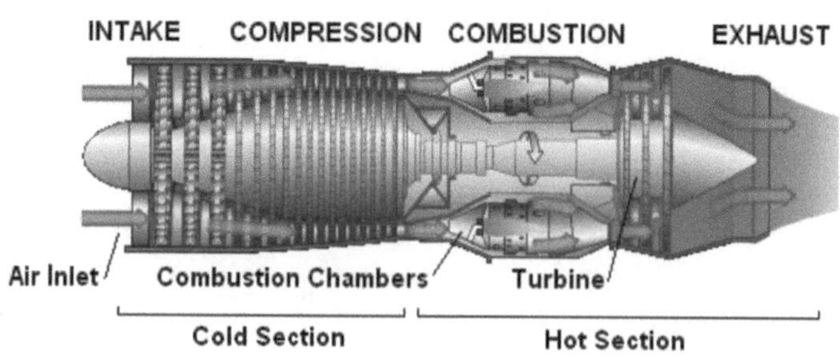

FIGURE 8.1 Gas turbine parts and stages [3]. (*Under CC-4.0 license*).

DOI: 10.1201/9781003542957-8

the compressor. Finally, during the Exhaust stage, the gases are ejected, which produces thrust. This section of the engine is called the Cold Section, made up of Intake and Compression, whereas the Hot Section is made up of Combustion and Exhaust [3]. Aerospace manufacturers seeking higher efficiency and performance standards are today raising operating temperatures to new levels. This trend can make it necessary to create new generations of TBCs. The next-generation TBCs should be able to accommodate higher temperatures, aggressive combustion situations, and additional thermal cycles [4]. It must possess higher thermal and electrical insulation, better protective characteristics against oxidation and corrosion, and enhanced resistance to mechanical and thermal loads at extreme temperatures. The issue here is that in such conditions the coatings need to stay protective and hold their structure to allow for lightweight and affordable solutions. Satisfying these requirements is vital to supporting the future progress of aero engine development and efficiency improvements [5].

Figure 8.2 illustrates the key properties of zirconia-based coatings. These coatings are thus emphasized for their ability to have low thermal conductivity, as this assists in a material losing heat. It also contains a high cation of thermal expansion, meaning that it can expand and contract without cracking. Oxidation prevents the chemical degradation caused by changes in the surface chemistry. Additionally, they have a high mechanical quality and thereby guarantee sturdiness and toughness. Thermal shock resistance enables them to support temperature gradients, while microstructure stability supports these structures under stress. These attributes render zirconia-based coatings suitable for high-temperature use and protective coatings [6].

This review focuses on trends in TBC material and structure in the last decade with the view to improving hot corrosion resistance for aero engine applications. Such developments not only make components last longer but also improve the general drive of the engine, leading to minimal maintenance costs and better fuel efficiency. Moreover, current research in the evolution of AM technologies is expanding the use of complex and inserted forms and composition of materials for the development of TBC systems that could redefine their remedial action against hot corrosion.

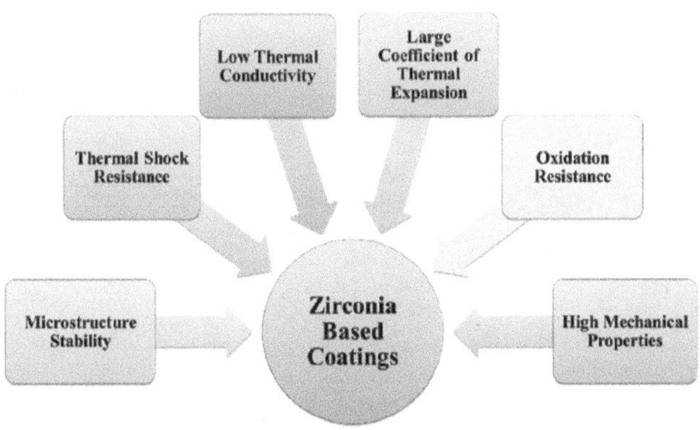

FIGURE 8.2 Schematic illustration of the properties of a zirconia-based coating [6].

8.2 STRUCTURAL DESIGN APPROACHES

8.2.1 Multi-Layer Coatings

Enhancing TBCs' hot corrosion performance is one of the main approaches in the application of multi-layer coating structures. Two-layer protective TBCs utilize corrosion-resistant, correlation-stabilized zirconia as a top layer, making them more advanced than single-layer yttria-stabilized zirconia (YSZ) coatings [7]. Multi-layer coatings are advanced materials designed to improve functionality across a range of applications spanning from (i) tribology, (ii) optics, and (iii) TBCs. These coatings consist of layers, each of which is endowed with unique characteristics, and they collectively form the coatings. The multi-layered structure also enables the use of different materials, enhancing durability, reducing friction, and providing the right optical properties [8]. When choosing and placing all those layers properly, the scientists and engineers will be able to develop the coatings that will fulfill certain applications, such as increasing wear resistance in mechanical elements, increasing the transparency of optical equipment, or increasing heat shields in aerospace engineering. For this reason, multi-layer coatings play a crucial role in developing technology in different sectors [9].

This has a new structure that utilizes the concept of YSZ's high thermal conductivity as the bottom layer, at the same time overcoming the shortcomings of the material in terms of corrosion resistance [10]. The incorporation of a custom upper layer is one way through which the general coating performance is boosted, for instance, using lanthanum zirconate ($La_2Zr_2O_7$). This combination may raise the temperature capacity by more than 100K in comparison with traditional YSZ coatings only [11]. The substantial increase in temperature tolerance associated with these advanced ceramics has paved the way for higher operating temperatures of engines with likely improvement in the efficiency and performance of aero engines. Two-hop systems with the creel of the intermediate coat between the bond coat (BC) and the topcoat have also been studied to enhance durability and corrosion resilience [8]. Figure 8.3 illustrates

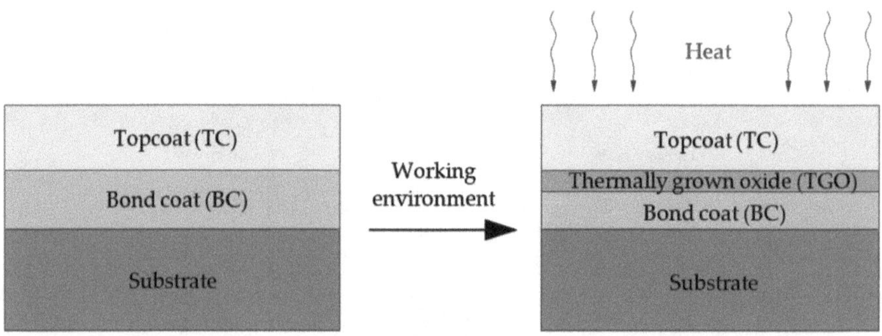

FIGURE 8.3 Layered systems consist of various components with distinct chemical characteristics. Application of this system on a substrate typically includes a bond coat (BC) and a ceramic topcoat (TC). Thermal exposure creates a third layer, known as thermally generated oxide (TGO), between the two layers [12]. (*Under CC-4.0 License*).

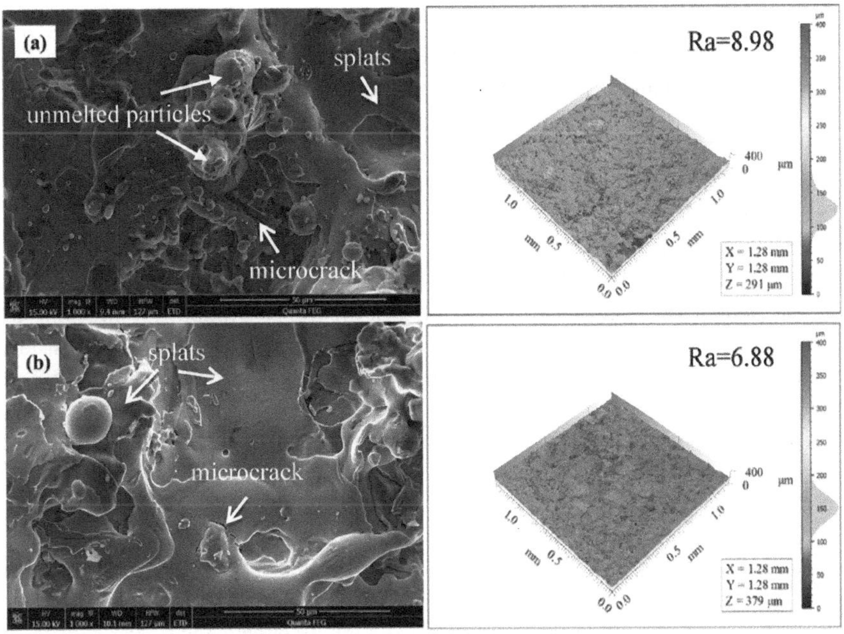

FIGURE 8.4 3D micrographs of surface roughness of bond coat Al–NiCr by (a) air cooling and (b) dry ice blasting [14].

the structure of TBCs, which consist of several layers. Initially, there are three layers: the topcoat (TC) for thermal insulation, the BC for adhesion, and the substrate [12].

In the working environment, heat results in the buildup of a thermally grown oxide (TGO) layer between the TC and the BC. This TGO layer is also the precursor to high thermal stability and oxidation protection, which improves the lifespan of the substrate. The layers cooperatively prevent various detrimental impacts on the substrate, such as thermal and oxidative degradation [13].

Figure 8.4 compares two surfaces analyzed using scanning electron microscopy (SEM) and 3D surface profilometry. The Left Panels (a and b) SEM images reveal surfaces with solid unmelted particles, splats, and microcrack morphology. Such features point to exponential imperfection and possible structural flaws in the molten ores. Surface (a) is rougher, having an Ra of 8.98 compared to the smoother surface and (b) having an Ra of 6.88. These analyses are important for the characterization of material properties and behavior of coatings in services demanding certain surface features [14].

Surface roughness patterns are clearly captured in high-resolution 3D micrographs that can impact the functionality and stability of the material. These patterns, developed during manufacturing processes, can inform the frictional characteristics, wear behavior, and even adhesion properties of coatings applied to those surfaces [15]. Knowledge of these patterns is therefore critical in materials selection and the processing strategies needed to optimize the overall performance.

8.2.2 Nanostructured Coatings

Nanostructured coatings are widely used in almost all industries and are critical elements in processes and products that impact solar energy and electronics. Coatings in solar cells are helpful in enhancing the ability of cells to accept light to convert it to useful energy, a significant step in the development of renewable energy technologies [16]. Also, in the electronics field, nanostructured coatings have improved conductivity, reduced friction, and enhanced thermal dissipation, which promote miniaturization and efficiency of devices in the electronics sector. Metallic nanostructured coatings include the deposition of metals at the atomic level and add-on properties like corrosion and wear resistance, which are essential in the transport industry, including cars and airplanes [17]. The advanced nanostructured coatings with hot corrosion resistance for aero engine applications are superior to conventional coatings. The distinguishing characteristic of these superior coatings is their nanoscale grain size, which establishes a qualitatively different material. The nanostructure makes the coating to be tighter in that it is much denser and does not have many cracks through which the corrosive agents can penetrate [18]. This leads to increased protection against molten salt infiltration, which is a principal mechanism of hot corrosion. Moreover, impulse-pulse treatment can develop fine grain boundaries in nanostructured TBCs and serve as a source of rejection of impurities, enhancing the corrosion feature. Altogether, all these properties enable a longer useful coating life along with enhanced protection of engine components in severe and high-temperature applications [19].

The application of nanostructured YSZ coatings could be considered as the new generation TBC materials for aero engines. These coatings have a fine-grained structure with grain sizes in the nanoscale range and have been identified to have the following advantages over the conventional YSZ coatings. The nanostructure also causes a marked improvement in the resistance to attack by molten salts, which is normally responsible for the deterioration of the coatings at high temperatures [20]. This improved barrier effect assists in preventing corrosive stuff from having its way through the bare metal surface. Also, YSZ coatings in the form of nanostructures revealed improved resistance to an undesirable phenomenon known as spallation, which is the process when the coating delaminates from the substrate. This means that the claimed angular contacting patch-bearing architecture results in longer component life and improved engine performance when exposed to high levels of operation [21].

Figure 8.5 presents SEM images of a coating material's microstructure. The images (a) and (b) reflect the coercive interaction of the YSZ layer with the BC and the substrate, which demonstrated the growth of microcracks and pores. Images (c) and (d) are concerned with nanoscale morphology and show spherical morphology of 101–139 nm. These images stress features of the ceramic coating at the microscale and macroscale [22].

Two layers consist of nanostructured and conventional structures to optimize the TBC for aero engines. This remarkable structure capitalizes on the properties displayed by both layers to improve general performance. The nanostructured layer often offers better resistance to penetration by the molten salt and is more stably durable, while the traditional layer has been validated to offer excellent thermal insulation capability [23]. When these layers are layered, the coating itself gives better

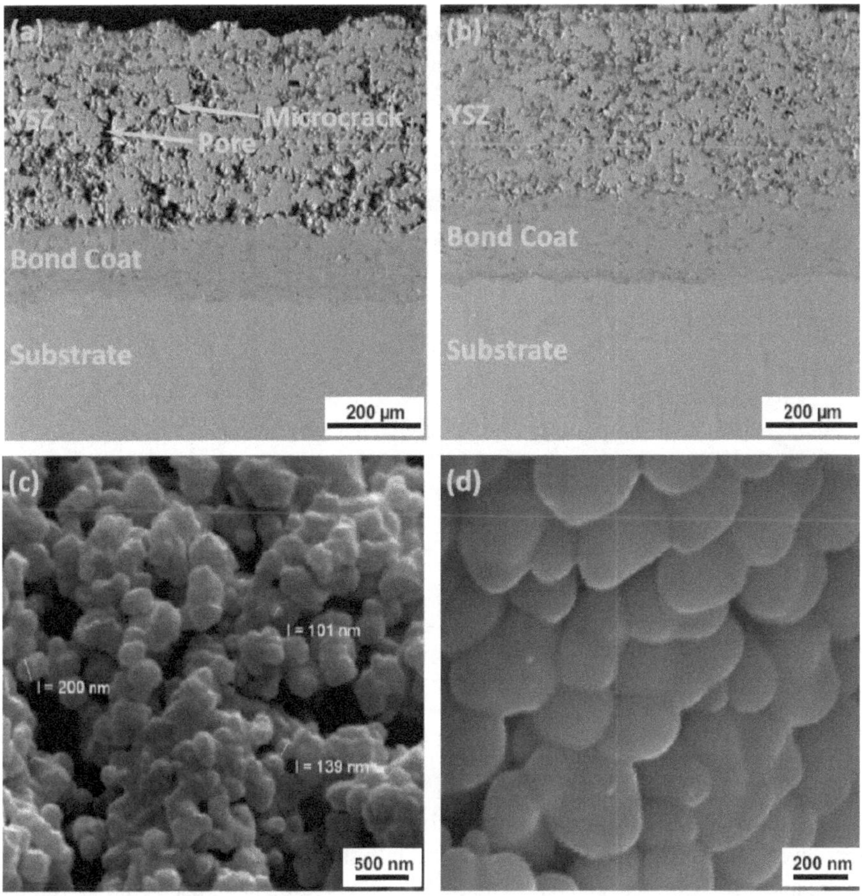

FIGURE 8.5 SEM micrographs of cross-sections (as-sprayed) of the: (a) conventional coating; (b) nanostructured coating; (c) nanostructured coating outer surface morphology; and (d) cross-section nanostructure [22]. SEM, scanning electron microscopy.

thermal protection and a somewhat longer life than the layers of either type alone. By doing this, the coating properties are designed to meet the needs of the specific engine and may provide for enhanced insulation efficiency and longer useful life of vulnerable engine components cycling in severe environments [24].

8.2.3 Segmented/Columnar Microstructures

A special type of TBC with segmented or columnar structures demonstrated considerable benefits for the aero engine part. These antecedent structures are characterized by specific columns or portions within the coating, which are divided by voids. Thus, the present architectural structure contributes to increased tolerance to the chosen type of strain: due to adjusting the size of individual columns within the architectural structure, the overall stress absorbed by the coating is minimized [25].

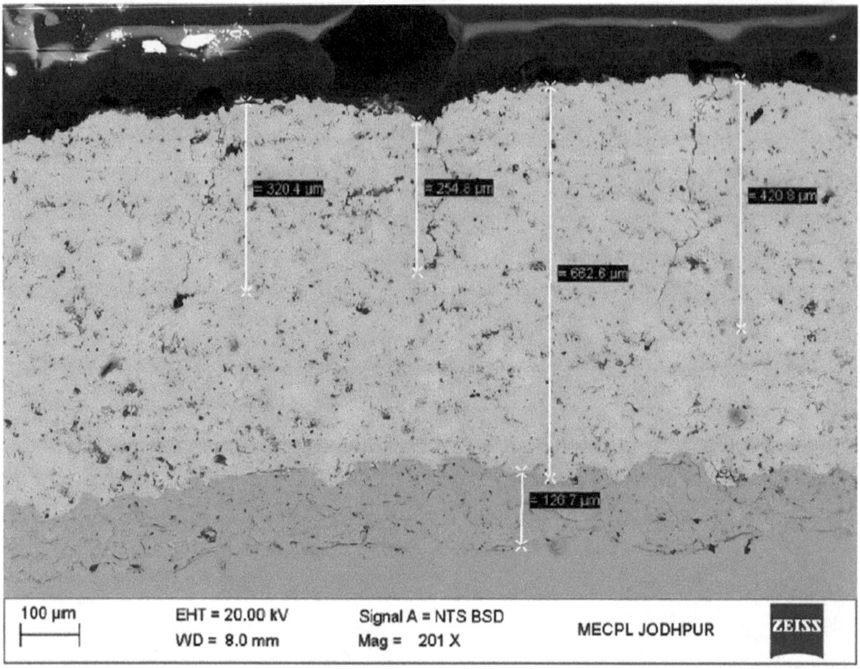

FIGURE 8.6 A cross-sectional picture of segmented YSZ coatings exhibiting varying crack lengths and depths [26].

Figure 8.6 is a SEM micrograph showing a cross-section of a material with distinct layers. It features five measured layers with thicknesses of 320.4, 254.8, 662.6, 420.8, and 126.7 µm, respectively. The SEM was operated at an accelerating voltage of 20.00 kV with a working distance of 8.0 mm and a magnification of 201×. The image provides detailed structural information about the material's composition, likely for analysis or quality control purposes, and was produced by MECP Jodhpur using a Zeiss microscope [26].

The segmented nature also impedes crack propagation and limits the infiltration of corrosive substances, improving corrosion resistance. Moreover, a columnar structure can improve thermal insulation characteristics somewhat, at least when it comes to metals. The advantages cause the enhancement of the durability and lifetimes of some engine parts working under severe conditions and enhance performance and reliability under high-temperature operation [27]. Figure 8.7 compares the thermal conductivity of different coatings. The first image shows a bar chart with coatings applied by atmospheric plasma spraying (APS), suspension plasma spraying (SPS), and electron beam-physical vapor deposition (EB-PVD) methods. Thermal conductivity values range from approximately 0.6 to 1.5 W/m·K. APS and SPS coatings have lower conductivities, while EB-PVD shows the highest. The second image is a recreated bar chart using similar data, displaying the same trend: APS and Coating C are lower in conductivity, while SPS and EB-PVD are higher. Each coating type is illustrated with microscopic images showing their structural differences [28].

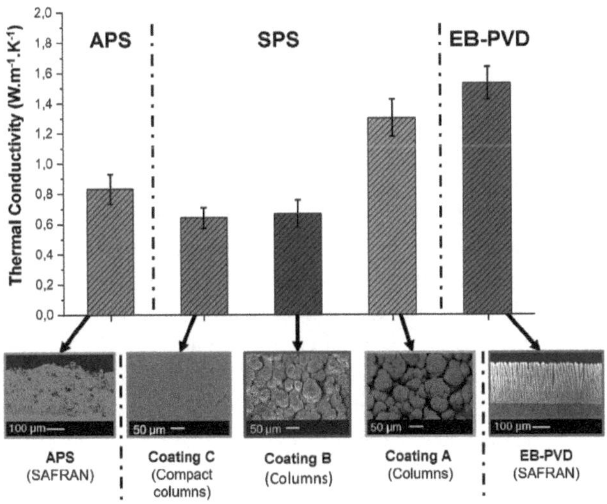

FIGURE 8.7 Thermal conductivities at 25°C of YSZ samples performed by SPS compared to YSZ obtained by APS and EB-PVD [28]. YSZ, yttria-stabilized zirconia; SPS, suspension plasma spraying; APS, atmospheric plasma spraying; EB-PVD, electron beam-physical vapor deposition.

EB-PVD coatings with a columnar grain structure, when applied in aero engine components, provide better hot corrosion resistance than the others because of this structure. These coatings have rather columnar crystals formed as separate columns with small inter-columnar spaces that proved to have several benefits [29]. The spaces between the columns are engineered to facilitate independent movement in response to thermal stresses. The columnar structure that the grain exhibits effectively prevents crack transmission across the entire coating. Segmentation of the coating design also influences thermal shock distribution. The structure prevents corrosive materials from penetrating through. Altogether, these properties make blunt coatings have longer Coating service life even under the most severe conditions [30].

This microstructure allows the EB-PVD coatings to sustain their protection characteristics in the severe environment in aero engines and provides a good application for hot corrosion protection. TBCs, which are segmented microstructures, have been developed using a low-cost plasma spray instead of the expensive EB-PVD [31]. Other more advanced plasma spray techniques can deposit coatings that have nearly equivalent microstructure to EB-PVD coatings as far as concerns for strain tolerance and corrosion resistance are concerned. The parameters of spraying must be managed, and the feedstock materials have to be selected carefully to obtain the desired segmentations through the process [32]. This approach integrates the benefits of columnar microstructures with lower cost and improved deposition rate of plasma spraying. Therefore, these techniques are useful to produce TBCs for aero engines of higher performance at comparatively less cost so that they can be effectively used in industrial chambers.

8.3 NOVEL MATERIALS

8.3.1 Rare Earth Zirconates

Rare earth zirconate ceramics have gained significant attention as potential replacements for traditional YSZ in TBCs for aero engines.

Some of the most well-known compositions are gadolinium zirconate ($Gd_2Zr_2O_7$) and lanthanum zirconate ($La_2Zr_2O_7$) [33]. These materials show quite a high performance in severe conditions, which can lead to increased working temperature of the engine and therefore efficiency. But issues like the lower fracture toughness than that in YSZ must be resolved for widespread application. Gadolinium fluoride and lanthanum fluoride emerged as potential ceramics with superior capabilities as TBC material compared to the conventional YSZ material of aero engines [34]. These rare earth zirconates reveal closely related compounds but significantly have less thermal conductivity and hence better insulating capacity, enabling more efficient operation at higher temperatures. Furthermore, it has much better phase stability at high temperatures and minimizes the probability of coating degradation and failure at high temperatures [35]. These relative stabilities add to longer life spans of coating as well as improved reliability in the engine's systems. These properties make $Gd_2Zr_2O_7$ and $La_2Zr_2O_7$ the potential candidates for next-generation TBCs and have the potential to advance the aero engine technology and performance [36].

Figure 8.8 shows two SEM micrographs of a TBC. The left image (Figure 8.8a) displays a cross-section of the coating, highlighting a layer of gadolinium zirconate ($Gd_2Zr_2O_7$) atop a CoNiCrAlY BC. The micrograph reveals columnar structures typical of such coatings. The right image (Figure 8.8b) presents the top surface of the $Gd_2Zr_2O_7$ layer, illustrating inter-columnar voids. These voids are characteristic of the as-sprayed surface, contributing to the coating's thermal insulation properties. Both images are labeled with details like SEM settings and were taken at Bartin University [34].

Figure 8.9 shows SEM micrographs. Figure 8.9a shows a cross-section of a coated material with layers of La Zr_2O_7 and CoNiCrAlY, featuring columnar grains.

FIGURE 8.8 Cross-sectional and top surface images of $Gd_2Zr_2O_7$ coating [34].

FIGURE 8.9 Cross-sectional and top surface images of $La_2Zr_2O_7$ coating [34].

It is captured at a magnification of 3,000×. Figure 8.9b depicts the as-sprayed top surface of La Zr_2O_7, highlighting inter-columnar voids and an as-sprayed texture, with a magnification of 200×. Both images are from Bartin University, using a SEM with 20.0 kV voltage [34].

Lanthanum zirconate is an example of a pyrochlore structure and is highly resistant to molten salts such as sulfate and vanadate, hence their importance in the development of TBCs for aero engines. This resistance is due to the nature of the crystal structure and the chemical stability of the compounds in the semiconductor. The pyrochlore organization gives a definite structure that will not yield to penetration and chemical attack by these dissolving salts, which are often associated with aviation fuels [37]. This property is vital to application at high temperatures when salt can bring about a quick deterioration of a normal coating. Thus, the increased stability of $La_2Zr_2O_7$ to these aggressive compounds means fewer frequent replacements of engine parts and increased service life of the aero engines under severe conditions.

8.3.2 Multi-Component Stabilized Zirconia

Enhancement of TBC of the aero engine through co-doping zirconia with multiple rare earth oxides is due to the improvement of phase steady state and corrosion behaviors. This technique forms a more compact crystal lattice that has enhanced heat resistance and does not permit the penetration of the material by corrosive agents. The product is an enhanced and longer-lasting coating for engine parts [38]. Yb_2O_3-Gd_2O_3-Y_2O_3 co-stabilized zirconia is a breakthrough for the advanced TBCs for aero engines, especially with respect to hot corrosion protection. It is evidenced here that this new material exhibits high-performance characteristics over $Na_2SO_4 + V_2O_5$ molten salts, which are widely used corrosive agents at high-temperature operations [39]. A more enhanced resistance is attributed to the harder microstructure and improved chemical resistance from the co-stabilization with several rare earth oxides. In comparison to standard YSZ, this material possesses superior thermal resistance, lower thermal coefficient, and enhanced performance in conditions of high stress [40]. These properties in unison enable the formulation of a more robust coating layer that can effectively weather aggressive corrosive stress factors in the operation of an engine and may even enhance the durability of the individual parts and consequently the performance of the engines.

8.4 FUTURE DIRECTIONS

8.4.1 Advanced Manufacturing Techniques

These numerical methods happen to be essential in the development of thermal spray coating processes, something discussed in a 2023 published review paper [41]. These techniques allow researchers and engineers to simulate and optimize different aspects of the coating process, such as particle behavior, heat transfer, and coating formation [42]. It is believed that advanced manufacturing techniques such as additive manufacturing, laser cladding, and spark plasma sintering will improve functionally graded material (FGM) manufacturing. These techniques provide high control on gradation of material and can result in better production control and economy [43].

Figure 8.10 outlines future directions for FGMs across six key areas. It focuses on the control of process parameters and the application of mixed methods to improve gradient formation. The subprocess of research and development is characterized by studying the influence of mixing various sizes to construct multiple gradient regions and the perspective of new nanomaterials. In this field, its aim is to create theoretical models and simulations that would allow to study gradient layer formation and its relationship with the matrix. Largely focused on corrosion, bending, fatigue, and toughness tests to capture broader ideas concerning FGM utilization, large-scale performance, and long-term testing. Is aimed at increasing the production rate of

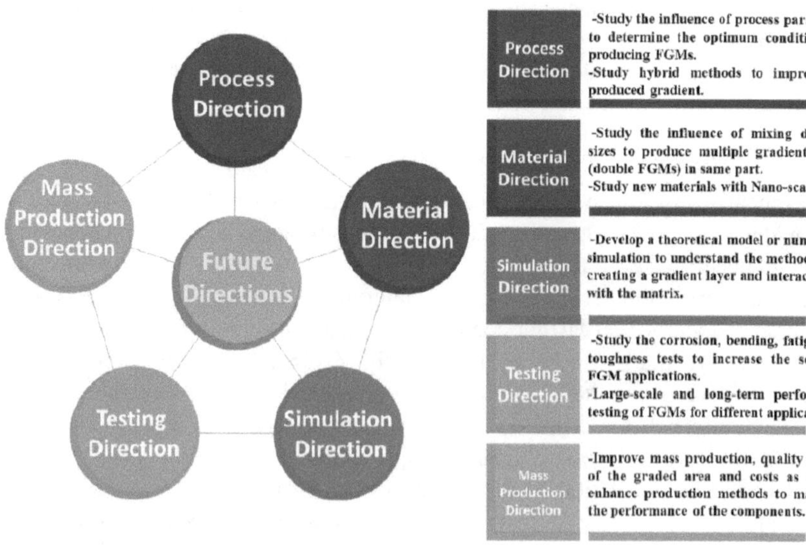

FIGURE 8.10 The most important expected future trends for FGMs [44]. FGMs, functionally graded material.

the mass production technique, quality of graded areas, minimizing costs, and fine-tuning production techniques that would yield improved performance of the new component. Each direction covers the concerns and directions in the further development and implementation of FGMs [44].

8.4.2 Environmental Impact and Sustainability

TBCs are indispensable for the gas turbine engines' environmental footprint and sustainability. Since sustainable aviation fuels (SAF) have been introduced into the aviation industry to minimize climate change effects, TBCs have some novelties. Higher hydrogen/carbon equivalent of SAFs results in higher levels of water vapor during combustion, possibly shortening the life of the TBCs and other hot section parts [45]. The higher rates of this kind of degradation could result in the more frequent replacement of parts and components with shorter useful lifetimes for the engines themselves. But TBCs also have a beneficial impact on sustainability by enhancing the efficiency of the engine and minimizing fuel consumption. Current work is aimed at achieving higher thermal stability of TBCs and lower thermal inertia of the coatings to address these environmental concerns and improve the engine's operation [46]. It is presumed that future research will investigate the possibilities concerning the minimization of FGM's environmental footprint. This entails reducing energy consumption using waste during production and the use of recyclable or biodegradable material on the FGMs. As for the biocompatibility of FGMs, they are currently actively studied to create multiphase materials for the medical industry, including for implants into bones and dentition [47]. These materials can reproduce in themselves the gradation of human tissue, which may help to decrease stress concentrations and increase the duration of implantation. Current computation models, including finite element analysis and molecular dynamics simulations of FGMs, are being developed to enhance the prediction of the composite material behavior under different circumstances [48]. This can lead to designs that are ventured out to undertake a particular task in the best way possible. Figure 8.11 shows the profile of a coating across the substrate [49].

The distributions of final residual stress σ_{xx} (in kPa) and equivalent plastic strain (PEEQ) after depositing two layers using two strategies: repeat and reverse. In the repeat strategy (a, b), the stress and strain are more concentrated along Path-3, with higher stress values indicated by red and orange areas. The PEEQ distribution shows significant plastic strain at the edges. In the reverse strategy (c, d), stress is more evenly distributed, with slightly lower peak values. The PEEQ distribution is similar but shows variations in strain patterns, suggesting different deformation behaviors between the two strategies [50].

The future for FGMs seems bright in all classes of application, with every industry appearing to hold the structures in high regard because of the compositional gradients that make them the ideal performance solutions. Given the foregoing analysis, it is believed that the application field of FGMs will expand rapidly with the continuous development of manufacturing technologies [51]. Engine performance with the corresponding flight mission profile is used to predict hot corrosion damage and develop a computational model for coating design [52–58]. The development of TBCs for improved hot corrosion performance in aero engines has focused on

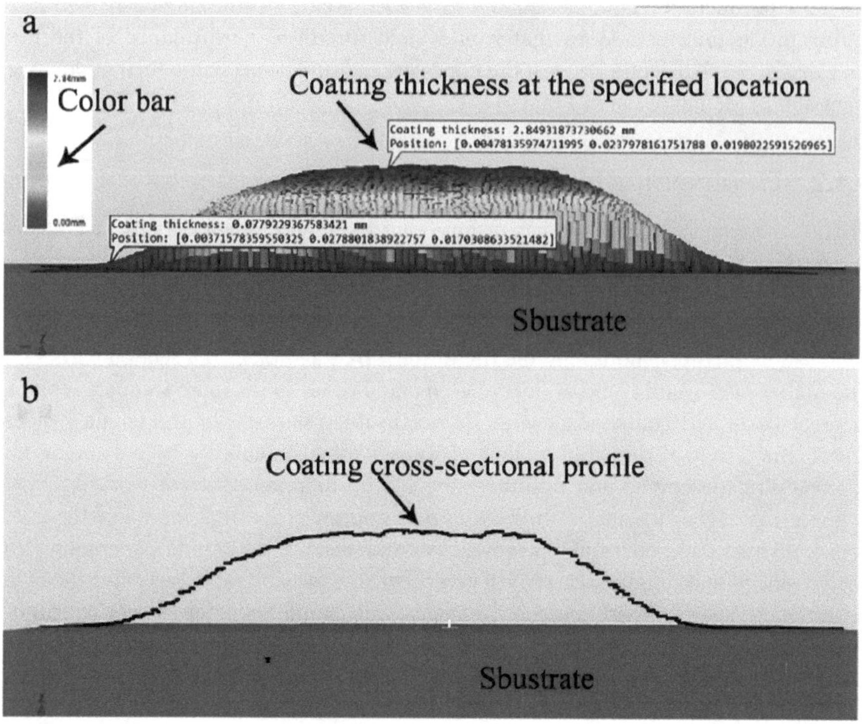

FIGURE 8.11 Determination and quantification of the coating thickness: (a) the thickness of the coating at the designated location; (b) the cross-sectional profile of the coating [49].

advanced structural designs and the integration of novel materials [59–63]. These coatings effectively withstand high temperatures and aggressive environments, protecting engine components from degradation [64–66]. Research highlights the use of deposition techniques such as thermal spraying to achieve dense and durable TBCs with enhanced thermal stability and corrosion resistance.

8.5 CONCLUSION

TBCs are integral anti-erosional and thermal insulating coatings required to be used as protective coatings on critical components in aero engines. New trends in the design of TBC structures and materials are focused on improving the hot corrosion protection. Three-layer structures with the primary layer of a metal or oxide, a middle layer of a metal or oxide having good phase stability, and the final layer having higher corrosion resistance can enhance the overall performance of the coating layer. A nanostructured coating with grain size ranging from a few nanometers to tens of nanometers provides better defense against molten salt and spallation. Such microstructures as planar or segmented-columnar ones can help to enhance strain tolerance, thermal insulation capability, as well as corrosion resistance. Some new materials, including rare earth zirconates ($Gd_2Zr_2O_7$, $La_2Zr_2O_7$, and other heat-resistant ones) and multi-component stabilized zirconia ((Yb_2O_3–Gd_2O_3)–Y_2O_3 co-stabilized zirconia), have received greater attention due to their low thermal conductivity, high phase stability, and hot corrosion resistance. Future work in TBC developments includes improved manufacturing technologies, the evaluation of the environmental effects on the process and the TBC, and the development of computational tools to analyze the coating behavior and design the TBC.

REFERENCES

1. Bakhsheshi-Rad H, Hamzah E, Ismail A, Daroonparvar M, Kasiri-Asgarani M, Jabbarzare S, Medraj M. Microstructural, mechanical properties and corrosion behavior of plasma sprayed NiCrAlY/nano-YSZ duplex coating on Mg–1.2 Ca–3Zn alloy. *Ceramics International.* 2015;41:15272–7.
2. Brossard S, Munroe P, Tran A, Hyland M. Study of the splat formation for plasma sprayed NiCr on aluminum substrate as a function of substrate condition. *Surface Coatings Technology.* 2010;204:2647–56.
3. Karadimas G, Salonitis KJAS. Ceramic matrix composites for aero engine applications—a review. *Applied Sciences.* 2023;13:3017.
4. Prashar G, Vasudev H, Mehta A. Application of bimodal stuructured thermal sprayed coatings in power plant components. *Advanced Ceramic Coatings for Energy Applications.* 2024:273–87. https://doi.org/10.1016/B978-0-323-99620-4.00012-9
5. Bakan E, Vaßen R. Crack healing mechanisms in atmospheric plasma sprayed Yb-silicate coatings during post-process heat treatment. *Journal of the European Ceramic Society.* 2023;43:3684–93.
6. Sathish M, Radhika N, Saleh B. A critical review on functionally graded coatings: methods, properties, and challenges. *Composites Part B: Engineering.* 2021;225:109278.
7. Bobzin K, Öte M, Knoch M. Surface pre-treatment for thermally sprayed ZnAl15 coatings. *Journal of Thermal Spray Technology.* 2017;26:464–72.
8. Chen H, Fan M, Zhu W, Norton A. High temperature oxidation behaviour of combustion flame sprayed CoNiCrAlY coatings. *Surface Coatings Technology.* 2020;385:125431.
9. Cinca N, Lima CRC, Guilemany JM. An overview of intermetallics research and application: status of thermal spray coatings. *Journal of Materials Research Technology.* 2013;2:75–86.
10. Bobzin K, Öte M, Linke T, Sommer J, Liao X. Influence of process parameter on grit blasting as a pretreatment process for thermal spraying. *Journal of Thermal Spray Technology.* 2016;25:3–11.
11. Buchwalder A, Zenker R. Pre-and post-surface treatments using electron beam technology for load-related application of thermochemical and PVD hard coatings on soft substrate materials. *Surface Coatings Technology.* 2019;375:920–32.
12. Bogdan M, Peter I. A comprehensive understanding of thermal barrier coatings (TBCs): applications, materials, coating design and failure mechanisms. *Metals.* 2024;14:575.
13. Daroonparvar M, Bakhsheshi-Rad HR, Saberi A, Razzaghi M, Kasar AK, Ramakrishna S, Menezes PL, Misra M, Ismail AF, Sharif S. Surface modification of magnesium alloys using thermal and solid-state cold spray processes: challenges and latest progresses. *Journal of Magnesium Alloys.* 2022;10:2025–61.
14. Liang P, Zeng J, Yang X, Zhang H, Dong S, Jiang J, Deng L, Zhou X, Cao X. Duplex and functionally graded Al@ NiCr/8YSZ thermal barrier coatings on aluminum substrates. *Journal of Alloys Compounds.* 2019;790:928–40.
15. Di Girolamo G, Blasi C, Brentari A, Schioppa M. Microstructural, mechanical and thermal characteristics of zirconia-based thermal barrier coatings deposited by plasma spraying. *Ceramics International.* 2015;41:11776–85.

16. Di Girolamo G, Blasi C, Pagnotta L, Schioppa M. Phase evolution and thermophysical properties of plasma sprayed thick zirconia coatings after annealing. *Ceramics International*. 2010;36:2273–80.
17. Di Girolamo G, Blasi C, Schioppa M, Tapfer L. Structure and thermal properties of heat treated plasma sprayed ceria–yttria co-stabilized zirconia coatings. *Ceramics International*. 2010;36:961–8.
18. Ding Y, Li H, Tian Y. Bristle blasting surface preparation in thermal spraying. *Journal of Thermal Spray Technology*. 2019;28:378–90.
19. Fan W, Bai Y. Review of suspension and solution precursor plasma sprayed thermal barrier coatings. *Ceramics International*. 2016;42:14299–312.
20. Fan X, Zou B, Gu L, Wang C, Wang Y, Huang W, Zhu L, Cao X. Investigation of the bond coats for thermal barrier coatings on Mg alloy. *Applied Surface Science*. 2013;265:264–73.
21. Fanicchia F, Axinte D, Kell J, McIntyre R, Brewster G, Norton A. Combustion flame spray of CoNiCrAlY & YSZ coatings. *Surface Coatings Technology*. 2017;315:546–57.
22. Keyvani A, Saremi M, Sohi MH, Valefi Z. A comparison on thermomechanical properties of plasma-sprayed conventional and nanostructured YSZ TBC coatings in thermal cycling. *Journal of Alloys Compounds*. 2012;541:488–94.
23. Fanicchia F, Axinte DA. Transient three-dimensional geometrical/thermal modelling of thermal spray: normal-impinging jet and single straight deposits. *International Journal of Heat Mass Transfer*. 2018;122:1327–42.
24. Fanicchia F, Maeder X, Ast J, Taylor A, Guo Y, Polyakov M, Michler J, Axinte D. Residual stress and adhesion of thermal spray coatings: microscopic view by solidification and crystallisation analysis in the epitaxial CoNiCrAlY single splat. *Materials Design*. 2018;153:36–46.
25. Fauchais PL, Heberlein JV, Boulos MI, Fauchais PL, Heberlein JV, Boulos MI. Overview of thermal spray. In: MI Boulos, PL Fauchais, JVR Heberlein (eds.), *Thermal Spray Fundamentals: From Powder to Part*. Springer International Publishing AG, Cham, 2014:17–72.
26. Tailor S, Modi A, Modi S. Effect of controlled segmentation on the thermal cycling behavior of plasma sprayed YSZ thick coatings. *Ceramics International*. 2018;44:6762–8.
27. Fauchais PL, Heberlein JV, Boulos MI, Fauchais PL, Heberlein JV, Boulos MI. Industrial applications of thermal spraying technology. In: MI Boulos, PL Fauchais, JVR Heberlein (eds.), *Thermal Spray Fundamentals: From Powder to Part*. Springer International Publishing AG, Cham, 2014:1401–566.
28. Bernard B, Quet A, Bianchi L, Joulia A, Malié A, Schick V, Rémy B. Thermal insulation properties of YSZ coatings: suspension plasma spraying (SPS) versus electron beam physical vapor deposition (EB-PVD) and atmospheric plasma spraying (APS). *Surface Coatings Technology*. 2017;318:122–8.
29. Chavez JJG, Naraparaju R, Mikulla C, Mechnich P, Kelm K, Ramana C, Schulz U. Comparative study of EB-PVD gadolinium-zirconate and yttria-rich zirconia coatings performance against Fe-containing calcium-magnesium-aluminosilicate (CMAS) infiltration. *Corrosion Science*. 2021;190:109660.
30. Dong H, Liang X, Wang Z, Luo N, Zhang J, Zhan Y, Zhang X. Enhancing the performances of EB-PVD TBCs via overlayer Al-modification. *Surface Coatings Technology*. 2023;473:130001.
31. Huang R, Yang W, Fan X, Lai C, Mao J, Liu M, Zhan Y, Dai H, Zhang X. Insights into the primitives and sequence deposition mechanism of 7YSZ TBCs respectively by EB-PVD and PS-PVD. *Surface Coatings Technology*. 2023;468:129673.
32. Kadam NR, Karthikeyan G, Kulkarni DM. Effect of substrate rotation on the microstructure of 8YSZ thermal barrier coatings by EB-PVD. In Kumar VS, Karthikeyan R, Upadhyaya P (Eds.), *Materials Today: Proceedings, 2nd International Conference on Recent Advances in Materials & Manufacturing Technologies*. 2020;28:678–83.

33. Jonnalagadda KP, Peng RL, Mahade S, Markocsan N, Nylén P, Björklund S, Curry N, Li X-H. Hot corrosion behavior of multi-layer suspension plasma sprayed $Gd_2Zr_2O_7$/YSZ thermal barrier coatings. *Interceram-International Ceramic Review.* 2017;66:180–4.
34. Karaoglanli AC, Doleker KM, Ozgurluk Y. Interface failure behavior of yttria stabilized zirconia (YSZ), $La_2Zr_2O_7$, $Gd_2Zr_2O_7$, YSZ/$La_2Zr_2O_7$ and YSZ/$Gd_2Zr_2O_7$ thermal barrier coatings (TBCs) in thermal cyclic exposure. *Materials Characterization.* 2020;159:110072.
35. Khan MJ, Moskal G, Iqbal A, Mikuśkiewicz M, Pawlik T, Olesik P. Hot corrosion behavior of single-layered $Gd_2Zr_2O_7$, $Sm_2Zr_2O_7$, and $Nd_2Zr_2O_7$ thermal barrier coatings exposed to $Na_2SO_{4+}MgSO_4$ environment. *Coatings.* 2023;13:1311.
36. Mahade S, Curry N, Björklund S, Markocsan N, Nylén P. Thermal conductivity and thermal cyclic fatigue of multilayered $Gd_2Zr_2O_7$/YSZ thermal barrier coatings processed by suspension plasma spray. *Surface Coatings Technology.* 2015;283:329–36.
37. Moskal G, Swadźba L, Hetmańczyk M, Witala B, Mendala B, Mendala J, Sosnowy P. Characterization of microstructure and thermal properties of $Gd_2Zr_2O_7$-type thermal barrier coating. *Journal of the European Ceramic Society.* 2012;32:2025–34.
38. Bahamirian M, Hadavi S, Farvizi M, Keyvani A, Rahimipour M. ZrO_2 $9.5Y_2O_3$ $5.6Yb_2O_3$ $5.2Gd_2O_3$; a promising TBC material with high resistance to hot corrosion. *Journal of Asian Ceramic Societies.* 2020;8:898–908.
39. Bobzin K, Zhao L, Öte M, Linke TF. Deposition and characterization of thermal barrier coatings of ZrO_2–4 mol.% Y_2O_3–1 mol.% Gd_2O_3–1 mol.% Yb_2O_3. *Surface Coatings Technology.* 2015;268:205–8.
40. Daroonparvar M, Yajid MAM, Yusof NM, Bakhsheshi-Rad HR, Hamzah E, Nazoktabar M. Investigation of three steps of hot corrosion process in Y_2O_3 stabilized ZrO_2 coatings including nano zones. *Journal of Rare Earths.* 2014;32:989–1002.
41. Mehta A, Vasudev H, Thakur L. Applications of numerical modelling techniques in thermal spray coatings: a comprehensive review. *International Journal on Interactive Design Manufacturing.* 2023;18:1–21.
42. Yedida VS, Mehta A, Vasudev H, Singh S. Role of numerical modeling in predicting the oxidation behavior of thermal barrier coatings. *International Journal on Interactive Design Manufacturing.* 2023;18:1–10.
43. Ding Z, Fang Z, Wang G, Wu C, Yang SJO, Technology L. Microstructure and properties of Mo-Si-B-Y_2O_3 composite coating on nickel-based alloy by laser cladding. *Optics Laser Technology.* 2023;164:109473.
44. Saleh B, Jiang J, Fathi R, Al-Hababi T, Xu Q, Wang L, Song D, Ma A. 30 Years of functionally graded materials: an overview of manufacturing methods, applications and future challenges. *Composites Part B: Engineering.* 2020;201:108376.
45. Fang H, Wang W, Yang Z, Yang T, Wang Y, Huang J, Ye D. Phase stability, thermal shock behavior and CMAS corrosion resistance of Yb_2O_3-Y_2O_3 co-stabilized zirconia thermal barrier coatings prepared by atmospheric plasma spraying. *Surface Coatings Technology.* 2021;427:127864.
46. Guo L, Xin H, Zhang Z, Zhang X, Ye F. Microstructure modification of Y_2O_3 stabilized ZrO_2 thermal barrier coatings by laser glazing and the effects on the hot corrosion resistance. *Journal of Advanced Ceramics.* 2020;9:232–42.
47. Guo L, Zhang C, Li M, Sun W, Zhang Z, Ye F. Hot corrosion evaluation of Gd_2O_3-Yb_2O_3 co-doped Y_2O_3 stabilized ZrO_2 thermal barrier oxides exposed to $Na_2SO_4+V_2O_5$ molten salt. *Ceramics International.* 2017;43:2780–5.
48. Li J, Luo X, Li G. Effect of Y_2O_3 on the sliding wear resistance of TiB/TiC-reinforced composite coatings fabricated by laser cladding. *Wear.* 2014;310:72–82.
49. Wu H, Xie X, Liu M, Chen C, Liao H, Zhang Y, Deng S. A new approach to simulate coating thickness in cold spray. *Surface Coatings Technology.* 2020;382:125151.

50. Lin E, Nault I, Ozdemir OC, Champagne VK, Nardi A, Müftü S. Thermo-mechanical deformation history and the residual stress distribution in cold spray. *Journal of Thermal Spray Technology*. 2020;29:1424–36.
51. Lima RS. Perspectives on thermal gradients in porous ZrO_2–7–8 wt.% Y_2O_3 (YSZ) thermal barrier coatings (TBCs) manufactured by air plasma spray (APS). *Coatings*. 2020;10:812.
52. Prashar G, Vasudev H, Thakur L. High-temperature oxidation and erosion resistance of Ni-based thermally-sprayed coatings used in power generation machinery: a review. *Surface Review and Letters*. 2022;29(03):2230003.
53. Singh G, Vasudev H, Bansal A, Vardhan S. Influence of heat treatment on the microstructure and corrosion properties of the Inconel-625 clad deposited by microwave heating. *Surface Topography: Metrology and Properties*. 2021;9(2):025019.
54. Prashar G, Vasudev H. Hot corrosion behavior of super alloys. *Materials Today: Proceedings*. 2020;26:1131–5.
55. Vasudev H, Prashar G, Thakur L, Bansal A. Electrochemical corrosion behavior and microstructural characterization of HVOF sprayed inconel718-Al_2O_3 composite coatings. *Surface Review and Letters*. 2022;29(02):2250017.
56. Singh P, Bansal A, Vasudev H, Singh P. In situ surface modification of stainless steel with hydroxyapatite using microwave heating. *Surface Topography: Metrology and Properties*. 2021;9(3):035053.
57. Vasudev H, Singh P, Thakur L, Bansal A. Mechanical and microstructural characterization of microwave post processed Alloy-718 coating. *Materials Research Express*, 2020;6(12):1265f5.
58. Prashar G, Vasudev H. Structure–property correlation of plasma-sprayed Inconel625-Al_2O_3 bimodal composite coatings for high-temperature oxidation protection. *Journal of Thermal Spray Technology*. 2022;31(8):2385–408.
59. Prashar G, Vasudev H. Surface topology analysis of plasma sprayed Inconel625-Al_2O_3 composite coating. *Materials Today: Proceedings*. 2022;50:607–11.
60. Prashar G, Vasudev H. A review on the influence of process parameters and heat treatment on the corrosion performance of Ni-based thermal spray coatings. *Surface Review and Letters*. 2022;29(01):2230001.
61. Yedida VS, Vasudev H. A review on the development of thermal barrier coatings by using thermal spray techniques. *Materials Today: Proceedings*. 2022;50:1458–64.
62. Majji BGR, Vasudev H, Bansal A. A review on the oxidation and wear behavior of the thermally sprayed high-entropy alloys. *Materials Today: Proceedings*. 2022;50:1447–51.
63. Sunitha K, Vasudev H. A short note on the various thermal spray coating processes and effect of post-treatment on Ni-based coatings. *Materials Today: Proceedings*. 2022;50:1452–7.
64. Vasudev H. Wear characteristics of Ni-WC powder deposited by using a microwave route on mild steel: microwave cladding of Ni-WC. *International Journal of Surface Engineering and Interdisciplinary Materials Science (IJSEIMS)*. 2020;8(1):44–54.
65. Prashar G, Vasudev H, Thakur L, Bansal A. Performance of thermally sprayed hydroxyapatite coatings for biomedical implants: a comprehensive review. *Surface Review and Letters*. 2023;30(01):2241001.
66. Parkash J, Saggu HS, Vasudev H. A short review on the performance of high velocity oxy-fuel coatings in boiler steel applications. *Materials Today: Proceedings*. 2022;50:1442–6.

9 Plasma-Sprayed Thermal Barrier Coatings for High-Temperature Applications

Mathanbabu Mariappan,
Thirumalaikumarasamy Duraisamy,
Guruprasad Balachandran,
Murugan Kaliyamoorty, and
Praveen Kandasamy

9.1 INTRODUCTION

Thermal barrier coatings (TBCs) have become significantly important for the design of modern turbines [1–4]. From the materials perspective, the superior performance of TBCs is more dependent on the following characteristics: a high melting for improved phase stability; a high thermal expansion coefficient (CTE) for enhanced fracture toughness; high coating density; coating soundness (no cracking); controlled porosity with high hardness; good thermal and chemical stability with high corrosion resistance; low sintering activity; low thermal diffusivity; low specific heat capacity; resistance to hot corrosion; calcium-magnesium-aluminosilicate (CMAS) attack; high strain tolerance; and high durability under elevated conditions [5–12].

To attain the aforementioned qualities, TBCs are produced using many techniques, including plasma spraying (PS), high velocity oxy-fuel spraying, detonation gun, plasma spray-physical vapour deposition (PS-PVD), and electron beam physical vapour deposition (EB-PVD) [10,13–17]. The PS technology is extensively utilised across several industries. It also generates coatings of exceptional quality in demanding working situations because of its superior flexibility, reduced cost, and capacity for complete automation [18]. The potential of PS coating might significantly benefit from the worldwide patenting of both the equipment utilised in this enhanced process and the technique itself. In comparison to other thermal spray counterparts, PS offers deposition characterised by superior bonding and hardness, accompanied by little porosity. In contrast to combustion techniques, this PS technology facilitates the spraying of components with exceptionally high melting points, including refractory metals, pyrochlores, and perovskite-based substances. An additional benefit is the application of an inert gas environment, which significantly reduces the

DOI: 10.1201/9781003542957-9

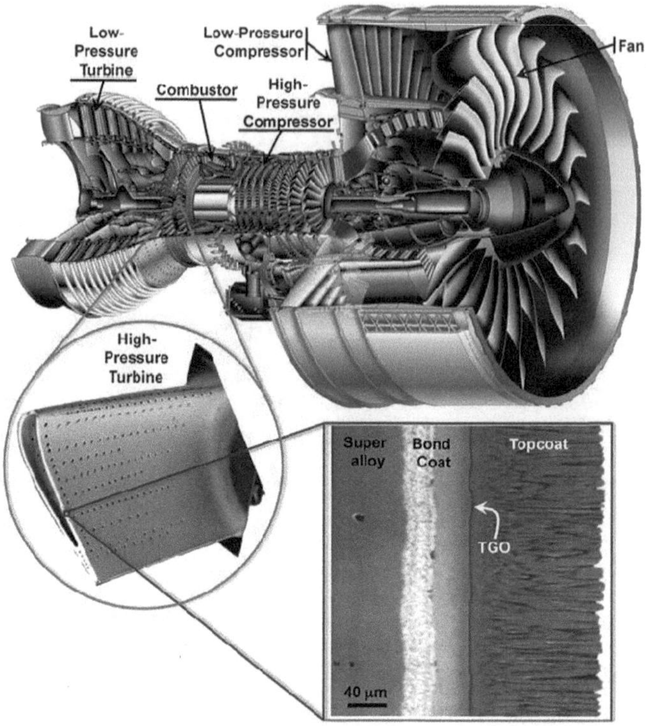

FIGURE 9.1 Sectional view of gas turbine engine with a high-pressure turbine protected with thermal barrier coating [9].

oxidation of molten particles in the atmosphere and its effect on the substrate [19–22]. A cross-sectional view of an aero engine with a high-pressure turbine coated with TBC is illustrated in Figure 9.1.

In addition to developing modern gas turbines, the operative temperature should be increased without affecting the characteristics of TBC. Researchers are paying a lot of attention to pyrochlore oxides in recent days, like lanthanum zirconate and gadolinium zirconate, because of their better properties than the traditional YSZ [23–25]. Moreover, pyrochlore oxides have a lower CTE and lower toughness than the YSZ. Pyrochlore oxides and YSZ are combined to maximise the benefits through various compositions with multilayers in order to enhance the coating life [26]. Figure 9.2 illustrates that spray settings are essential in determining the required coating quality. The management of process control in the PS technique is challenging because of the impact of several process variables, since the intricate morphology of PS coating is linked to multiple factors. This may be modified by altering the process settings through the optimisation of process parameters utilising the response surface methodology approach [27–30]. Research comparing PS to alternative thermal spray methods on pyrochlore-based TBCs has consistently intensified [31–44].

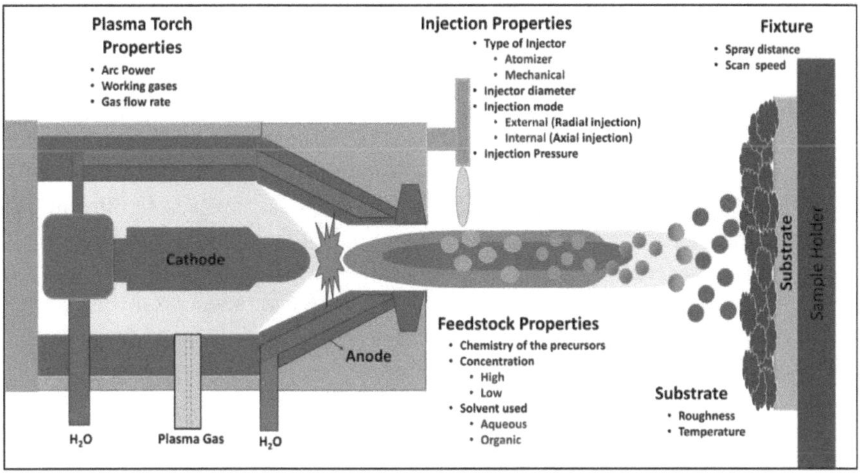

FIGURE 9.2 Schematic diagram of the plasma spray process with spray parameter [27]. *(Under CC-4.0 license).*

9.2 PLASMA SPRAYING TECHNIQUE

PS facilitates the application of various coatings, including ceramic-based coatings, refractory materials, metals, polymers, and other substances, while minimising the heat effect on the substrate materials.

A conventional PS technique has the following stages: The powdered materials are continuously supplied into the plasma jet generated by electric arc combustion through a gas flow, which may include Ar, N_2, H_2, and He gases. The ions engage with the electrons in the plasma produced by the torch, generating an enormous energy that elevates the gas temperature over 20,000°C. The qualities of the gas impact it and induce fast growth. The plasma flame liquefies the feedstocks, and the molten droplets are projected to the surface at velocities reaching 300 m/s before rapidly hardening. The coating is influenced by the feed rate, angle, and SOD [24].

Zhang et al. applied all the coatings utilising the air plasma spray (APS) method. Haynes 188 served as the standard substrate material. The bond coat comprised NiCrAlY powder. The LZ and 8YSZ served as TC [36]. To guarantee the optimal porosity of the top coat, the parameters were suitably modified [24]. Guo et al. employed PS to examine the LZ/8YSZ-based functionally graded composite TBC. The double layers produced with LZO/8YSZ exhibit superior durability compared to the other coatings during service [31]. Di Girolamo et al. examined the APS LZO and discovered that the splat grains consist of some porosity structures and fractures. The thermal conductivity of the PS deposition is lesser than that of the EB-PVD due to the splat structure. The findings indicate that altering the spray parameters enables precise control over the porosity, resulting in modifications to its thermal and mechanical characteristics [25]. Additionally, the LZ TBC was examined using the suspension PS technique. The optimal parameters are 30 wt.% coating substance with 1 wt.% polyethylene glycol in water, a standoff distance of 40 mm, and a power output of 40 kW. The porosity ranges from 6% to 10%.

Ramachandran et al. employed PS to fabricate multilayered coatings of LZ and YSZ. The double YSZ-based FGM coating offers exceptional high-temperature stability [28,45]. The plasma method is a multifaceted thermal spray technique; however, the liquefaction of powders, compaction, and solidification of PS powder deposits on a substrate are exceedingly challenging [27]. This process involves significant fluctuations in the dynamic and temperature conditions of the melting materials, which depend on several factors. In PS, few oxides will consistently be available in atmospheric PS coatings. In order to overcome the above factor, some promising deposition processes using allied plasma techniques, coating structures, and properties are needed as described in the following chapters.

9.3 ADVANCED PS TECHNIQUES

In a standard plasma technique, certain oxides are invariably present in atmospheric PS deposition. To mitigate the disadvantages, environmentally controlled spraying has been implemented [13]. The existence of different oxides in the microstructure might be alleviated by spraying in an inert atmosphere or utilising an inert gas veil [3]. It can eliminate the presence of oxides in the structure. In these methods, air is excluded, and coating oxide is removed by specific methodologies [18]. The plasma gas contributes to a reduction in oxide content compared to conventional thermal spray techniques by mitigating the effects of ambient air mixing.

The APS methods have been refined into two effective ways, namely, suspension (SPS) and solution precursor (SPPS), which produce a top layer exhibiting reduced thermal conductivity and superior strain resistance compared to conventional PS methods [14]. Currently, the plasma approach combined with physical vapour deposition is among the most innovative techniques in TBCs [46–48].

9.3.1 Comparison of Various Techniques

This chapter concentrates on the architecture, characteristics, drawbacks, and future research trajectories for bi-layer and multilayer PS-based TBC and associated methods such as APS, SPS, SPPS, PS-PVD, and EB-PVD described as follows.

The APS technique is easy to configure and employs a horizontal splat architecture to create the coating, augmented by a highly textured bond coat surface. It offers coatings with low heat conductivity but a limited lifespan. The feedstock characteristics of APS powders and microstructural inadequacies result in limited thermal conductivity for high-temperature applications. Porous layers could be tailored to specific uses. APS provides an extensive selection of foundational materials for the coating structure, encompassing ceramics, glassware, and composite materials [13].

APS is more cost-effective than the EB-PVD method. Nonetheless, challenges in the spraying technique arise from numerous parameters, complicating the deposition of intricate shapes and surfaces; difficulties in splash measurement due to diverse interfacing boundaries; line-of-sight measurement issues; obstacles in covering complex shapes and surfaces; and the frequent necessity for plasma anode support to maintain coating integrity [17]. APS necessitates the preservation of coating quality and requires extensive maintenance of the PS gun electrode. Elevated particle velocities and deformation upon impact lead to significant adhesion strengths in base materials [18,19,49,50]. The elevated drop/substrate attachment is achieved because of the significant molecular velocity and distortion occurring during oscillation. The inactive gas plasma stream, considering the effects of surrounding air integration, reduces oxide content more effectively than previous thermal spray methods [51].

SPS is an advanced and effective PS technique that employs liquid feedstock containing submicron ranges. In the SPS technique, fine powder is initially combined with water or ethanol to create a suspension, which is subsequently transferred to the plasma vessel. The development of the coating by mechanical bonding is augmented by a highly textured bond coat surface. The features of SPS coating provide enhanced strain tolerance, resulting in superior heat cycle performance and an extended lifespan. However, managing liquid feedstock requires further expenditure and is exceedingly complex. Furthermore, optimised parameters have yet to be defined [52–62].

The SPPS microstructure serves as a solution precursor that induces vertical cracking, hence enhancing strain tolerance. The expense is more analogous to that of the SPS approach, which necessitates a solution precursor. The presence of insoluble solution precursors resulting from contaminant phases in the coating induces sintering and re-melting [63–67].

The PS-PVD microstructure consists of coating powders measuring 25 μm or exhibiting a columnar microstructure. Coating characteristics encompass enhanced strain tolerance, the capacity to cover complex structures, and increased deposition speeds. Nonetheless, it is costlier than the conventional technique [13]. EP-PVD exhibits an ingot microstructure and strain tolerance deposition, characterised by uniform columnar deposition rates of 12 μm/minute. Filament deterioration results in inconsistent evaporation, rendering it unsuitable for some metals [68–79]. Figure 9.3 illustrates a comparison of several PS-TBCs.

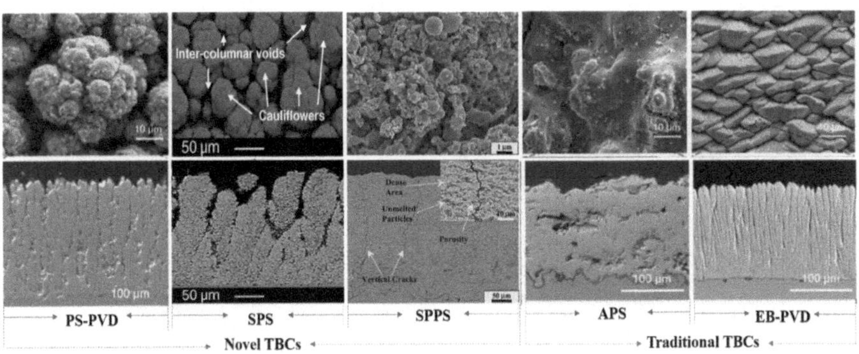

FIGURE 9.3 Comparison of various plasma-sprayed TBCs [13,66,67,69]. *TBCs*, thermal barrier coatings.

Wang et al. [80] investigated the SPS technique for depositing an LZ coating using a solution comprising 30% LZO granules in 99.5% ethanol. Additionally, a 1% polyethylene glycol electrostatic dispersion was included, resulting in a morphology characterised by many holes and fissures. Weber et al. examined spray pyrolysis to produce the LZ coating using raw powder dissolved in deionised water. The precursor solution is then sprayed at 2,400°C with a flow rate of 1 m/minute [81].

9.3.2 Synthesis of Powder and Process Parameters of TBCs

A numerical spray parameter affects the PS process. The correlation between microstructure and parameters remains enigmatic concerning processing control and is very challenging to regulate. Multiple studies were previously undertaken to examine the technique, microstructure, and characteristics to mitigate this disadvantage. Conversely, the processing maps are constrained to the specific spray gun, feedstocks, and several other parameters, rendering the transition of process mapping from one gun to another complex [5,82–84]. Figure 9.4 illustrates the most prevalent control settings. Consequently, process parameters must be improved by controlling microstructural features to enhance the longevity of TBC. The factors, including powder feed rate, power, and standoff, are critical parameters influencing the coating structure and are adjusted to regulate porosity [27–30,85].

Various ways exist for the production of feedstock powder, including solid-state reaction (SSR) procedures, co-precipitation processes, and sol-gel techniques. The majority of powders were synthesised with the SSR method for PS. Ramachandran et al. [82] synthesised LZ and produced a bulk quantity by the tap melting process.

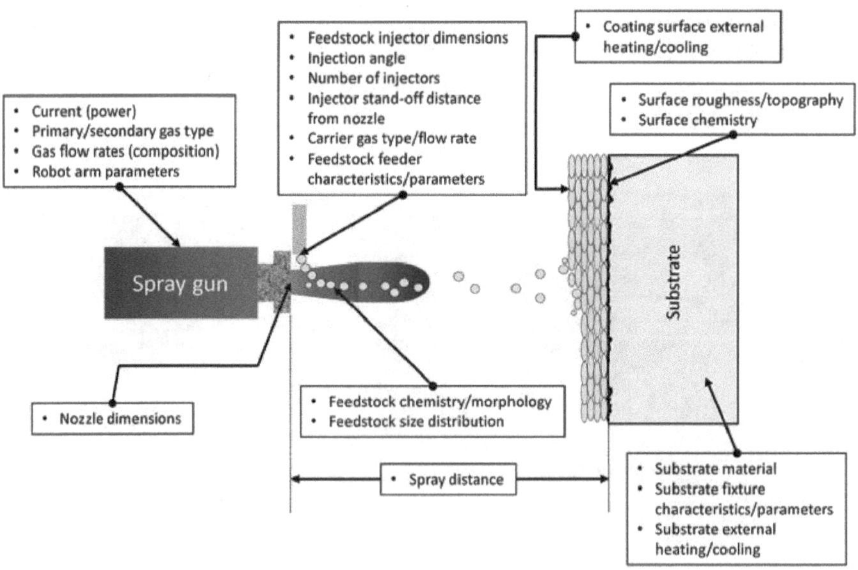

FIGURE 9.4 Processing parameters of plasma spray techniques [28]. (*Under CC-4.0 license*).

9.4 MICROSTRUCTURES OF TBC

A thin layer applied by PS coatings is fundamentally distinct in structure from a thick base material. The mechanical characteristics of coated substrates differ because of their unique architectures. The internal structure of the layer is heterogeneous, including multilaterally bonded multigrains. The thin layer coating originated from the molten substrate and signifies coatings with a lamellar structure. The manufacturing of coatings in liquid and solid forms differs minimally from that of dense substrates. The use of an inert gas or vacuum chamber produces a more refined structure with less porosity and improved adhesion among layers and substrates. The lamellar structure transforms into grains during annealing at the recrystallisation temperature of the substrate. The layer's architecture lacks insertions or oxides, in contrast to PS coatings in the environment. Consequently, a fundamental comprehension of microstructural assets is essential for the implementation of microstructural control.

Numerous elongated fissures, including horizontal deformations and vertical fractures, are evident, alongside conspicuous round, oval, and elliptical pores. Coating delamination may transpire either longitudinally within the inter-layer or at the edges. Minimal adhesion between the laminas or splines leads to vertical microcracks. Their production results in a low adhesive limit and/or initial defects in the layers of the deposited surface. When substantial biaxial tensile residual stresses are induced in the coated surface, these are believed to arise from quenching and subsequent stress relaxation during the cooling process [1,86].

The microstructure of APS coatings is significantly influenced by several elements like porous formations, delaminations, and fractures. This impact is intensified due to the inadequate spray factors, resulting in particles that are either unaltered or partially liquefied. Consequently, imperfect structures are generated [87–89].

Delamination effects arise from inadequate bonding between progressively coated particles, often occurring during consecutive passes of the spray guns. Cracks occur because of the internal tension generated in the deposited particles when the coated substrate is maintained at a low temperature. The rapid cooling leads to increased shrinkage, which causes the tensile strains in the base material to impede the shrinking. The tension is not constrained by the formation of fractures in the bond covering. In the correct spraying process, vertical fissures in the surface deposition may extend to segmentation fractures. The particular splats possess a columnar structure formed by the solidification of the grains. This microstructure is often observable on a fractured section of the coated surface, as illustrated. The microstructures and particle sizes are influenced by processing variables during spray and may differ significantly under service conditions, impacting the coating parameters [13].

Conversely, vertical cracks enhance the thermal conductivity, hence augmenting the heat transfer rate to the base material [6]. To attain low thermal conductivity and enhanced strain tolerance in TBCs with extended longevity, optimisation of microstructural features is essential [90,91].

The properties and durability of TBC are primarily influenced by the shape of the topcoat surface. The coating will exhibit enhanced insulating properties if many pores and fissures are present. Conversely, these horizontal fissures may expand due to heat and mechanical stresses during operation, leading to the spallation of the coating [18]. Segmental fissures facilitate the alleviation of residual tensions inside the coating, hence enhancing its flexibility [36]. The inherent intricacy of the PS technique and coating designs renders the establishment of processes, microstructures, and property connections exceedingly challenging.

9.4.1 Formation of Coating in PS

Particles liquefy during deposition in PS coatings and, upon contact with the base material, flatten, rapidly solidify, and form splats. The high-energy plasma impacting particles provide a rough and undulating texture that differs from EB-PVD. Moreover, the existence of microstructural defects in ceramics, like porosity, leads to long-term consistency issues, resulting in sintering under prolonged exposure to elevated temperatures, hence enhancing the elastic stiffness of the coating [92,93].

The feedstock powders absorb the thermal energy from the plasma, subsequently melting and striking the substrate at high velocities. The molten particles rapidly solidify upon the substrate, resulting in the formation of splats [94]. The individual splats possess a thickness of a few microns. These splats coalesce to form a brick-like substructure, as seen in Figure 9.5 [95,96]. A figure illustrating the splat formation during different processes is presented in Figure 9.6.

Porosities and flaws resulting from minute fractures are a fundamental characteristic of the APS deposited layer. The existence of these defects diminishes the heat conductivity of the coating. Among EB-PVD techniques, APS methods are frequently utilised. Due to the performance of the APS sprayed coating, thicker coatings may be easily produced. The surface morphology of an APS coating consists of a porous lamellar structure characterised by a complex porous configuration [97].

The formation of the thermally grown oxide (TGO) led to the de-cohesion of the bond coat, resulting in the final fracture of the TBC. Numerous experiments were undertaken to evaluate the coating's efficacy in a challenging operational setting.

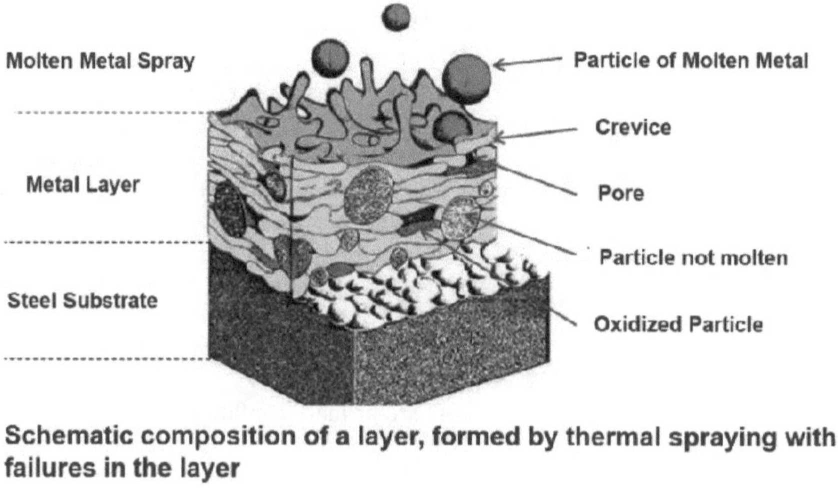

FIGURE 9.5 Plasma-sprayed coating exhibiting lamellar structure [23]. (*Under CC-4.0 license*).

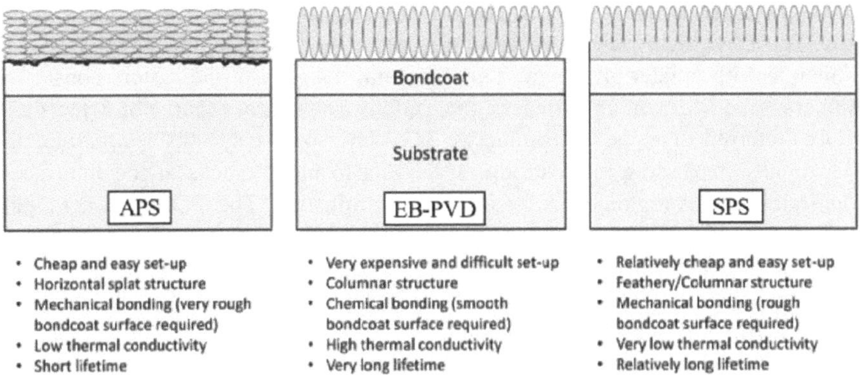

FIGURE 9.6 Formation of splat on the various plasma spray [28,62]. *(Under CC-4.0 license).*

The efficacy of the coating is predominantly assessed by the deterioration of surface morphology [31]. Consequently, modifying the surface shape of the bond layer may contribute to improved durability. Altering the microstructure mostly entails controlling porosity, achievable through the modification of coating methods [59].

The interfacial and microstructural characteristics of TBC related to the coating process for each layer significantly affect the longevity and failure assessment of PS-TBCs. Among various PS techniques, APS is a conventional coating method frequently employed for ceramic coatings on metallic structural elements [98,99].

9.4.2 Effect of Microstructure on Coating Thickness

The performance of the coating was enhanced with an increase in TBC thickness, a finding corroborated in relatively thin TBCs. The TBC performance demonstrated a proportional association up to a certain thickness, as the attributes are augmented by a comparable ratio. The combination of multilayer and buffer layer TBC technologies enhances TBC performance by augmenting the top coat thickness.

Jung et al. [100] investigated the effects of TBC composition, structure, and thickness on TBC performance. The author said that the total thicknesses of the TBC were established at 250 and 1,000 μm, with the thickness ratio between the top and bond coats modified accordingly. The author ultimately determined that the thickness of the top coat influenced TBC performance and sought to identify the optimal coating thickness ratio [101].

9.4.3 Effect of TGO Layer on TBCs

TGO is an oxide layer that forms between the bond and top coat due to the infiltration of oxygen molecules from the top coat, which reacts with the components of the bond coat. Oxidation happens when attaining 700°C, leading to the production of TGO. The TGO primarily consists of a thin layer of thick Al_2O_3 oxides, measuring less than 10 μm, and is generated at elevated temperatures [31,34,36]. It serves as a barrier to inhibit cations and oxygen, functioning as a safeguard against more oxidation and hot corrosion by enhancing good adhesion, referred to as the protective TGO.

It is a gradually expanding entity, the fundamental element in complying with the TBC. The TGO facilitates the mitigation of substrate corrosion, enabling the engine to operate with greater efficiency. The structural integrity of the system constrains its operational lifespan; nevertheless, the TGO is a significant source of defects and failures referred to as the non-protective TGO layer (above 1,000°C), which exhibits a rapidly expanding microstructure, leading to many cracks at the interfaces. Consequently, this region necessitates more examination. The TGO thickness, pattern of thermal cycling behaviours, and thermal stresses are considered crucial in influencing the longevity of TBCs [1,9–12,33].

Takahashi et al. [102] studied the TGO layer that has been formed through heat treatment in an atmosphere and vacuum above 800°C, 1,100°C, and 1,150°C for 4 hours. According to the author's findings, surface finish rises with increasing temperature, and the oxide layer in the vacuum treatments is greater than in the atmospheric air. The microstructure analysis demonstrated that the specimen at 1,100°C attained the utmost acceptable oxide layer thickness, and XRD investigations demonstrated the development of higher amounts of aluminium oxides. A cross-sectional morphology of TGO layers formed on the bond coat is shown in Figure 9.7.

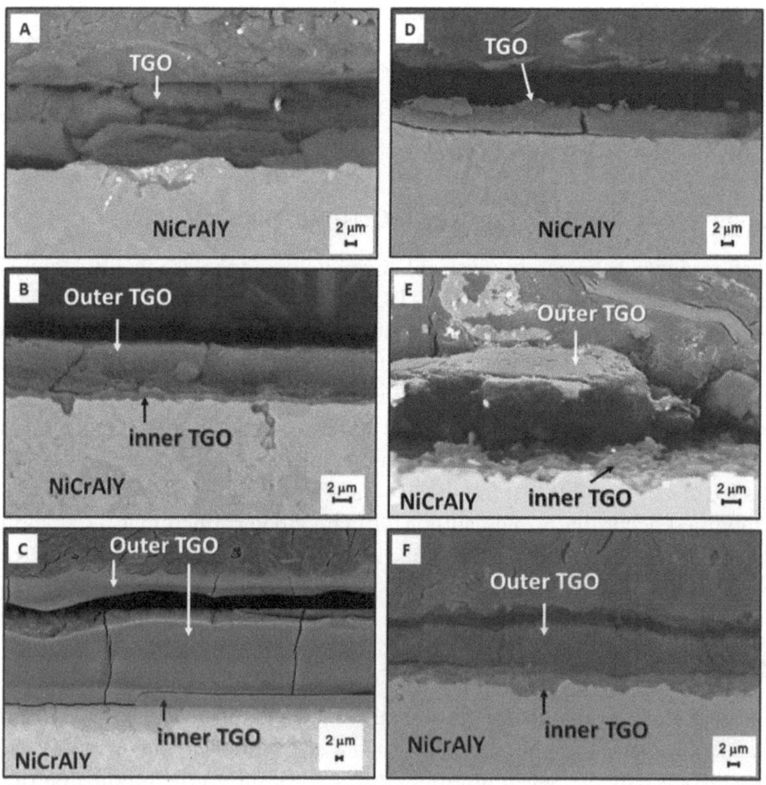

FIGURE 9.7 Cross-section micrograph showing TGO layers formed on NiCrAlY [102]. *TGO*, thermally grown oxide. (*Under CC-4.0 license*).

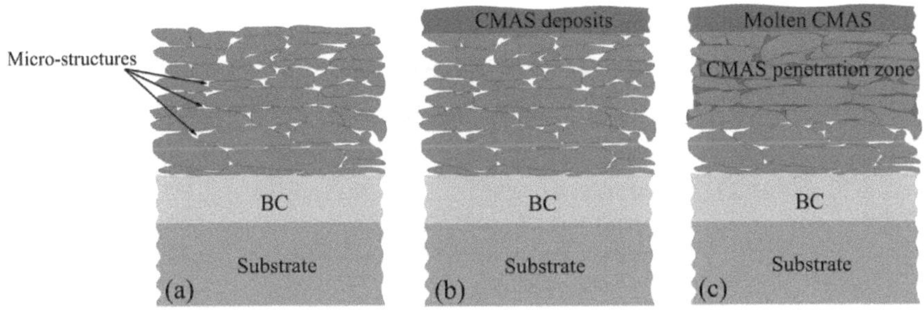

FIGURE 9.8 Schematic of the penetration process of CMAS by APS-TBCs [117]. CMAS, calcium-magnesium-aluminosilicate; APS-TBCs, air plasma-sprayed thermal barrier coatings. *(Under CC-4.0 license).*

9.5 PROPERTIES OF TBCs

Specific qualities must be assessed and considered while examining the characterisation, performance, and durability of TBCs. The surface morphology and features of TBCs are significantly affected by processing parameters, and the measurement procedures can also influence these characteristics [3]. The thermal, mechanical, and microstructural properties of TBCs are briefly discussed in a previously published article [26].

Thermal conductivity is the most essential attribute of TBC materials [1,24,34,36]. CMAS tests are generally carried out for gas turbine engine parts for TBCs. The CMAS test was carried out based on 33% of CaO, 9% of MgO, 13% of Al_2O_3, and 45% of SiO_2 [103–105]. The schematic diagram of the penetration process of CMAS into APS-TBCs is shown in Figure 9.8.

Hot corrosion has a significant impact on the lifetime of TBCs, and the degradation is accelerated significantly owing to the existence of molten deposits. Dehkharghani et al. [106] studied the APS approach utilising lanthanum cerate and 8YSZ-based composite coatings developed, and their hot corrosion behaviour was examined. The results showed that coatings with 75% lanthanum cerate and 25% 8YSZ have the strongest and lowest hot corrosion behaviour due to the highest percentage of tetragonal to monoclinic phase and the development of cavities and cracks. The properties of LZ and YSZ are listed in Table 9.1.

Porosity is a significant structural indicator of coatings. A denser film structure results in larger liquid droplets of the sprayed metal and accelerates their flow towards the substrate. Porosity is influenced by spray factors such as current, feed rate, and standoff distance. A denser coating can be achieved by increasing torch power, increasing the feed rate, or reducing the standoff distance. The porosity of the APS coating generally ranges from 15% to 25%. To ensure sufficient performance, the porosity must be adjusted to between 8% and 19% by modifying process parameters according to specific applications [107,108].

TABLE 9.1
Properties of LZ- and YSZ-Based PS-TBCs

S. No.	Nominal Materials Properties	LZ	8YSZ
1.	Melting point (°C)	2,300	2,680
2.	Maximum operating temperature (°C)	2,300	1,425
3.	Thermal conductivity (K) (W/m-K) (@800°C)	1.60	2.10
4.	Thermal expansion coefficient (CTE) ($\times 10^{-6}$/K) (@1,000°C)	9.1	11.0
5.	Density (g/m^3)	6,000	6,070
6.	Specific heat (J/kg K) (@1,000°C)	540	640
7.	Elastic modulus (GPa)	63	45

LZ, lanthanum zirconate; YSZ, yttria-stabilized zirconia; PS-TBCs, plasma-sprayed thermal barrier coatings.

9.6 POSTTREATMENTS OF PS-TBCs

On top of ceramic coats, technologies for eliminating and restoring damaged TBCs exist. However, there is no uniform procedure that can be applied to all TBCs [109]. The coating clean-up and repair method needs to be tailored to the damaged TBCs. Posttreatment processes included grit blasting with an abrasive medium, blasting with solid CO_2, chemical stripping, non-abrasive water jet, and laser glazing. Among these techniques, laser glazing has been proposed as a viable approach for densifying the surfaces of TBCs [110–116].

Guo et al. [110] investigated the laser glazing layers, in which certain vertical fractures were created by adjusting the laser settings. The modified coating was then subjected to hot corrosion tests in V_2O_5 molten salt for 4 hours at 700°C and 1,000°C, and the results showed that the altered one provided better corrosion protection. The glaze film retained structural integrity during heat corrosion, with no signs of breakdown. The vertical fractures in the glazed layer, however, served as pathways for penetrating the molten salt, speeding the corrosion of the non-modified coating.

Zhong et al. [115] utilised an uninterrupted CO_2 laser to glaze, and the results revealed that laser glazing significantly improved the hot corrosion performance of the as-sprayed 8YSZ coating. The as-sprayed coating suffered from severe cracking and spalling, but the as-glazed coating displayed superior hot corrosion behaviour and hence had a longer lifespan.

According to a previous study, laser glazing can create a network of vertical fissures, which improves the corrosion resistance, prevents the creation of massive macroscopic fractures, and produces a smooth surface and densified top coat, but it is prone to cracking due to non-uniform cooling following the point-like introduction. As a result, the author suggested that much effort has been expended in developing methods to reduce molten salt corrosion. Future studies will necessitate additional optimization of the glazed layer TBCs collapse due to the destruction of coating morphologies and glazing layer thickness. Finally, the investigation into how to completely leverage the benefits of the laser modification approach to increase TBC hot corrosion performance is recommended.

9.7 CONCLUSIONS AND FUTURE SCOPES

The surface-modified coating techniques for gas turbine components on PS-TBC were studied and reported in depth in this article. The following are the study's major findings:

1. The research of many characteristics such as crack morphology, residual stress, high-temperature performance, hot corrosion, and TGO layer efficiency of TBCs is highly important in the advancement of novel coating materials and TBC structures.
2. A life prediction model for newly developed coatings must be built before they are deployed in actual working environments. Before using next-generation advanced TBCs, it is critical to understand the failure mechanism and process-induced stresses.
3. TBCs produced via APS, SPS, and SPPS have demonstrated efficacy in addressing both internal and external failure modes and might serve as a viable substitute for the current YSZ, contingent upon further refinement to meet industrial requirements.
4. In addition, the PS-PVD method should be investigated for the development of double-layer and multilayer TBC systems. If further refined to the industry's needs, it could be an effective replacement for the existing YSZ. It is apparent that the manufacture of multilayered TBCs by SPS, SPPS, and PS-PVD is all promising in their own right.
5. Consequently, the use of FGM and multilayered composite coating in industry continues to pose challenges. Altering the microstructure mostly entails regulating porosity and hardness, achievable through the modification of coating methods.
6. Laser glazing is a potential approach for increasing the corrosion performance of TBC in the presence of molten salt. Nevertheless, additional optimization of glazing parameters and coating surface morphology is required to fully utilize the benefits of this technique.
7. Although there is no standardized technique applicable to all coating systems, surface posttreatments must be selected individually for damaged TBC, considering their composition, nature of damage, and available resources.

REFERENCES

1. Guo, X., Lu, Z., Jung, Y. G., & Zhang, J. (2020). *Novel Lanthanum Zirconate-based Thermal Barrier Coatings for Energy Applications.* Springer International Publishing AG.
2. Hardwicke, C. U., & Lau, Y.-C. (2013). Advances in thermal spray coatings for gas turbines and energy generation: a review, *J. Therm. Spray Technol.*, 22(5), 564–576.

3. Thakare, J. G., Pandey, C., Mahapatra, M. M., & Mulik, R. S. (2020). Thermal barrier coatings—a state of the art review, *Metals Mater. Int.*, 27, 1–22.
4. Vaßen, R., Jarligo, M. O., Steinke, T., Mack, D. E., & Stöver, D. (2010). Overview on Advanced thermal barrier coatings, *Surf. Coat. Technol.*, 205(4), 938–942.
5. Viswanathan, V., Dwivedi, G., & Sampath, S. (2015). Multilayer, multi material thermal barrier coating systems: design, synthesis, and performance assessment, *J. Am. Ceram. Soc.,* 98(6), 1769–1777.
6. Vaßen, R., Bakan, E., Mack, D. E., & Guillon, O. (2022). A perspective on thermally sprayed thermal barrier coatings: current status and trends, *J. Thermal Spray Technol.*, 31, 1–14.
7. Darolia, R. (2013). Thermal barrier coatings technology: critical review, progress update, remaining challenges and prospects, *Int. Mater. Rev.*, 58(6), 315–348.
8. Cao, X. Q., Vassen, R., & Stoever, D. (2004). Ceramic materials for thermal barrier coatings, *J. Eur. Ceramic Soc.*, 24(1), 1–10.
9. Clarke, D. R., Oechsner, M., & Padture, N. P. (2012). Thermal-barrier coatings for more efficient gas-turbine engines, *MRS Bull.*, 37, 891–898.
10. Schulz, U., Saruhan, B., Fritscher, K., & Leyens, C. (2004). Review on advanced EB-PVD ceramic topcoats for TBC applications, *Int. J. Appl. Ceram. Tec.*, 1(4), 302–315.
11. Mittal, G., & Paul, S. (2022). Suspension and solution precursor plasma and HVOF spray: a review, *J. Thermal Spray Technol.*, 31, 1–33.
12. Miller, R.A. (1987). Current status of thermal barrier coatings: an overview, *Surf. Coat. Tech.*, 30(1), 1–11.
13. Lashmi, P. G., Ananthapadmanabhan, P. V., Unnikrishnan, G., & Aruna, S. T. (2020). Present status and future prospects of plasma sprayed multilayered thermal barrier coating systems, *J. Eur. Ceramic Soc.*, 40(8), 2731–2745.
14. Ashokkumar, M., Thirumalaikumarasamy, D., Thirumal, P., &Barathiraja, R. (2021). Influences of mechanical, corrosion, erosion and tribological performance of cold sprayed coatings a review, *Mater. Today Proc.*, 46, 7581–7587.
15. Singh, H., Kaur, M., & Bala, N. (2016). High velocity oxy-fuel spraying and surface finish. In: *Reference Module in Materials Science and Materials Engineering. Comprehensive Materials Finishing* (pp. 207–219). Elsevier, 3.
16. Vuoristo, P. (2014). Thermal spray coating processes. In: MSJ Hashmi (ed.), *Comprehensive Materials Processing, 1st edition Volume* 4*: Coatings and Films* (pp. 229–276). Elsevier.
17. Feuerstein, A., Knapp, J., Taylor, T., Ashary, A., Bolcavage, A., & Hitchman, N. (2008). Technical and economical aspects of current thermal barrier coating systems for gas turbine engines by thermal spray and EBPVD: a review, *J. Therm. Spray Technol.*, 17(2), 199–213.
18. Ružbarský, J., & Panda, A. (2017). *Plasma and Thermal Spraying*. Springer International Publishing.
19. Bakan, E., & Vaßen, R. (2017). Ceramic top coats of plasma-sprayed thermal barrier coatings: materials, processes, and properties, *J. Therm. Spray Technol.*, 26(6), 992–1010.
20. Vaßen, R., Kerkhoff, G., & Stover, D. (2001). Development of a micromechanical life prediction model for plasma sprayed thermal barrier coatings, *Mater. Sci. Eng. A*, 303, 100–109.
21. Anupam, A., Kottada, R. S., Kashyap, S., Meghwal, A., Murty, B. S., Berndt, C. C., & Ang, A. S. M. (2020). Understanding the microstructural evolution of high entropy alloy coatings manufactured by atmospheric plasma spray processing, *Appl. Surf. Sci.*, 505, 144117.

22. Mauer, G., Schlegel, N., Guignard, A., Jarligo, M. O., Rezanka, S., Hospach, A., & Vassen, R. (2014). Plasma spraying of ceramics with particular difficulties in processing, *J. Therm. Spray Technol.*, 24, 30–37.
23. Boronenkov, V., & Korobov, Y. (2016). *Fundamentals of Arc Spraying: Physical and Chemical Regularities.* Springer.
24. Zhang, J., Guo, X., Jung, Y. G., Li, L., & Knapp, J. (2017). Lanthanum zirconate based thermal barrier coatings: a review, *Surf. Coat. Technol.*, 323, 18–29.
25. Di Girolamo, G., Marra, F., Schioppa, M., Blasi, C., Pulci, G., & Valente, T. (2015). Evolution of microstructural and mechanical properties of lanthanum zirconate thermal barrier coatings at high temperature, *Surf. Coat. Technol.*, 268, 298–302.
26. Mathanbabu, M., Thirumalaikumarasamy, D., Thirumal, P., & Ashokkumar, M. (2021). Study on thermal, mechanical, microstructural properties and failure analyses of lanthanum zirconate based thermal barrier coatings: a review, *Mater. Today. Proc.*, 46, 7948–7954.
27. Unabia, R., Candidato, R., & Pawłowski, L. (2018). Current progress in solution precursor plasma spraying of cermets: a review, *Metals*, 8(6), 420.
28. Gupta M. (2015). *Design of Thermal Barrier Coatings, a Modeling Approach.* Springer, London.
29. Ramachandran, C. S., Balasubramanian, V., & Ananthapadmanabhan, P. V. (2011). Multi objective optimization of atmospheric plasma spray process parameters to deposit yttria-stabilized zirconia coatings using response surface methodology, *J. Therm. Spray Technol.*, 20(3), 590–608.
30. Keshavamurthy, R., Naveena, B. E., Ramesh, T., & Shashikumara, N. K. (2021). Optimization of process parameters for plasma sprayed lanthanum zirconate TBCs on nickel based superalloy, *Adv. Sci. Technol.*, 106, 90–96.
31. Guo, X., Lu, Z., Park, H. Y., Li, L., Knapp, J., Jung, Y. G., & Zhang, J. (2019). Thermal properties of $La_2Zr_2O_7$ double-layer thermal barrier coatings, *Adv. Appl. Ceram.*, 118(3), 91–97.
32. Wang, L., Di, Y., Wang, H., Li, X., Dong, L., & Liu, T. (2019). Effect of lanthanum zirconate on high temperature resistance of thermal barrier coatings, *Trans. Indian Ceram. Soc.*, 78(4), 212–218.
33. Ramachandran, C. S., Balasubramanian, V., Ananthapadmanabhan, P. V., & Viswabaskaran, V. (2012). Influence of the intermixed interfacial layers on the thermal cycling behaviour of atmospheric plasma sprayed lanthanum zirconate based coatings, *Ceram. Int.*, 38(5), 4081–4096.
34. Guo, X., Lu, Z., Jung, Y. G., Li, L., Knapp, J., & Zhang, J. (2016). Thermo mechanical properties of bilayer $La_2Zr_2O_7$ thermal barrier coatings, *Energy Technol.*, 155–160.
35. Lyu, G., Kim, I. S., Song, D., Park, H. M., Kim, J. S., Song, T., Myoung, S., Jung, Y. G., & Zhang, J. (2020). Sintering behavior and phase transformation of YSZ-LZ composite coatings, *Ceram. Int.*, 46(2), 1307–1313.
36. Guo, X., Lu, Z., Jung, Y. G., Li, L., Knapp, J., & Zhang, J. (2016). Thermal properties, thermal shock, and thermal cycling behavior of lanthanum zirconate-based thermal barrier coatings, *Metall. Mater. Trans. E*, 3(2), 64–70.
37. Song, D., Paik, U., Guo, X., Zhang, J., Woo, T. K., Lu, Z., Jung, S. H., Lee, J. H., & Jung, Y. G. (2016). Microstructure design for blended feedstock and its thermal durability in lanthanum zirconate based thermal barrier coatings, *Surf. Coat. Technol.*, 308, 40–49.
38. Yugeswaran, S., Kobayashi, A., & Ananthapadmanabhan, P. V. (2012). Initial phase hot corrosion mechanism of gas tunnel type plasma sprayed thermal barrier coatings, *Mater. Sci. Eng.*, 177(7), 536–542.
39. Liu, X., Wang, X., Javed, A., Zhu, C., & Liang, G. (2016). The effect of sintering temperature on the microstructure and phase transformation in tetragonal YSZ and LZ/YSZ composites, *J. Ceram. Int.*, 42(2), 2456–2465.

40. Zhang, C., Zhao, J., Yang, L., Zhou, Y., Wang, Q., Chen, H., Yang, G., Gao, Y., & Liu, B. (2020). Preparation and corrosion resistance of nonstoichiometric lanthanum zirconate coatings, *J. Eur. Ceram. Soc.*, 40, 3122–3128.
41. Zhang, J., Guo, X., Jung, Y.-G., Li, L., & Knapp, J. (2014). Microstructural non-uniformity and mechanical property of air plasma-sprayed dense lanthanum zirconate thermal barrier coating, *Mater. Today Proc.*, 1(1), 11–16.
42. Myoung, S.-W., Kim, J.-H., Lee, W.-R., Jung, Y.-G., Lee, K.-S., & Paik, U. (2010). Microstructure design and mechanical properties of thermal barrier coatings with layered top and bond coats, *Surf. Coat. Technol.*, 205(5), 1229–1235.
43. Praveen, K., Alroy, R. J., Rao, D. S., & Sivakumar, G. (2022). Hot corrosion behavior of hybrid double-layered LZ/LZC + YSZ thermal barrier coatings made using powder and solution precursor feedstock, *Surf. Coat. Technol.*, 436, 128260.
44. Chellaganesh, D., Khan, M. A., & Jappes, J. W. (2021). Thermal barrier coatings for high temperature applications–a short review, *Mater. Today Proc.*, 45, 1529–1534.
45. Zhong, X., Zhao, H., Liu, C., Wang, L., Shao, F., Zhou, X., Tao, S., & Ding, C. (2015). Improvement in thermal shock resistance of gadolinium zirconate coating by addition of nanostructured yttria partially-stabilized zirconia, *Ceram. Int.*, 41(6), 7318–7324.
46. Fan, W., & Bai, Y. (2016). Review of suspension and solution precursor plasma sprayed thermal barrier coatings, *Ceram. Int.*, 42, 14299–14312.
47. Shinozawa, A., Eguchi, K., Kambara, M., & Yoshlda, T. (2009). Feather-like structured YSZ coatings at fast rates by plasma spray physical vapor deposition, *Proc. Int. Therm. Spray Conf.*, 19, 1140–1145.
48. Góral, M., Poreba, M., Kubaszek, T., Ochał, K., & Drajewicz, M. (2022). Structure and thickness of Y_2O_3 coatings deposited by plasma spray physical vapour deposition (PS-PvD) method on graphite, *Metalurgija*, 61(3–4), 677–680.
49. Tejero-Martin, D., Rezvani Rad, M., McDonald, A., & Hussain, T. (2019). Beyond traditional coatings: a review on thermal-sprayed functional and smart coatings, *J. Therm. Spray Technol.*, 28(4), 598–644.
50. Fauchais, P., Vardelle, M., Vardelle, A., & Bianchi, L. (1996). Plasma spray: study of the coating generation, *Ceram. Int.*, 22, 295–303.
51. Shen, Z., Liu, Z., Mu, R., He, L., Liu, G., & Huang, Z. (2021). LaGdZrO/YSZ thermal barrier coatings by EB-PVD: microstructure, thermal properties and failure mechanism, *Chem. Eng. J. Adv.*, 5, 100073.
52. Ganvir, A., Calinas, R. F., Markocsan, N., Curry, N., & Joshi, S. (2019). Experimental visualization of microstructure evolution during suspension plasma spraying of thermal barrier coatings, *J. Eur. Ceram. Soc.*, 39, 470–481.
53. Farrokhpanah, A., Coyle, T. W., & Mostaghimi, J. (2017). Numerical study of suspension plasma spraying, *J. Therm. Spray Technol.*, 26, 12–36.
54. Vanevery, K., Krane, J. M. M., Trice, R. W., Wang, H., Porter, W., Besser, M., Sordelet, D., Ilavsky, J., & Almer, J. (2011). Column formation in suspension plasma-sprayed coatings and resultant thermal properties, *J. Therm. Spray Technol.*, 20 (2011) 817–828.
55. Delbos, C., Fazilleau, J., Rat, V., Coudert, J. F., Fauchais, P., & Pateyron, B. (2006). Phenomena involved in suspension plasma spraying part 2: zirconia particle treatment and coating formation, plasma them, *Plasma Process,* 26, 393–414.
56. Kassner, H., Siegert, R., Hathiramani, D., Vaßen, R., & Stoever, D. (2008). Application of suspension plasma spraying (SPS) for manufacture of ceramic coatings, *J. Therm. Spray Technol.*, 17, 115–123.
57. Chidambaram, R., Seshadri, Dwivedi, G., Viswanathan, V., & Sampath, S. (2016). Characterizing suspension plasma spray coating formation dynamics through curvature measurements, *J. Therm. Spray Technol.*, 25, 1666–1683.

58. Zhong, X., Li, D., Shao, F., Tao, S., Zhao, H., Wang, L., Zhao, Y., & Liu, C. (2014). Thermal shock behaviors of YSZ thick thermal barrier coatings fabricated by suspension and atmospheric plasma spraying, *Surf. Coat. Technol.*, 249, 48–55.
59. Mauer, G., Schlegel, N., Guignard, A., Jarligo, M. O., Rezanka, S., Hospach, A., & Vaßen, R. (2014). Plasma spraying of ceramics with particular difficulties in processing, *J. Therm. Spray Technol.*, 24, 30–37.
60. Sokołowski, P., Pawłowski, L., Dietrich, D., Lampke, T., & Jech, D. (2106). Advanced microscopic study of suspension plasma-sprayed zirconia coatings with different microstructures, *J. Therm. Spray Technol.*, 25, 94–103.
61. Sokołowski, P., Kozerski, S., Pawłowski, L., & Ambroziak, A. (2014). The key process parameters influencing formation of columnar microstructure in suspension plasma sprayed zirconia coatings, *Surf. Coat. Technol.*, 260, 97–106.
62. Łatka, L. (2018). Thermal barrier coatings manufactured by suspension plasma spraying-a review, *Adv. Mater. Sci.*, 18(3), 95–117.
63. Padture, N. P., Schlichting, K. W., Bhatia, T., Ozturk, A., Cetegen, B., Jordan, E. H., Gell, M., Xiao, T. D., Strutt, P. R., Garci, E., Miranzo, P., & Osendi, M. I. (2001). Towards durable thermal barrier coatings with novel microstructures deposited by solution precursor plasma spray, *Acta Mater.*, 49, 2251–2257.
64. Govindarajan, S., Dusane, R. O., & Joshi, S. V. (2011). In situ particle generation and splat formation during solution precursor plasma spraying of yttria-stabilized zirconia coatings, *J. Am. Ceram. Soc.*, 94, 4191–4199.
65. Sivakumar, G., Banerjee, S., Raja, V. S., & Joshi, S. V. (2018). Hot corrosion behavior of plasma sprayed powder-solution precursor hybrid thermal barrier coatings, *Surf. Coat. Technol.*, 349, 452–461.
66. Ma, X., Xie, L., Chen, D., Jordan, E. H., Cetegen, B. M., Wu, F., Ozturk, A., & Gell, M. (2006). Formation of vertical cracks in solution-precursor plasma-sprayed thermal barrier coatings, *Surf. Coat. Technol.*, 201, 1058–1064.
67. Gell, M., Jordan, E. H., Teicholz, M., Cetegen, B. M., Padture, N. P., Xie, L., Chen, D., Ma, X., & Roth, J. (2008). CHEN spray process, *J. Therm. Spray Technol.*, 17, 124–135.
68. Anwaar, A., Wei, L., Guo, Q., Zhang, B., & Guo, H. (2017). Novel prospects for plasma spray–physical vapor deposition of columnar thermal barrier coatings, *J. Therm. Spray Technol.*, 26, 1810–1822.
69. Zhang, B., Song, W., Wei, L., Xiu, Y., Xu, H., Dingwell, D. B., & Guo, H. (2019). Novel thermal barrier coatings repel and resist molten silicate deposits, *Scr. Mater.*, 163, 71–76.
70. Dorier, J. L., Gindrat, M., Hollenstein, C., Loch, M., Refke, A., Sailto, A., & Barbezat, G. (2001). Plasma jet properties in a new spraying process at low pressure for large area thin film deposition, *Proceedings of the International Thermal Spray Conference*, May 28–30, (Singapore, Singapore), pp. 759–764.
71. Zhang, B., Wei, L., Gao, L., Guo, H., & Xu, H. (2017). Microstructural characterization of PSPVD ceramic thermal barrier coatings with quasi-columnar structures, *Surf. Coat. Technol.*, 311, 199–205.
72. Rezanka, S., Mauer, G., & Vaßen, R. (2014). Improved thermal cycling durability of thermal barrier coatings manufactured by PS-PVD, *J. Therm. Spray Technol.*, 23, 182–189.
73. Mauer, G., Jarligo, M. O., Rezanka, S., Hospach, A., & Vaßen, R. (2015). Novel opportunities for thermal spray by PS-PVD, *Surf. Coat. Technol.*, 268, 52–57.
74. Deng, Z., Zhang, X., Zhou, K., Liu, M., Deng, C., Mao, J., & Chen, Z. (2018). 7YSZ coating prepared by PS-PVD based on heterogeneous nucleation, *Chin. J. Aeronaut.*, 31, 820–825.

75. Rezanka, S., Mauer, G., & Vaßen, R. (2014). Improved thermal cycling durability of thermal barrier coatings manufactured by PS-PVD, *J. Therm. Spray Technol.*, 23, 182–189.
76. Mauer, G., Jarligo, M. O., Rezanka, S., Hospach, A., & Vaßen, R. (2015). Novel opportunities for thermal spray by PS-PVD, *Surf. Coat. Technol.*, 268, 52–57.
77. Zhao, H., Zhang, X., Deng, C., Deng, Z., & Chen, X. (2022). Performance evaluation and thermal shock behavior of PS-PVD (Gd0.9Yb0.1) $2Zr_2O_7$/YSZ thermal barrier coatings, *Coatings*, 12(3), 323.
78. He, W., Mauer, G., Gindrat, M., Wäger, R., & Vaßen, R. (2017). Investigations on the nature of ceramic deposits in plasma spray–physical vapor deposition, *J. Therm. Spray Technol.*, 26, 83–92.
79. Yang, J., Zhao, H., Zhong, X., Ni, J., Zhuang, Y., Wang, L., & Tao, S. (2018). Evolution of residual stresses in PS-PVD thermal barrier coatings on thermal cycling, *J. Therm. Spray Technol.*, 27, 914–923.
80. Wang, C., Wang, Y., Wang, L., Hao, G., Sun, X., Shan, F., & Zou, Z. (2014). Nanocomposite lanthanum zirconate thermal barrier coating deposited by suspension plasma spray process, *J. Therm. Spray Technol.*, 23(7), 1030–1036.
81. Weber, S. B., Lein, H. L., Grande, T., & Einarsrud, M. A. (2013). Influence of the precursor solution chemistry on the deposition of thick coatings by spray pyrolysis, *Surf. Coat. Technol.*, 221, 53–58.
82. Ramachandran, C. S., Balasubramanian, V., & Ananthapadmanabhan, P. V. (2012). Synthesis, spheroidization and spray deposition of lanthanum zirconate using thermal plasma process, *Surf. Coat. Technol.*, 206(13), 3017–3035.
83. Ganesan, B., Hariharan, P., & Dhinasekaran, D. (2021). Hot corrosion studies of nanostructured gadolinium zirconate thermal barrier coatings, *Ceram. Int.,* 47(18), 25959–25972.
84. Satpathy, R., Mishra, B. R., Mallick, B., & Mishra, B. K. (2019). Synthesis and application of nano structured Bi layer YSZ LZ thermal barrier coating, *Defence Sci. J.*, 69(2), 185.
85. Wu, S., Zhao, Y., Li, W., Liu, W., Wu, Y., & Liu, F. (2021). Optimization of process parameters for thermal shock resistance and thermal barrier effects of 8YSZ thermal barrier coatings through orthogonal test, *AIP Adv.*, 11(6), 065215.
86. Weng, W. X., Zheng, Z. H., & Li, Q. (2020). Cracking evolution of atmospheric plasma-sprayed YSZ thermal barrier coatings subjected to isothermal heat treatment, *Surf. Coat. Technol.*, 402, 125924.
87. Montavon, G., Sampath, S., Berndt, C. C., & Coddet, C. (1995). Effects of vacuum plasma spray processing parameters on splat, *Morphology,* 4, 5–6.
88. Foroushani, M. H., Shamanian, M., Salehi, M., & Davar, F. (2016). Porosity analysis and oxidation behavior of plasma sprayed YSZ and YSZ/$LaPO_4$ abradable thermal barrier coatings, *Ceram. Int.*, 42, 15868–15875.
89. Rokni, M. R., Nutt, S. R., Widener, C. A., Champagne, V. K., & Hrabe, R. H. (2017). Review of relationship between particle deformation, coating microstructure, and properties in high-pressure cold spray, *J. Therm. Spray Technol.* 26, 1308–1355 (2017).
90. Fauchais, P., Montavon, G., Vardelle, M., & Cedelle, J. (2006). Developments in direct current plasma spraying, *Surf. Coat. Technol.*, 201, 1908–1921.
91. Sampath, S., Jiang, X. Y., Matejicek, J., Prchlik, L., Kulkarni, A., & Vaidya, A. (2004). Role of thermal spray processing method on the microstructure, residual stress and properties of coatings: an integrated study of Ni-5 wt% Al bond coats, *Mater. Sci. Eng. A.*, 364, 216–231.
92. Kim, Y., Kim, J., Han, J. W., & Choi, J. (2022). Multiscale mechanics of yttria film formation during plasma spray coating, *Appl. Surf. Sci.*, 572, 151416.

93. Vardelle, A., Moreau, C., Themelis, N. J., &Chazelas, C. (2015). A perspective on plasma spray technology, *Plasma Chem. Plasma Proces.*, 35(3), 491–509.
94. Mostaghimi, J., Pasandideh-Fard, M., & Chandra, S. (2002). Dynamics of splat formation in plasma spray coating process, *Plasma Chem. Plasma Proces.*, 22(1), 59–84.
95. Sampath, S., & Jiang, X. (2001). Splat formation and microstructure development during plasma spraying: deposition temperature effects, *Mater. Sci. Eng.* A, 304, 144–150.
96. Vassen, R., Kassner, H., Stuke, A., Mack, D. E., Jarligo, M. O. D., & Stöver, D. (2010). Functionally graded thermal barrier coatings with improved reflectivity and high-temperature capability. In: *Materials Science Forum* (Vol. 631, pp. 73–78). Trans Tech Publications Ltd.
97. Curry, N., Leitner, M., & Körner, K. (2020). High-porosity thermal barrier coatings from high-power plasma spray equipment—processing, performance and economics, *Coatings*, 10(10), 957.
98. Chen, L. (2011). Processing, microstructures and properties of thermal barrier coatings (TBCs) by plasma spraying (PS). In: Huibin Xu and Hongbo Guo (eds.), *Thermal Barrier Coatings* (pp. 132–160). Woodhead Publishing.
99. Fauchais, P., Vardelle, M., Vardelle, A., & Bianchi, L. (1996). Plasma spray: study of the coating generation, *Ceram. Int.*, 22(4), 295–303.
100. Jung, S. H., Jeon, S. H., Lee, J. H., Jung, Y. G., Kim, I. S., & Choi, B. G. (2016). Effects of composition, structure design, and coating thickness of thermal barrier coatings on thermal barrier performance, *J. Korean Ceram. Soc.*, 53(6), 689–699.
101. Nayak, H., Krishnamurthy, N., & Shailesh, R. A. (2018). Studies on plasma sprayed thermal barrier coating with increase in coating thickness, *Tribol. Ind.*, 40(3), 420.
102. Takahashi, R. J., Assis, J. M. K., Neto, F. P., & Reis, D. A. P. (2020). Heat treatment for TGO growth on NiCrAlY for TBC application, *Mater. Res. Express*, 6(12), 126442.
103. Cai, Z., Hong, H., Peng, D., Zhao, X., Wang, W., Liu, Y., & Cao, Z. (2020). Stress evolution in ceramic top coat of air plasma-sprayed thermal barrier coatings due to CMAS penetration under thermal cycle loading, *Surf. Coat. Technol.*, 381, 125–146.
104. Hitchman, N. J., Li, L., & Knapp, J. (2010). Failure of thermal barrier coatings subjected to CMAS attack, *J. Therm. Spray. Technol.*, 19, 148–155.
105. Gok, M. G., & Goller, G. (2017). Microstructural characterization of GZ/CYSZ thermal barrier coatings after thermal shock and CMAS+hot corrosion test, *J. Eur. Ceram. Soc.*, 37(6), 2501–2508.
106. Dehkharghani, A. M. F., Rahimipour, M. R., & Zakeri, M. (2021). Hot corrosion behavior of plasma sprayed $La_2Ce_2O_7$/YSZ thermal barrier composite coating in the presence of sulfate and vanadate molten salts, *Corrosion Sci.,* 183, 109349.
107. Mahade, S., Venkat, A., Curry, N., Leitner, M., & Joshi, S. (2021). Erosion performance of atmospheric plasma sprayed thermal barrier coatings with diverse porosity levels, *Coatings*, 11(1), 86.
108. Li, Y., Huang, J., Wang, W., Ye, D., Fang, H., Gao, D., & Yu, Z. (2021). Control of the pore structure of plasma-sprayed thermal barrier coatings through the addition of unmelted porous YSZ particles, *Coatings*, 11(3), 360.
109. Yang, X., Zhang, J., Lu, Z., Park, H. Y., Jung, Y. G., Park, H., … & Zhang, J. (2020). Removal and repair techniques for thermal barrier coatings: a review, *Trans. IMF*, 98(3), 121–128.
110. Guo, L., Xin, H., Zhang, Z., Zhang, X., & Ye, F. (2020). Microstructure modification of Y_2O_3 stabilized ZrO_2 thermal barrier coatings by laser glazing and the effects on the hot corrosion resistance, *J. Adv. Ceram.*, 9(2), 9.
111. Zhang, P., Sun, L., Zhang, X., Wang, Y., Zhang, Q., Yao, J., & Chan g, F. (2021). Thermal cycling behavior of selective laser-remelted thermal barrier coatings with different laser dot distances, *J. Therm. Spray Technol.*, 30(4), 1038–1048.

112. Li, W., Li, Z., An, G., Cheng, B., Song, Q., Sun, J., &Goransky, G. (2022). Isothermal oxidation TGO growth behaviors of laser-remolten LZO/YSZ thermal barrier coatings, *Coatings*, 12(2), 107.
113. Huang, K., Li, W., Pan, K., Lin, X., & Wang, A. (2021). Microstructure and corrosion properties of $La_2Zr_2O_7$/NiCoAlY thermal barrier coatings deposited on Inconel 718 superalloy by laser cladding, *Coatings*, 11(1), 101.
114. Arshad, A., Yajid, M. A. M., & Idris, M. H. (2021). Microstructural characterization of modified plasma spray LZ/YSZ thermal barrier coating by laser glazing, *Mater. Today Proc.*, 39, 941–946.
115. Zhong, X., Wang, Y., Xu, Z., Zhang, Y., Zhang, J., & Cao, X. (2009). Influence of laser-glazing on hot corrosion resistance of yttria-stabilized zirconia TBC in molten salt mixture of V_2O_5 and Na_2SO_4, *Mater. Corrosion*, 60(11), 882–888.
116. Yan, Z., Guo, L., Li, Z., Yu, Y., & He, Q. (2019). Effects of laser glazing on CMAS corrosion behavior of Y_2O_3 stabilized ZrO_2 thermal barrier coatings, *Corrosion Sci.*, 157, 450–461.
117. Wu, S., Zhao, Y., Li, W., Liu, W., Wu, Y., & Liu, F. (2021). Research progresses on ceramic materials of thermal barrier coatings on gas turbine, *Coatings*, 11(1), 79.

10 Artificial Intelligence in High-Temperature Tribology

Umesh Ghorpade, Shubhangi Suryawanshi, Digvijay Bhosale, Ganesh Babar, Vanita Kshirsagar, and Manpreet Singh

10.1 INTRODUCTION

The basic study of surfaces that interact in relative motion, or tribology, has drawn interest from scholars because of its applications in a variety of industries, including food, automotive, aerospace, biomedicine, and energy [1]. Consequently, numerous research studies in tribology have explored key aspects such as dynamics and surface characteristics like friction, wear, adhesion, and indentation [2]. For instance, excessive energy consumption, mechanical equipment failure, industrial process failure, and increased transportation costs due to inevitable wear and friction have prompted scientists to investigate the tribological phenomenon at various time and length scales [3]. In recent years, the majority of these studies have concentrated on developing and applying state-of-the-art experimental techniques to reduce wear and corrosion [4]. In particular, the study of tribological systems has led to the development of sophisticated experimental methods like spectroscopy and digital image correlation. The finite element method has made significant strides in the study of tribology at the nano level and micro level [5]. Lubrication is one of the most efficient ways to reduce and manage wear and friction [6]. In recent decades, a great deal of research has been done to develop novel lubricants that will enhance the lubrication of machinery's moving parts. These include innovative types of lubricants like gel composites and diamond-like carbon. Furthermore, lubrication in oil/polymer-based liquids is improved by the addition of solid-state reinforcements such as fly ash, graphite, short carbon fibers, and titanium oxide particles [7]. Contemporary data-driven machine learning (ML) techniques have surfaced as novel means of comprehending and progressing intricate processes in tribological research. This is due to the fact that higher-dimensional issues and datasets can be easily and affordably modeled using ML, providing them with improved adaptability to changing conditions [8]. Several studies have used ML in conjunction with modeling and design of experiments (DOE) to expedite the creation of tribological substances with the required features [9].

The ability of thermal coating processes to enhance the surface qualities of materials, like greater wear resistance, enhanced heat insulation, and improved resistance

DOI: 10.1201/9781003542957-10

to corrosion, has drawn a lot of interest from a variety of industries [10]. These coatings are essential for extending the life and functionality of important parts in the automotive, aerospace, energy, and other industries. However, attaining the best possible coating quality and appearance necessitates a thorough comprehension of the intricate relationships that exist between the coating properties, material characteristics, and course parameters [11]. Thermal spray techniques are used to apply coatings to surfaces using heat [12]. These methods are used in variety of industries for wear resistance, corrosion protection, and thermal barrier coatings. Dense, strongly bonded coatings with low porosity and high adhesion are produced by high-velocity oxy-fuel spraying (HVOF). Coating wires are melted via an electric arc in arc spraying, while coating particles are applied to a substrate surface via cold spraying [13–14]. The technique chosen depends on the required application, substrate and coating materials, and desired properties. Every method has particular advantages and uses in various industries [15]. The most popular surface texturing techniques for numerical analysis involve the deposition of thin coating layers onto substrates to enhance their properties [16]. The difficulties were primarily attributed to the intricate and nonlinear surface coating methods and process variables. For the purpose of modeling and process parameter optimization in deposition processes, ML techniques were presented as an alternative to traditional and statistical approaches [17].

10.2 EFFECTIVENESS OF TRADITIONAL COATINGS AT HIGH TEMPERATURES

Coatings generally improved the substrate properties like hardness, toughness, and porosity, resulting in enhanced tribological properties of the specimen. It was discovered that coated specimens had greater erosion/wear resistance than bare samples. This section covers studies related to the performance of coatings at high temperatures. Praveen et al. noticed that at higher temperatures and a 30° impact angle, there was greater resistance to erosion by composite coating because of the alumina particles' pinning and shielding effects deposited. The erosion resistance of coatings is 2.5 times higher than that of the substrate. For an impact angle of 30°, the primary mechanisms of erosion were micro-cutting and deformation. Conversely, at 90° impact angles, craters and lips formed [18]. It was demonstrated in other studies also that coatings exhibit higher erosion resistance than the substrate that is not coated [19–20].

On a gray cast iron substrate using the HVOF method, the erosion pattern of coated samples was investigated at high temperatures (800°C). An alloy-718 bi-layer coating exhibits superior corrosion resistance compared to its substrate. For a 30° impact angle, micro-cutting and deformation were the primary erosion mechanisms. However, lips and craters formed at 90° impact angles. High hardness and the surface development of stable oxides (Al_2O_3 and Cr_2O_3) are linked to enhanced erosion resistance [21]. But Sharma [22] demonstrates that with rises in temperature, the rate of erosion also rises because high temperatures cause materials to soften, which lowers coating strength. Substrate erosion wear was four times less than coating erosion wear.

Wang et al. [23] deposited self-developed FeTi/CrB metal matrix composite on boiler steels by arc spraying and compared its performance with locally available boiler material $Fe_{17}Cr_5NiB$, which was used commonly for developing metal matrix composite of circulating fluidized bed (CFB) combustion boiler in China. According to the authors, the self-developed coating outperforms its counterpart in terms of porosity, bond strength, and fracture toughness across a wide temperature range and impact angle [23]. The experimental data on high-temperature erosive wear that examines the effects of Al_2O_3 micrometer-sized reinforcement particles (10, 20, and 30 wt.%) in the Inconel-718 matrix show that the 30 wt.% coating was found to exhibit superior resistance due to its high hardness, which helped to limit plastic deformation [24]. Vicenzi et al. carried out erosive wear experiments on three different types of coatings (WC–12Co, Cr_3C_2 NiCr, and WC CrC Ni) on SAE 1020 steel by using the HVOF process with the main objective to combat erosion at higher temperatures (300°C). The WC–12Co was found to wear out 18 times quicker than the substrate [25]. Zhang et al. also studied the erosion behavior at three different temperature ranges: 400°C, 500°C, and 600°C of the Cr_3C_2–NiCr coating deposited on a nickel base superalloy by the HVOF technique. As the temperature rises to 600°C, the rate of erosion decreases because the plastic properties of the coating also increase at higher temperatures. As a result, erosion can only occur through micro-cutting at a normal angle [26]. The effects of extreme temperatures, solid particle erosion (800°C), and impact angles (30°, 60°, 90°) on SiC-WC composites were examined in a different study [27].

10.3 FAILURE MECHANISMS AND CAUSES

The main failure mechanisms and their causes at high temperatures are summarized in Table 10.1.

TABLE 10.1
Boiler Tube Failure Mechanisms and Causes

Sr. No.	Failure Mechanisms	Cause
1.	Lack of quality control and monitoring	• Damage during maintenance cleaning • Material defects • Defects in welding • Metal loss
2.	Erosion	• Impact of fly ash • Damage from falling slag • Erosion by soot blowers • Abrasion by fuel particles (coal, peat)
3.	Corrosion: fire side	• Hot corrosion (Types 1 and 2) • Erosion caused by ash deposits
4.	Corrosion: steam/water side	Pitting • Stress corrosion cracking • Hydrogen damage • Oxygen attack • Caustic corrosion
5.	Fatigue	• Fatigue caused by vibrations • Thermal fatigue • Creep fatigue • Fatigue due to corrosion

(Continued)

TABLE 10.1 (*Continued*)
Boiler Tube Failure Mechanisms and Causes

Sr. No.	Failure Mechanisms	Cause
6.	Stress rupture	• Overheating in the short term • High-temperature creep (including long-term overheating) • Stress accumulation due to dissimilar metal welds

10.4 EFFECTIVENESS OF NANOSTRUCTURED COATINGS AT HIGH TEMPERATURES

When comparing nano-coatings to conventional coatings, they have superior properties due to their unique structure and nano-scale features, which are typically in the range of 1–100 nm. The nano-scale particles produce a more uniform and dense coating layer. A substantial improvement in hardness, wear resistance, and corrosion resistance is achieved by nano-coating thanks to this structural refinement. In addition, nano-coatings have better thermal stability, making them more appropriate for high-temperature applications; they can also form tightly bonded, self-assembling layers, which improve their barrier qualities against chemical and environmental deterioration; and when compared to conventional coatings, nano-coatings have better durability, smoother surfaces, and stronger resistance to mechanical stress, making them perfect for advanced engineering applications where traditional coatings might not be able to meet the high-performance requirements. The NGC-v4 nano-composite coating, which has a ceramic and metal matrix, was chosen. The results indicate that coatings save fuel between 0.5% and 2%. Compared to uncoated blades, which showed a 4% chord loss, coated blades showed only a 0.2% chord loss under the same testing conditions. In the test, an abrasive medium called Arizona Road Dust was applied at a pressure of 80 PSI and an impact angle of 20° to the tip of the rotor blade. The result of this rough action was chord loss. To guarantee nano-scale attachment to the substrate, a high-vacuum technique was used to apply the coating material. For optimal performance, the coating thickness must stay higher than 10 µm. In an experiment, the eighth-stage compressor rotor blade of the CFM56 engine was coated with NGC V4 using physical vapor deposition. The results of the testing also showed that the applied coatings could tolerate fatigue, vibration, corrosion, and erosion. These coatings have to go through and pass a fatigue test as per FAA regulations [28]. Wire arc spraying was used to apply the nano-composite coatings, and impact angles of 30° and 60° were used in high-temperature erosion testing at 300°C, 450°C, and 600°C. According to research, erosion rates increased by 20%–30% between 300°C and 450°C and by 4%–12% at 600°C [29]. In contrast, the erosion rate increased by about 220% when the Alpha 1800 alloy was heated from 450°C to 600°C. These findings emphasize how crucial the glass content of SHS717 is in reducing erosion, especially in hot weather [30]. Experiments on solid particle erosion show that as the temperature rises, the erosion rate falls. AISI 1045 steel was coated with two iron-based amorphous/nanocrystalline coatings using a wire arc spraying technique. Traditional coatings show a homogenous structure free of cracks. The three coatings were tested in a jet air erosion tester with impact angles of 30°, 60°, and 90° and a temperature of 900°C. The maximum erosion for substrate and nanostructured coatings occurs at an oblique angle, indicating their ductile nature. Traditional coatings behave in a semi-ductile manner. Nano-coatings outperformed substrates and traditional TiAlN coatings in terms of performance [31]. The nano-composite's erosion resistance remained superior to that of conventional and uncoated substrates even after undergoing an erosion test at 550°C. The rate of erosion was higher at 30° angles for materials exhibiting ductile behavior [32]. The behavior of 7YSZ nanostructured coatings eroded by solid particles at high temperatures, which were applied to the NiCrAlY bond coat using the APS process, was investigated. A bond coat was applied to the steel substrate using the HVOF method. Alumina particles of varying masses (3, 6, and 9 g) were used in high-temperature erosion experiments, which were conducted at 900°C and an impact angle of 80°C. The particle masses before and after every erosion test are denoted by m_1 and m_2. This method works well for a number of failure modes, including sintering, oxidation, and corrosion [33].

10.5 APPLICATIONS OF ML IN TRIBOLOGY

Several studies have demonstrated that a comprehensive understanding of structure characteristics relationships in intricate higher-dimensional tribological problems can be achieved through the application of ML [34,35]. Generally speaking, probability theory's stochastic approach and algorithmic theory are what drive the creation of ML models. The generalized structure of ML is illustrated in Figure 10.1.

Through the evaluation and improvement of the marked output for a given input, the model learns over time through the feature-property relationship that underpins the supervised learning approach. The process of successfully building a supervised learning model involves determining and characterizing a vector of distinct features that are expected to have a major impact on the result. Supervised learning techniques encompass a variety of methods, such as the Naive Bayes model, decision

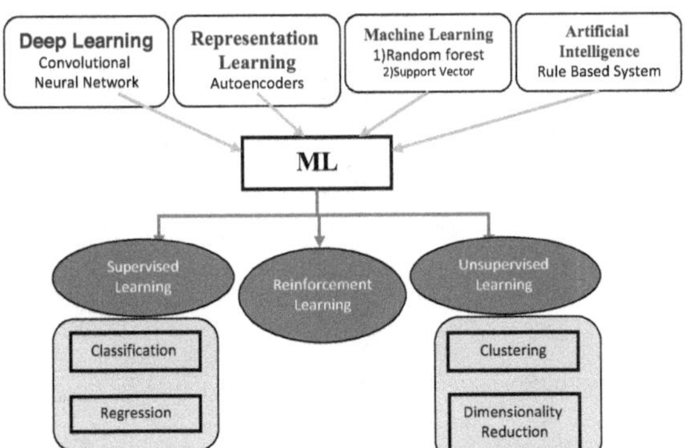

FIGURE 10.1 Applications of machine learning.

tree, random forest, and support vector machine (SVM) [36]. However, an unsupervised learning method analyzes and clusters data according to shared features using an unlabeled dataset. Reinforcement learning is a significant subset of ML models. It falls under the category of supervised learning and uses a reward-punishment mechanism [37]. In tribology, the performance of materials under test is influenced by their chemical and physical characteristics as well as the external temperature, pressure, and humidity of the testing environment. Therefore, it is essential to comprehend how various experimental parameters affect a material's performance [38]. It is possible to distinguish between the impact of each variable thanks to the orthogonality of this DOE. Apart from the fewer experiments, Taguchi's method is among the most reliable, effective, and methodical ways to offer the parameters design for the experiments. When working with a set of variables where only a few have significance and are generally small, Taguchi's method is especially helpful [39].

10.5.1 Artificial Neural Networks

Applications of artificial neural networks (ANNs) have been shown to be valuable in a variety of research domains for the study, creation, and forecasting of material formulations with improved properties (Figure 10.2). Their capacity for adaptive learning, generalizability, independence from earlier models, and logical reasoning is credited for this. The application of ANNs has been the subject of groundbreaking research, with promising results in diagnosing faulty rolling bearings, determining fretting wear damage, tool wear, roughness of the surface, and disk brake performance. ANNs have also been widely used to optimize low coefficients of friction, high lubrication, and/or minimum wear rates. Notably, the incorporation of ANNs into lubricant design has yielded insightful information about the composition and interrelationships between the input and outcome parameters, in addition to facilitating the creation of unique lubricants with a variety of additives. A thorough grasp of the state of tribology research today can be gained from current research on the application of ML in the research and development of lubricants, particularly in metal matrix composites (MMCs) and structural components. The function of ANNs in lubricants and materials is covered in detail in the sections that follow [40].

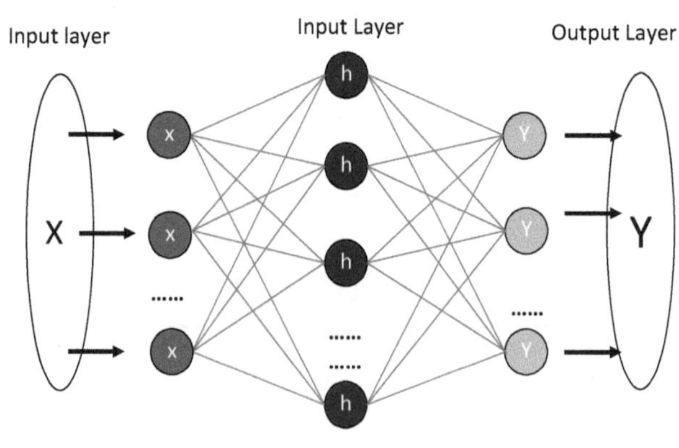

FIGURE 10.2 Extreme learning machine.

10.5.1.1 Lubricants Based on Oil

The transportation and manufacturing industries use lubricants extensively. The research on lubricants currently focuses on lubricants that are based on petroleum, composites, and carbon. However, petroleum-based lubricants have drawn a lot of attention over the years because of their increased viscosity, additive susceptibility, and thermal stability [41]. In addition to highlighting the impact of six input variables and material system specifications, such as sliding speed and distance, they demonstrated how well the ANN model predicted the wear rate. RNNs are able to learn more effectively from previous instances because they recycle both new inputs and previous outputs [42]. An ANN model was trained using data points gathered from experiments conducted using lubricant produced by differing levels of the oils mentioned. Only the coefficient of friction (CoF) and wear scar diameter (WSD) were examined, regardless of the testing [43].

10.5.1.2 Designing Materials with a Composite Basis for Tribology

Composites are interesting candidates for engineered materials because they combine several heterogeneous components to achieve desired chemical or physical properties. Composites can be identified from their constituent elements due to their distinct characteristics, which include greater mechanical strength, reduced cost and density, and longer life cycles. Three primary categories can be used to classify composites: polymer matrix composite, MMC, and ceramic matrix composite [44].

10.5.2 Artificial Neural Networks: From Early Development to Back-Propagation Success

The first logical connection between neural events in the nervous system was established by Warren McCulloch and Walter Pitts. Using the concepts of dynamic programming, Kelley and Bryson developed continuous back-propagation in 1960 and 1961, respectively (Figure 10.3).

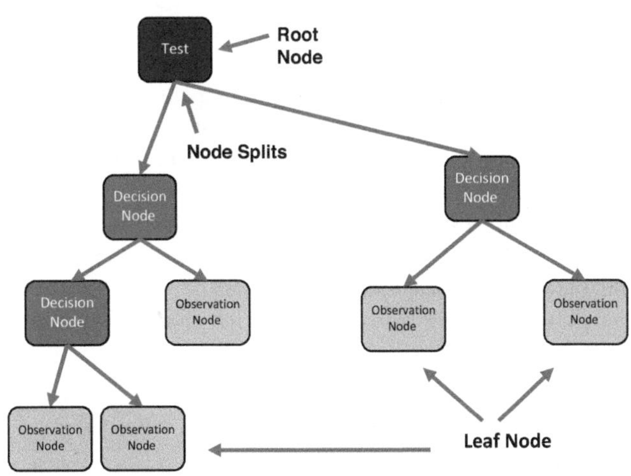

FIGURE 10.3 Decision tree.

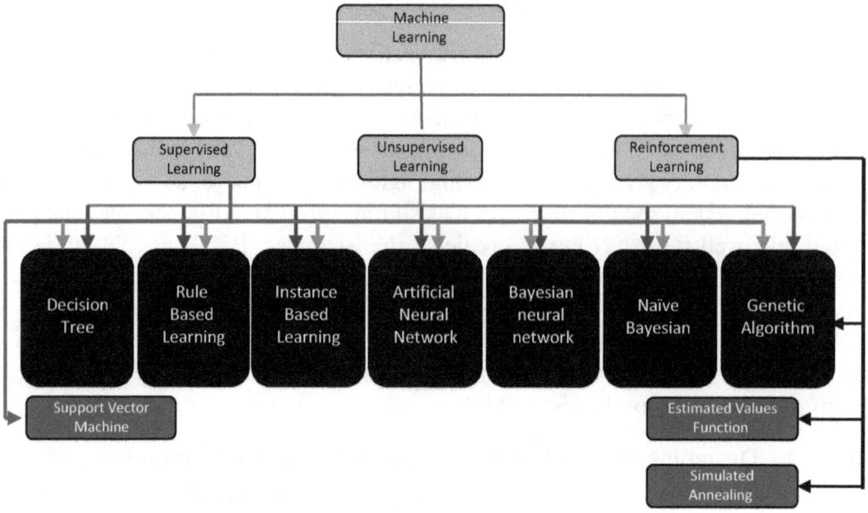

FIGURE 10.4 Categorization of machine learning methods.

In the back-propagation technique, multi-layer networks could be trained. ANNs became more and more well-liked between 2009 and 2012 due to human-like performance in ML. When it came to solving linear and nonlinear problems, ANN models performed better than traditional empirical models [45] (Figure 10.4).

10.5.3 ANNs' Function in Surface Coatings

ANNs are essential for surface coating process optimization because they improve the creation and accuracy of control predictions. Given input parameters such as the composition of the material, the processing parameters, and the performance, ANNs can predict properties and optimize coating formulations. They evaluate data from sensors that track variables like temperature, viscosity, and coating thickness, managing real-time operations. ANNs also aid in quality assurance (inset Figures 10.5 and 10.6). Their ability to forecast coating performance in various scenarios allows for the early identification of possible problems and raises the quality of the final product.

Through surface characterization techniques, coating formulations can be optimized for better functionality. Finally, by evaluating data from sensors tracking the performance and health of the equipment, ANNs can forecast maintenance requirements or equipment failures in coating processes. All things considered, ANNs are an effective instrument for the surface coatings sector, providing capacities for control, prediction, quality assurance, and optimization at different phases of the coating lifecycle. In the surface coating sector, ANNs have grown in significance over the last few years [46] (Figure 10.7).

10.6 ROLE OF MACHINE LEARNING IN TRIBOLOGY

Robust tribo-engineering components require a thorough understanding of material tribological behavior. Tribological datasets are used by ML algorithms to leverage examples or prior knowledge to improve task performance. Tribological properties

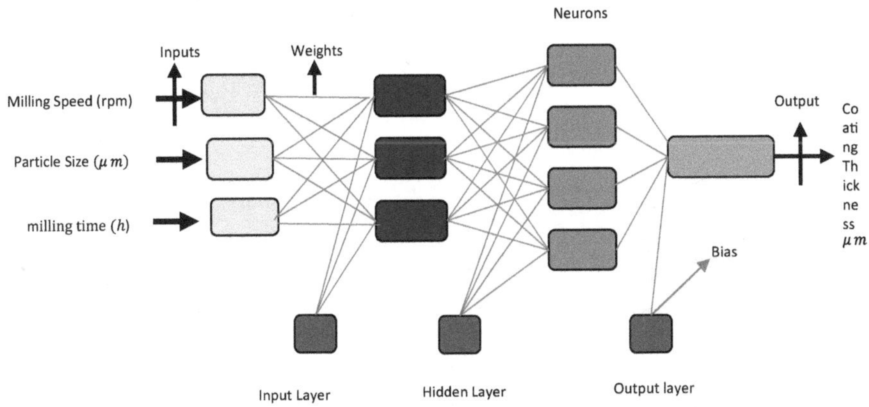

FIGURE 10.5 ANN architecture of prediction model for coating thickness. ANN, artificial neural network.

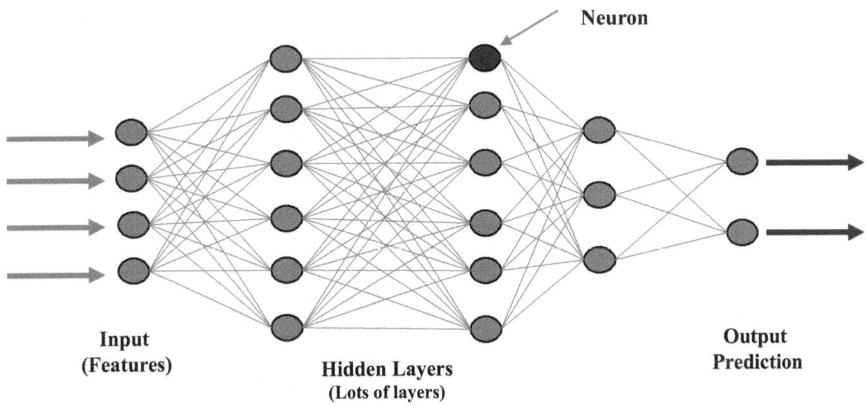

FIGURE 10.6 Typical architecture for an ANN. The inward and outward flow of neurons shown by arrow. ANN, artificial neural network.

can be affected by a variable's operational parameters and mechanical and material properties. However, figuring out how these factors affect performance is essential for the product's tribological characterization. The temperature, load, distance, sliding velocity, and sliding speed all negatively impact a material's tribological characteristics. Given that ML techniques seem to handle the complexity of data, they might be the key to solving this problem. Early in the 1980s, attempts were made to integrate ML into the field of tribology and develop an expert system and computerized database for tribological design fault diagnosis and decision-making. Numerous comparable studies have been carried out to show the ML algorithm's superior performance in predicting the tribological properties of materials [47] (Figure 10.8).

The numerous uses of ML exist in the field of tribology, such as condition monitoring and diagnosis. Nonetheless, it is essential in order to forecast and maximize behavior (tribological). The friction can be found in both nature and industry, where it is useful for preserving energy and essential for daily living. Because of this, researchers

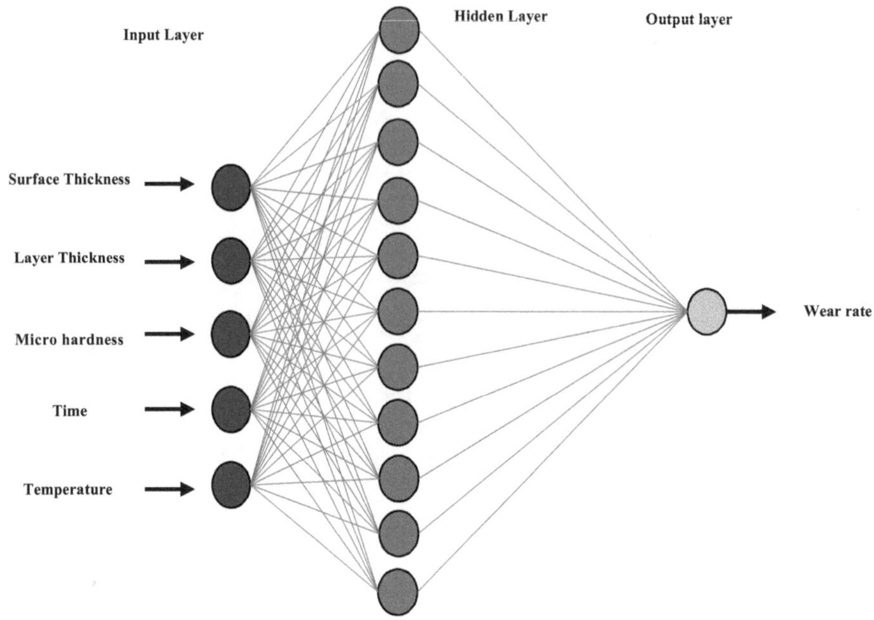

FIGURE 10.7 ANN schematic models for 316L stainless steel wear rate prediction. ANN, artificial neural network.

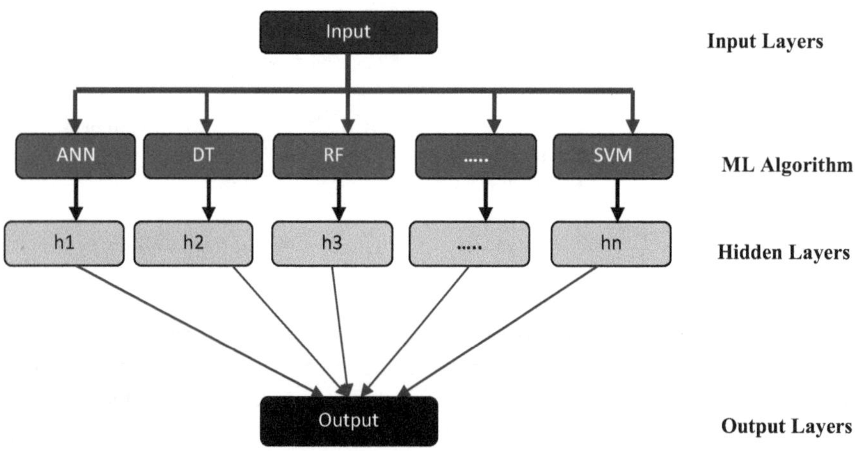

FIGURE 10.8 Collective learning technique.

have long worked to comprehend the causes of friction and have even tried to show that it is possible to achieve a completely frictionless state, with some significant advancements made in recent years. The adhesion and deformation make up the friction force. The friction coefficients for adhesion and deformation are therefore typically thought to be independent of one another. Frictional forces are greatly impacted by lubricants, surface coatings, and impurities. The collection and publication of experimental friction coefficients by researchers from a variety of fields have persisted in spite of the variable's uncertainty and complexity. Adverse events can happen when engineers solve specific tribology problems with tabular friction coefficients. Thus, identifying the component's tribological behavior and how operating conditions affect its performance requires modeling the process and parameter optimization. The output variable for such nonlinear processes can be quickly understood and predicted by ML algorithms, as previously mentioned in the literature. Using a Kriging surrogate model, the parameters influencing a wet clutch's friction coefficient were optimized. An R^2 value of 0.997 indicated the model's strong predictive ability in estimating the friction factor with fuzzy modeling, showing outstanding results at the 0.05 significance level. To predict the friction coefficient of a car brake system, ANNs are employed. By including inputs like specific pressure during molding, temperature during molding, duration of the molding process, temperature during the heat treatment, and heat treatment time, the author was able to accurately model the process and predict the material's friction [48]. Table 10.2 summarizes studies using ML techniques.

TABLE 10.2
Summarizing Studies Related to High-Temperature Wear of Coatings Using ML Techniques [48]

Sr. No.	Aspect	Details
1	Independence of friction coefficients	Adhesion and deformation friction coefficients are typically considered independent
2	Factors impacting friction forces	Lubricants, surface coatings, impurities have a significant effect on frictional forces
3	Challenges with friction coefficients	Experimental friction coefficient data are complex and uncertain, causing adverse events when engineers rely on tabular coefficients for solving tribology issues
4	Need for modeling	Identifying tribological behavior and understanding how operating conditions affect performance require process modeling and parameter optimization
5	Machine learning for nonlinear processes	Machine learning algorithms effectively predict output variables for nonlinear processes, as noted in the literature
6	Kriging surrogate model	Used to optimize parameters influencing a wet clutch's friction coefficient. Model achieved an R^2 value of 0.997, indicating strong predictive ability
7	Fuzzy modeling	Applied to predict the friction factor, showing outstanding results at the 0.05 significance level
8	Artificial neural networks (ANN)	Used to predict the friction coefficient in a car brake system. Input variables for modeling included: • Specific pressure during molding • Temperature during molding • Duration of the molding process • Temperature during heat treatment • Heat treatment time

ML, machine learning.

10.7 FUTURE RESEARCH DIRECTION

In order to model and optimize variables, this seeks to steer future research directions and persuade scientists to favor computational intelligence over conventional and statistical approaches. To determine the best course for future tribology research, 20 years of research have been analyzed. ANNs are effective for tribological applications due to their ability to handle complex, nonlinear problems. Other ML algorithms like SVM, decision trees, GA, fuzzy logic, RNN, random forests, and ensemble learning can also be used effectively. It is evident from the results that more study and application are possible. Artificial intelligence (AI) techniques could be used in future research to predict outcomes on material compositions and test conditions, enabling researchers to predict complex material systems, accelerate formulation discovery, and optimize performance under different conditions. The problem is that a ML strategy by itself may not always be able to resolve a specific tribological issue, which makes algorithm selection crucial. Therefore, it is advised that the research article include data repositories that include the underlying databases and related models. Furthermore, a large amount of data needs to be automatically optimized—something that is currently mostly done by hand—in order to take advantage of the tribology and ML knowledge that is currently available across several industries. Moreover, the use of various nanoparticles has expanded the field of tribology. AI is transforming high-temperature tribology by enabling the analysis of wear, friction, and lubrication data to improve material performance under extreme conditions [49–55]. Research has shown that ML models can assist in designing coatings and surface treatments for high-temperature applications, including Ni-based alloys and composite coatings [56–61]. These approaches help in understanding complex tribological behaviors and advancing the development of durable materials [62,63].

10.8 SUMMARY

The purpose of this chapter is to steer future directions and convince scientists that computational intelligence is a better option for modeling and variable optimization than traditional statistical methods. The lack of adequate and comparable datasets and inadequate uncertainty management with regard to test conditions and variations are the current obstacles. Therefore, it is advised that the research article include supplements or archives of data with the underlying databases and related models. The principal aim of this chapter is to perform a comprehensive analysis of the function of ML in tribological studies as well as to provide future researchers with an identifiable research path. The wider use of AI and different ML techniques in tribology, including the description and design of intricate tribological systems, has been highlighted as one of the main areas for prospective studies and advancements. Furthermore, interdisciplinary research can be carried out to better understand resource conservation and efficient use by integrating ML techniques with tribological experimental data.

REFERENCES

1. B. Bhushan, *Introduction to Tribology*. John Wiley & Sons, 2013.
2. H. Tang, J. Sun, J. He and P. Wu, Research progress of interface conditions and tribological reactions: a review. *J. Ind. Eng. Chem.,* 2021, **94**, 105–121.
3. D. Berman, S. A. Deshmukh, S. K. R. S. Sankaranarayanan, A. Erdemir and A. V. Sumant, Friction. Macroscale superlubricity enabled by graphene nanoscroll formation. *Science*, 2015, **348**, 1118–1122.
4. K. Holmberg and A. Erdemir, Influence of tribology on global energy consumption, costs and emissions. *Friction*, 2017, **5**, 263–284.
5. D. J. Whitehouse, Fractal or fiction. *Wear*, 2001, **249**, 345–353.
6. B. J. Hamrock, *Fundamentals of Fluid Film Lubrication*. McGraw-Hill Science, Engineering & Mathematics, 1994.
7. P. L. Menezes, C. J. Reeves, P. K. Rohatgi and M. R. Lovell, Fundamentals of engineering surfaces, in *Tribology for Scientists and Engineers: From Basics to Advanced Concepts*, eds. P. L. Menezes, M. Nosonovsky, S. P. Ingole, S. V. Kailas and M. R. Lovell, Springer, New York, 2013, pp. 341–389.
8. A. Rosenkranz, M. Marian, F. J. Profito, N. Aragon and R. Shah, The use of artificial intelligence in tribology—a perspective. *Lubricants*, 2020, **9**, 2.
9. M. Marian and S. Tremmel, Current trends and applications of machine learning in tribology—a review. *Lubricants*, 2021, **9**, 86.
10. H. Guo, (ed.), *Thermal Barrier Coatings*. Woodhead Publishing, 2023.
11. W. Tillmann, C. Schaak, L. Hagen, G. Mauer and G. Matthäus, Internal diameter coating processes for bond coat (HVOF) and thermal barrier coating (APS) systems. *J. Therm. Spray Technol.,* 2019, **28**, 233–241.
12. S. Samal, Thermal plasma technology: the prospective future in material processing. *J. Clean. Prod.*, 2017, **142**, 3131–3150.
13. D. Tejero-Martin, M. R. Rad, A. McDonald and T. Hussain, Beyond traditional coatings: a review on thermal-sprayed functional and smart coatings. *J. Therm. Spray Technol.*, 2019, **28**, 598–644.
14. J. Henao, C. Poblano-Salas, M. Monsalve, J. Corona-Castuera and O. Barceinas-Sanchez. Bio-active glass coatings manufactured by thermal spray: a status report. *Journal of Materials Research and Technology*, 2019, **8**(5), 4965–4984.
15. A. Rahimi,, M. Hojjati, A. Dolatabadi and C. Moreau, Thermal spray coating on polymeric composite for de-icing and anti-icing applications. *J. Manuf. Sci. Eng.*, 2021, **143**(10), 101008.
16. N. Noble, N. Radhika and M. Sathishkumar, Characterisation and property evaluation of high entropy alloy coating on 316L steel via thermal spray synthesis. *Tribol. Int.*, 2023, **185**, 108525.
17. C. Xu, W. Nie, Z. Liu, H. Peng, S. Yang and Q. Liu, Multi-factor numerical simulation study on spray dust suppression device in coal mining process. *Energy*, 2019, **182**, 544–558.
18. A. S. Praveen, J. Sarangan, S. Suresh and J. S. Subramanian, Erosion wear behaviour of plasma sprayed NiCrSiB/Al_2O_3 composite coating. *Int. J. Refract. Metals Hard Mater.*, 2015, **52**, 209–218.
19. E. Edward Anand and S. Natarajan, High-temperature erosion behaviour of plasmasprayed NiCrBSi–graphite coatings. *Surf. Eng.*, 2018, **34**(10), 783–790.
20. H. S. Nithin, V. Desai and M. R. Ramesh, Elevated temperature solid particle erosion behaviour of carbide reinforced CoCrAlY composite coatings. *Mater. Res. Express*, 2018, **5**(6), 066529.

21. H. Vasudev, L. Thakur, A. Bansal, H. Singh and S. Zafar, High temperature oxidation and erosion behaviour of HVOF sprayed bi-layer Alloy-718/NiCrAlY coating. *Surf. Coating. Technol.*, 2019, **362**, 366–380.
22. S. Sharma, High temperature erosive wear study of NiCrFeSiB flame sprayed coatings. *J. Inst. Eng.*, 2012, **93**, 7–12. Series D.
23. Y. Wang, C. Sun, J. Sun, W. Zhao, L. Dong, L. Li and F. Meng, Erosion behavior of arc sprayed FeTi/CrB MMC coating at elevated temperature. *Surf. Coating. Technol.*, 2015, **262**, 141–147.
24. H. Vasudev, L. Thakur, H. Singh and A. Bansal, Effect of addition of Al_2O_3 on the high-temperature solid particle erosion behaviour of HVOF sprayed Inconel-718 coatings. *Mater. Today Commun.*, 2022, **30**, 103017.
25. J. Vicenzi, D. L. Villanova, M. D. Lima, A. S. Takimi, C. M. Marques and C. P. Bergmann, HVOF-coatings against high temperature erosion (~ 300C) by coal fly ash in thermoelectric power plant. *Mater. Des.*, 2006, **27**(3), 236–242.
26. H. Zhang, X. Dong and S. Chen, Solid particle erosion-wear behaviour of Cr_3C_2–NiCr coating on Ni-based superalloy. *Adv. Mech. Eng.*, 2017, **9**(3), 1687814017694580.
27. S. K. Sharma, B. V. M. Kumar and Y. W. Kim, Effect of impingement angle and WC content on high temperature erosion behavior of SiC-WC composites. *Int. J. Refract. Metals Hard Mater.*, 2017, **68**, 166–171.
28. D. Alman and D. Marcio, Erosion-resistant Nanocoatings for Improved Energy Efficiency in Gas Turbines (No. NETL-PUB-911). National Energy Technology Lab (NETL), Pittsburgh, PA and Morgantown, WV (United States). In-house Research.
29. D. J. Branagan, M. Breitsameter, B. E. Meacham and V. Belashchenko, High performance nanoscale composite coatings for boiler applications. *J. Therm. Spray Technol.*, 2005, **14**, 196–204.
30. S. Dallaire, Hard arc-sprayed coating with enhanced erosion and abrasion wear resistance. *J. Therm. Spray Technol.*, 2001, **10**, 511–519.
31. J. S. Grewal, B. S. Sidhu and S. Prakash, High temperature erosion performance of nanostructured and conventional TiAlN coatings on AISI-304 boiler steel substrate. *Trans. Indian Inst. Met.*, 2014, **67**, 889–902.
32. B. Thiyagarajan and V. Senthilkumar, Experimental studies on fly-ash erosion behavior of Ni-Cr based nanostructured thermal spray coating in boiler tubes. *Mater. Manuf. Process.*, 2017, **32**(11), 1209–1217.
33. Z. Qu, X. Cheng, J. Wu, R. He, Y. Pei and D. Fang, An investigation on erosion behavior of nanostructured 7YSZ coatings at elevated temperature. *Surf. Coating. Technol.*, 2016, **299**, 129–134.
34. Y. Yin, J. Bao and L. Yang, Wear performance and its online monitoring of the semimetal brake lining for automobiles. *Ind. Lubr. Tribol.*, 2014, **66**, 100–105.
35. T. Wuest, D. Weimer, C. Irgens and K.-D. Thoben, Machine learning in manufacturing: advantages, challenges, and applications. *Prod. Manuf. Res.*, 2016, **4**, 23–45.
36. V. Nasteski, An overview of the supervised machine learning methods. *Horizons. B*, 2017, **4**, 51–62.
37. A. Olaode, G. Naghdy and C. Todd, Unsupervised classification of images: a review. *Int. J. Image Data Fusion*, 2014, **8**, 325–342.
38. G. Taguchi, *System of Experimental Design; Engineering Methods to Optimize Quality and Minimize Costs.* American Suppliers Institute, 1987, 653–655.
39. M. N. Islam and A. Pramanik, Comparison of design of experiments via traditional and taguchi method. *J. Adv. Manuf. Syst.*, 2016, **15**, 151–160.
40. C. S. Vui, G. K. Soon, C. K. On, R. Alfred and P. Anthony, A review of stock market prediction with Artificial neural network (ANN). *2013 IEEE International Conference on Control System, Computing and Engineering*, Penang, Malaysia, 2013, pp. 477–482.

41. E. E. Klaus, E. J. Tewksbury and M. R. Fenske, Preparation, properties, and some applications of super-refined mineral oils. *A S L E Trans.*, 1962, **5**, 115–125.
42. S. P. Jones, R. Jansen and R. L. Fusaro, Preliminary investigation of neural network techniques to predict tribological properties. *Tribol. Trans.*, 1997, **40**, 312–320.
43. S. Bhaumik and M. Kamaraj, Artificial neural network and multi-criterion decision making approach of designing a blend of biodegradable lubricants and investigating its tribological properties. *Proc. Inst. Mech. Eng. Pt. J. J. Eng. Tribol.*, 2021, **235**, 1575–1589.
44. C. Humelnicu, S. Ciortan and V. Amortila, Artificial neural network-based analysis of the tribological behavior of vegetable oil–diesel fuel mixtures. *Lubricants*, 2019, **7**, 32.
45. Y. Tikhamarine, A. Malik, D. Souag-Gamane and O. Kisi, Artificial intelligence models versus empirical equations for modelling monthly reference evapotranspiration. *Environ Sci Pollut Res.,* 2020, **27**, 30001–30019. https://doi.org/10.1007/s11356-020-08792-3
46. R. Rasouli, A. Barhoum and H. Uludag, A review of nanostructured surfaces and materials for dental implants: surface coating, patterning and functionalization for improved performance. *Biomater. Sci.*, 2018, **6**(6), 1312–1338.
47. T. E. Tallian, Tribological design decisions using computerized databases. *J. Tribol.*, 1986, **109**, 381–386. https://doi.org/10.1115/1.3261450
48. G. Binnig, C. F. Quate and C. Gerber, Atomic force microscope. *Phys. Rev. Lett.*, 1986, **56**, 930–933. https://doi.org/10.1103/PhysRevLett.56.930
49. P. Singh, A. Bansal, H. Vasudev and P. Singh, In situ surface modification of stainless steel with hydroxyapatite using microwave heating. *Surf. Topogr. Metrol. Prop.*, 2021, **9**(3), 035053.
50. K. Sunitha and H. Vasudev, A short note on the various thermal spray coating processes and effect of post-treatment on Ni-based coatings. *Mater. Today Proc.*, 2022, **50**, 1452–1457.
51. G. Prashar and H. Vasudev, Structure–property correlation of plasma-sprayed Inconel625-Al_2O_3 bimodal composite coatings for high-temperature oxidation protection. *J. Thermal Spray Technol.*, 2022, **31**(8), 2385–2408.
52. H. Vasudev, Wear characteristics of Ni-WC powder deposited by using a microwave route on mild steel: microwave cladding of Ni-WC. *Int. J. Surf. Eng. Interdiscip. Mater. Sci.*, 2020, **8**(1), 44–54.
53. B. G. R. Majji, H. Vasudev and A. Bansal, A review on the oxidation and wear behavior of the thermally sprayed high-entropy alloys. *Mater. Today Proc.*, 2022, **50**, 1447–1451.
54. G. Singh, H. Vasudev, A. Bansal and S. Vardhan, Influence of heat treatment on the microstructure and corrosion properties of the Inconel-625 clad deposited by microwave heating. *Surf. Topogr. Metrol. Prop.*, 2021, **9**(2), 025019.
55. G. Prashar and H. Vasudev, Surface topology analysis of plasma sprayed Inconel625-Al_2O_3 composite coating. *Mater. Today Proc.*, 2022, **50**, 607–611.
56. J. Parkash, H. S. Saggu and H. Vasudev, A short review on the performance of high velocity oxy-fuel coatings in boiler steel applications. *Mater. Today Proc.*, 2022, **50**, 1442–1446.
57. V. S. Yedida and H. Vasudev, A review on the development of thermal barrier coatings by using thermal spray techniques. *Mater. Today Proc.*, 2022, **50**, 1458–1464.
58. G. Prashar, H. Vasudev and L. Thakur, High-temperature oxidation and erosion resistance of Ni-based thermally-sprayed coatings used in power generation machinery: a review. *Surf. Rev. Lett.*, 2022, **29**(03), 2230003.
59. G. Prashar and H. Vasudev, Hot corrosion behavior of super alloys. *Mater. Today Proc.*, 2020, **26**, 1131–1135.
60. H. Vasudev, G. Prashar, L. Thakur and A. Bansal, Electrochemical corrosion behavior and microstructural characterization of HVOF sprayed inconel718-Al_2O_3 composite coatings. *Surf. Rev. Lett.*, 2022, **29**(02), 2250017.

61. G. Prashar and H. Vasudev, A review on the influence of process parameters and heat treatment on the corrosion performance of Ni-based thermal spray coatings. *Surf. Rev. Lett.*, 2022, **29**(01), 2230001.
62. H. Vasudev, P. Singh, L. Thakur and A. Bansal, Mechanical and microstructural characterization of microwave post processed Alloy-718 coating. *Mater. Res. Express*, 2020, **6**(12), 1265f5.
63. G. Prashar, H. Vasudev, L. Thakur and A. Bansal, Performance of thermally sprayed hydroxyapatite coatings for biomedical implants: a comprehensive review. *Surf. Rev. Lett.*, 2023, **30**(01), 2241001.

11 Fabrication of Components with Cold Spray Additive Manufacturing

Praveen Kumar Verma, Hitesh Vasudev, and Rajwinder Singh

11.1 INTRODUCTION

Additive manufacturing (AM), or 3D printing, has revolutionized the manufacturing landscape with its flexibility in creating complex geometries and customized components. Of all the techniques in AM, Cold Spray Additive Manufacturing (CSAM) is one of the most unique, with a solid-state process that is drastically different from conventional thermal-based methods. CSAM was invented in the late 1980s. However, in recent years, it has attracted much attention because of the deposition of a wide variety of materials with minimal thermal degradation, making it quite suitable for industries requiring high-performance and durable components [1].

CSAM is a process where fine powder particles are accelerated to supersonic speeds using a high-pressure gas stream and then directed onto a substrate. The particles adhere to both the substrate and each other due to plastic deformation at higher speeds. This is a different mechanism of solid-state bonding, which is in many ways different from other modes of traditional AM by considering melting and solidification phenomena for particle adhesion. This removes the high-temperature phase transition in CSAM that provides some advantages, such as retention of inherent feedstock material characteristics, lesser oxidation and thermal stresses, and the handling of temperature-sensitive materials for processing [2].

The most basic advantage of CSAM technology is that it interacts with a wide range of materials, including metals, polymers, and composite materials. The most common metals deposited are aluminum, copper, titanium, and their alloys. The ductility and impact resistance of these metals help facilitate the effective bonding of the particles. Polymers are less common but still successfully deposited by this method, thus widening its range of applications. Importantly, the capability of creating metal matrix composites (MMCs) and polymer matrix composites (PMCs) using CSAM introduces new ways in which material properties are designed by combining the beneficial attributes of constituents to improve their performance [3].

Low input heat is specifically advantageous when applying CSAM, especially because it is useful for retaining the material properties of layers that are deposited.

DOI: 10.1201/9781003542957-11

In traditional AM processes, the high temperatures required for melting lead to undesirable microstructural changes, such as grain growth and phase transformations, which can deteriorate the mechanical properties of the final product. CSAM avoids these issues by operating at lower temperatures, thereby preserving the microstructure of the feedstock material. This characteristic is critical in applications in aerospace, biomedical, and electronics industries, where material integrity is most critical. Although CSAM offers many advantages, some of the challenges it is subjected to must be overpowered to fully exploit the benefits. One of these issues is the need to very carefully prepare the substrate's surface so that it can strongly adhere to particles. Any contamination or irregularity on the substrate surface can impede the bonding process, thus causing defects and low-quality coatings. In addition, the process parameters, including gas pressure, temperature, and particle velocity, must be well controlled to optimize the quality of deposition. Any changes in these parameters may cause inconsistent coatings with varying thicknesses and densities [4,5]. Figure 11.1 shows the suggested framework for modeling artificial neural networks with data efficiency aimed at predicting a single-track profile in CSAM.

Brittle material being one more challenge; whenever deposition tends to get fractured, impact velocity can become substantial and highly challenging, which is related to CSAM based on the nature of the materials. Other advanced techniques that are evolved will be done in terms of equipment designing, and the further work related to CSAM will be easily facilitated. Other than that equipment design, large

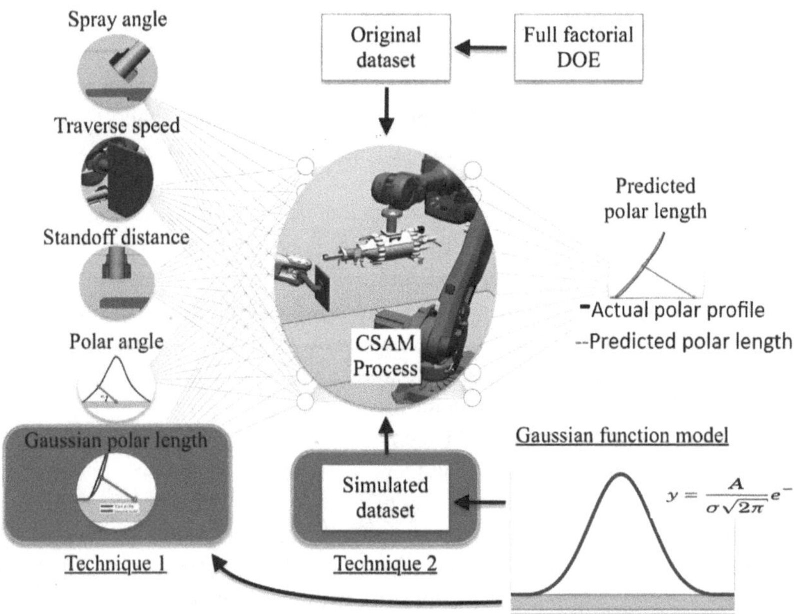

FIGURE 11.1 The proposed data-efficient artificial neural network modeling framework for the prediction of a single-track profile in cold spray additive manufacturing [6]. (CC By 4.0 Attribution.)

initial and operating costs that appear associated with CSAM equipment themselves become hindrances toward wider applications of CSAM. This technology still has to be improved further and more extensively developed in terms of cost-effectiveness and availability [7].

Applications of CSAM are very wide and increasingly diversified due to its unmatched capabilities. In the aerospace area, CSAM is used for the repair and production of high-performance components, including turbine blades and structural components, where the capability for durable coating deposition and the ability to build up the material rapidly is very attractive. In the automobile, CSAM is an input for the production of lightweight, strength-intensive components and the replacement of eroded parts, increasing service life. In biomedicine, CSAM is used to fabricate customized implants and prosthetics with tailored properties toward improved patient outcomes. Conductive and insulating layer depositions in electronics use CSAM to help enable miniaturization and integration [8].

The recent advancements in CSAM focus on the optimization of process parameters, the development of new material systems, and the integration of CSAM with other manufacturing techniques to leverage the strengths of their combined use. Computational modeling and simulation have also played a significant role in predicting and enhancing the outcomes of the process by providing valuable insights into the behavior of particles during deposition. Improved deposition quality, greater material compatibility, and expansion of application areas are some of the results of these efforts [9].

11.2 PRINCIPLES OF COLD SPRAY ADDITIVE MANUFACTURING

CSAM is a solid-state deposition process for building components that accelerates powdered feedstock materials to very high velocities through a high-velocity jet of gas, mostly helium and nitrogen. Such a technique is fundamentally different from the classical thermal-based techniques of AM because it depends on melting and solidification. The science behind CSAM is about acceleration dynamics, impacts, and bonding at temperatures below the material's melting point [10].

Particles of size from a few to tens of micrometers are injected into the gas stream at high velocities. Thus, the particles achieve speeds in the range of 300–1200 m/s, depending upon material and process parameters. To create the gas stream, a converging-diverging nozzle is used. Thereby, the expanding, high-pressure gas transfers some of its kinetic energy through the nozzle to accelerate particles to supersonic speeds. The high-speed particles deform plastically at the time of impact to the substrate. On impact, the particles get flattened and stuck onto the substrate without melting. In CSAM, mechanical interlocking is the dominant mechanism of bonding and metallurgical bonding. A large contact area is generated between the particle and substrate, and a lot of plastic deformation energy is taken from the kinetic energy. This deformation breaks up the surface oxides and contaminants, thereby allowing for intimate contact and bonding. The process, being solid-state in nature, preserves the inherent properties of the feedstock material, thus avoiding thermal degradation and oxidation [11]. Figure 11.2 shows the cold spraying gun.

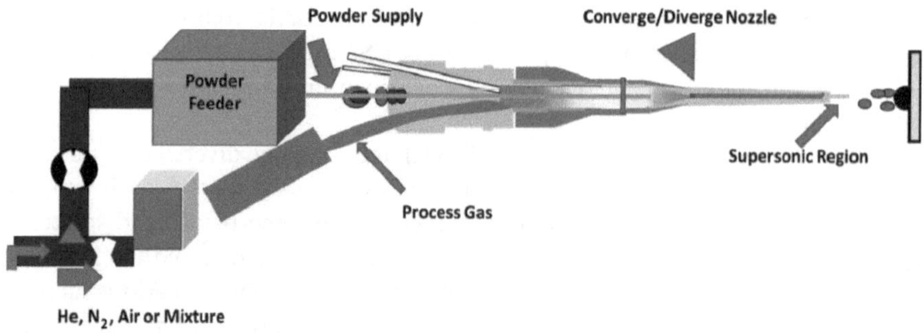

FIGURE 11.2 Schematic diagram of a cold spray gun [12]. (CC By 4.0 Attribution.)

Several critical parameters affect the efficiency and quality of the CSAM process, including gas pressure, gas temperature, particle velocity, and substrate condition. Both high gas pressure and temperature are necessary for achieving the desired velocities in the particles and adequate kinetic energy for bonding. The substrate surface must be prepared with appropriate care to remove contaminants and produce a suitable roughness that favors particle adhesion. The parameters should be optimized in such a way that particle rebound, insufficient bonding, and voids or cracks should not occur [13]. CSAM can be used with a wide range of materials, from metals to polymers to composites. Metals like aluminum, copper, and titanium are commonly used because they are ductile and impact-resistant, which ensures that the particles bond effectively. The process can deposit temperature-sensitive materials and create MMCs by co-depositing various types of particles, allowing the creation of materials with properties tailored to specific applications [14].

11.3 MATERIALS USED IN CSAM

CSAM is a versatile technology that accommodates a wide range of materials, each with distinct properties that make them suitable for various applications. The primary categories of materials used for CSAM are metals, polymers, and composites. Each category holds unique advantages and contributes toward the growing capabilities of CSAM in manufacturing high-performance components.

11.3.1 Metals

Metals are the most widely used materials in CSAM compared to other forms. These metals exhibit high ductility, impact resistance, and superior mechanical strength inherent within their properties. Alloys of aluminum, copper, titanium, and suchlike are the primary types of metals used in the composition of CSAM. Aluminum and its alloys are preferred because of their lightweight nature, excellent corrosion resistance, and high thermal and electrical conductivity. These properties make aluminum suitable for applications in the aerospace, automotive, and electronics industries. In CSAM, aluminum powders can be efficiently deposited to create coatings, repair worn-out parts, and fabricate complex geometries without compromising material integrity [15].

Copper exhibits excellent electrical and thermal conductivities, making it suitable for use in electrical and thermal management systems. CSAM also employs the use of copper powders in the manufacture of conductive tracks, thermal interface materials, and those components with superior heat dissipation qualities. The feature of not oxidizing copper and thermic degradation during CSAM deposition can ensure quality coatings as well as component quality. Titanium and its alloys are valued for their strength-to-weight ratio, superior corrosion resistance, and biocompatibility. Such characteristics make titanium ideal for aerospace, biomedical, and marine industries. CSAM allows titanium to be deposited with minimum thermal influence on its microstructure and mechanical properties. This is very important in the manufacture of aerospace parts, medical implants, and marine structures, where high performance and durability are demanded [16].

11.3.2 Polymers

Polymers are used much less often in CSAM than metals but offer specific benefits for specific applications. Deposition of polymers by CSAM could unlock new possibilities for lightweight, corrosion-resistant, and electrically insulating components. Some polymers such as polyether ether ketone (PEEK) and polytetrafluoroethylene (PTFE) can be deposited using CSAM. PEEK shows excellent mechanical strength, high thermal stability, and excellent chemical resistance, which makes it suitable for applications in aerospace, automotive, and biomedical fields. However, PTFE displays excellent chemical resistance and very low friction properties. This makes it suitable for some applications where non-stick surfaces and lubricious surfaces are required [17]. Polymer deposition with CSAM is helpful in complex geometries and custom components where certain material properties are desired. Polymers are used for insulating coatings, lightweight structural parts, and chemical environments where severe chemical exposures exist. Due to the solid-state nature of CSAM, the inherent properties of polymers are preserved during deposition, and therefore, components deposited by CSAM exhibit good quality and long durability [18].

11.3.3 Composites

Composites are an area of great interest in CSAM because they may hold the properties of different materials, thus resulting in improvement in mechanical, thermal, and electrical performance. Two types of composites most commonly used in CSAM are MM Composites and PM Composites. MMCs consist of a metal matrix with dispersed ceramic or metallic particles, fibers, or whiskers. In CSAM, MMCs are produced by co-deposition of metal powders and reinforcement material. The composite thus produced is much harder and more wear-resistant as well as thermally stable than the base metal. MMCs are applied wherever strength and durability are needed. Applications include aerospace components, automobile parts, and wear-resistant coatings [19].

PMCs are polymers filled with fibers or particles to enhance their mechanical properties. In CSAM, PMCs are deposited by mixing polymer powders with reinforcement materials. The final composites possess lightweight, corrosion-resistant, and high-strength properties, thus making them applicable in aerospace, automotive, and marine applications. Tailoring the properties of PMCs through CSAM allows the creation of customized components that can be made to perform under specific characteristics [20].

11.4 ADVANTAGES OF CSAM

CSAM is a technology that will offer quite distinctive advantages over traditional manufacturing and most other types of AM. These advantages are based directly on the novel nature of the cold spray process itself. This includes the operation based on solid-state processes with no melting and, therefore, no thermal input, coupled with very high deposition rates [21].

11.4.1 Low Thermal Input

The greatest strength of CSAM is low thermal input. Indeed, one of the differences between conventional AM processes like laser sintering or electron beam melting and CSAM is that CSAM does not melt the feedstock material at all. Instead, it utilizes a high-velocity gas stream to accelerate particles in an attempt to deform and then bond them to the substrate upon impact. The advantage of such a process in solid states is that it ensures no temperature rises beyond the melting points of the materials used. For this reason, CSAM eliminates the high-temperature decomposition, oxidation, and destructive phase transitions characteristic of these processes. Therefore, properties of the material can be preserved under applications wherein retention of intrinsic properties, such as, ductility, toughness, and resistance to corrosion, should be sustained [22].

The high deposition rates in this process drastically enhance manufacturing efficiency. That way, the material can pile up very fast, resulting in the manufacture of fairly large components within a remarkably short time. The resulting buildup or coating turns out to be dense with the impact of the velocity of the particles, and further reductions in multiple passes also cut down the production time. This high throughput capability makes it particularly advantageous in those industries where time-to-market is critical, such as aerospace, automotive, and defense. The ability to rapidly produce or repair components with very minimal post-processing requirements equates to lower production costs and faster turnaround times [23]. Figure 11.3 shows the selective laser sintering process.

11.4.2 Material Versatility

The other notable advantage of CSAM is its material versatility. The process can accommodate a wide range of materials, including metals, polymers, and composites. Aluminum, copper, and titanium are the metals most in use due to their great ductility and impact resistance. Polymers, though less common, are also processed for particular uses, including the advantages of being lightweight and corrosion-resistant. CSAM is also quite notable in the production of MMCs and PMCs. Different materials are combined for better mechanical properties. This versatility

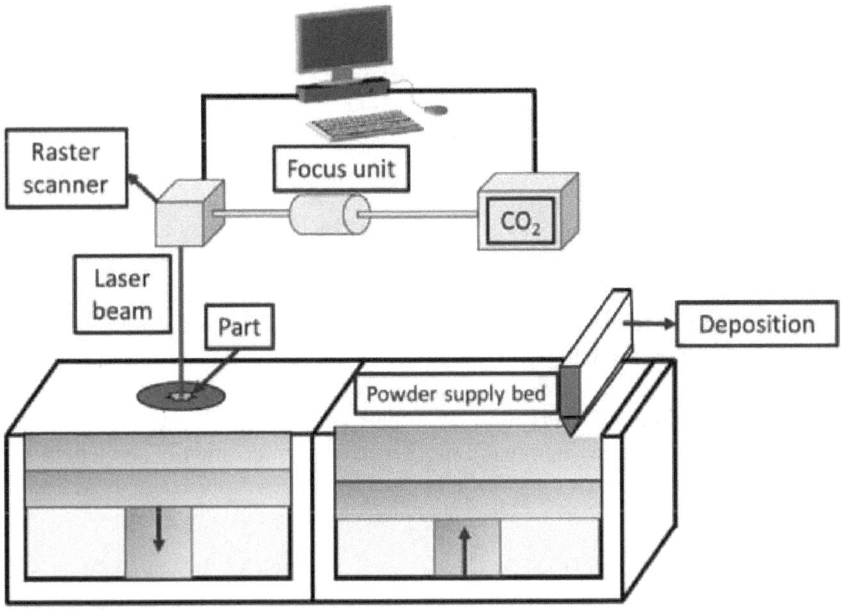

FIGURE 11.3 Schematic diagram illustrating the selective laser sintering process [24]. (CC By 4.0 Attribution.)

helps tailor material properties to needs in specific applications, so CSAM is a highly useful tool in various industries [25].

11.4.3 Repairing Abilities

CSAM is highly effective for repairing damaged parts. It extends the service life of parts and saves money by not replacing them. Low input of thermal energy and very high precision make it appropriate for repairing expensive components, which are difficult or too costly to replace. For instance, in aerospace, it can be used to weld turbine blades, landing gears, and other high-reliability components with complete retention of structural integrity without compromising their structural integrity. Material is added where required, and the repairs will be efficient and cost-effective. Furthermore, due to its solid-state nature, the repaired areas retain their original material properties and do not degrade in performance, thus ensuring that the component continues to perform well even in service [26].

11.5 CHALLENGES OF COLD SPRAY ADDITIVE MANUFACTURING

11.5.1 Surface Preparation

The first problem with CSAM is the surface preparation of the substrate. The adhesion strength of the deposited particles is very sensitive to the condition of the substrate surface. Poor preparation of the surface may lead to weak and unreliable coatings or structures due to poor bonding. Surface preparation generally involves cleaning to remove contaminants, roughening to increase surface area, and sometimes preheating to optimize particle adhesion. Poor surface preparation may severely degrade the quality and performance of the final product [27].

11.5.2 Process Control

In CSAM, process parameters must be controlled with accuracy to optimize deposition quality. The critical parameters are gas temperature and pressure, particle velocity, stand-off distance, and substrate temperature. Small differences in these parameters lead to failures such as inadequate bonding, high porosity, and undesirable microstructural changes. This balance can be reached only by developing deep insights into the interaction of the particles with the substrate. These parameters require continuous observation and adjustment to achieve the desired quality and uniformity of the deposited material; this process is technically challenging and time-consuming [28].

11.5.3 Material Limitations

Material limitations are the other limitations of CSAM. Even though the process is very efficient for ductile metals such as aluminum, copper, and their alloys, it is not that apt for brittle materials. Several ceramics and certain high-strength alloys break or fragment upon impact of a high velocity and cannot, therefore, be easily processed through CSAM. This limitation limits the range of materials that can be used effectively, which could be a major disadvantage when it comes to applications where unique material properties are desired, which cannot be achieved through more ductile materials [29].

11.5.4 Equipment Costs

High acquisition and operating costs make this type of CSAM equipment difficult to deploy on a larger scale. This expensive equipment can be required in the CSAM to produce high-pressure gas systems, sophisticated nozzles, and precise control systems, which are costly to purchase and maintain. Additional costs include the operating costs involved in the consumption of high-pressure gas and the wearing and tearing of equipment parts. These financial factors will discourage small to medium-sized enterprises from adopting CSAM, and it would only be used by large companies or in high-value applications where the value gained is more than the costs incurred [30].

11.6 APPLICATIONS OF CSAM

CSAM has been established as one of the most versatile technologies, finding applications across different fields and industries by exploiting the unique capabilities it offers toward enhancing manufacturing and repair processes.

CSAM has been widely used in the aerospace industry to repair and fabricate high-performance components. It has the ability to deposit material at low temperatures so that the properties of the substrate are kept intact. It is thereby suitable for the repair of critical parts like turbine blades, landing gear, and structural parts. Aerospace parts require a high strength combined with the capability of resistance in extreme conditions. CSAM allows for the usage of advanced materials like titanium and nickel alloys without thermal degradation. Furthermore, CSAM allows for the creation of entirely new parts with complex geometries that would enhance the efficiency and performance of aerospace systems [31].

CSAM can benefit the automotive industry as lightweight, high-strength components can be produced through it. Aluminum and magnesium alloys may be used in large extents for this purpose to reduce the total weight of automobiles. In turn, the efficiency of the vehicle and its overall performance improve. Besides effective repair of worn or damaged parts, such as engine blocks and transmission components, CSAM can simply restore parts to their original dimensions and properties, thus elongating service life while avoiding expensive replacements. This is extremely useful in competitive racing applications and high-performance automobiles, where component reliability and performance are of prime importance [32].

Biomedical sciences will be the primary gainers since CSAM can synthesize implants and prosthetic systems with desired properties. Medical implants may require intricate and well-controlled complexity of high precision, which in turn demands biocompatible materials like titanium alloys with good mechanical properties for such fine features to be processed effectively. This leads to implants that are better fitted and function properly in the human body. Additionally, low thermal input ensures that materials remain biocompatible and have mechanical integrity, both of which are very crucial in medical applications [33].

In electronics, CSAM is utilized for the deposition of conducting and insulating layers in electronic devices. The process can deposit thin, uniform coatings of materials like copper, silver, and other conductive metals, which are important for creating electrical pathways in devices such as printed circuit boards and sensors. CSAM's ability to work at low temperatures is particularly beneficial for electronics, as it prevents damage to sensitive components and substrates that could occur with traditional thermal processes. This capability allows the creation of more efficient and reliable electronic devices with enhanced performance characteristics [34].

11.7 RECENT ADVANCEMENTS IN CSAM

CSAM is an emerging technology, with current research concentrated on several fronts aimed at pushing the state of the art and extending the application areas. The main focus has been on process optimization, materials development, hybrid processes, and modeling and simulation. Process optimization is one of the new research areas for CSAM. Advanced control systems would allow the better adjustment of process parameters concerning a better quality of depositions. Methods of gas pressure, temperature, particle velocity, and stand-off distance for uniform and better-quality coatings are researched in this area. Sensor technology innovations with real-time monitoring systems are employed for the spraying process. Feedback is available on the instant basis of spraying, ensuring the parameters that are necessary for low-defect formation, high-strength bonding, and deposition of uniform material. CSAM produces high-quality products under advanced process control, making the whole process efficient and repeatable [35].

Material development has been another concern of prime focus in research in the CSAM field in recent times. Scientists are studying new material systems, which can be processed efficiently using CSAM, namely, functionally graded materials and nanocomposites. Functionally graded materials are the type of material that contains gradual variations in composition and properties, which are used to enhance performance characteristics such as thermal resistance and mechanical strength. The goal in developing nanocomposites, which contain nanoparticles in a matrix material, is to enhance properties such as hardness, wear resistance, and thermal conductivity. In this area of research, it is increasingly making more material options fit into CSAM, thereby enabling components with desired properties suitable for various applications [36].

Hybridizing CSAM with another process has also gained significant focus and momentum as hybrid processes, and their main aim includes exploiting the capabilities of combining CSAM with machining techniques or other cladding-based laser processing methods, which also incorporate the capabilities from the family of AM. It incorporates a selective laser melting process, among other technologies. For example, in a hybrid process, the rapid building up of material could be done by CSAM followed by precision machining for achieving the final geometry and surface finish. This process may lead to a reduction in the manufacturing time and cost but with an improvement in the overall quality and performance of the finished product. Current research is focused on the development of seamless integration methods as well as the interactions of different processes for the optimization of hybrid manufacturing workflows [37].

Modeling and simulation play an increasingly important role in the development of CSAM. The computational models predict and optimize process outcomes that reduce the need for lengthy experimental trials. These models provide insights into the complex interactions between particles and substrates, the effects of various process parameters on material properties, and microstructures that develop during deposition. Advanced simulation tools are now under development that can simulate the thermal and mechanical behavior of materials for CSAM and can be used for the design and optimization. These models can thus significantly help improve the efficiency and reliability of CSAM, leading to better-quality control and novel applications [38].

11.8 FUTURE DIRECTIONS

CSAM already has much potential in diverse industries, and yet unlocking its full strength and more applications will require overcoming many challenges. These encompass improving compatibility with materials, scaling of the processes involved, environmental impact, and advancing applications. One of the largest opportunities for CSAM lies in the ability to significantly widen the range of material it can process to full economic efficiency. Where the material has proven compatibility with CSAM, it is an excellent material – think aluminum, copper, and these alloys. Many more ceramics, or any such hard-to-deposit-high-strength material, present a challenge. Future developments, one hopes, will improve this compatibility. This may include optimizing particle size and shape, modification of substrate surfaces, or new bonding mechanisms. Hybrid material systems where brittle materials are combined with ductile matrices could open the possibility of depositing substances previously incompatible. An increase in material compatibility would allow CSAM to be applied to a greater range of industrial and technological challenges [39].

Scalability is another important area for future CSAM research. The process should be developed to be scalable both with respect to the volume and scale of components to be capable of meeting industrial applications. Current CSAM equipment as well as the costs incurred in operations can be inhibitive, especially for small and medium enterprises. Research should aim to generate more cost-effective systems while maintaining high precision and quality. This can be addressed through nozzle design improvement and gas dynamics as well as by automation to streamline the entire process and cut costs in the process. It also needs to scale the process without sacrificing the material properties and integrity of the components. The victories over these challenges will make CSAM possible and practically feasible for large-scale industrial use [40].

Understanding the environmental effects that can be associated with CSAM is critical for developing responsible, sustainable manufacturing practices. While CSAM has some environmental advantages since it consumes less energy than traditional thermal processes and reduces material waste, much more work remains to be done. Quantification of the environmental footprint of CSAM in terms of energy consumption, emissions, and material usage through the entire lifecycle of the process is still a need for future research. The three main strategies for reducing waste, recycling materials, and energy consumption are waste minimization, recycling of materials, and energy consumption reduction. There is also a need to study using environment-friendly materials and processes for better sustainability of CSAM. Thus, it will achieve global targets of sustainability and become an even more preferred material for green industries [41].

There are also a lot of opportunities for advanced applications of CSAM. Beyond the aerospace, automotive, biomedical, and electronics industries, CSAM will have an important role in emerging areas such as energy storage and advanced electronics. It could be used to make battery components, supercapacitors, and fuel cells where material properties and microstructures need to be well controlled for performance. Advanced electronics will allow for complex multi-material structures for next-generation devices. This exploration of all these new application areas demands interdisciplinary research and collaboration involving the specific benefits of CSAM in addressing particular technological challenges [42]. CSAM is an effective technique for fabricating components with excellent mechanical properties and minimal thermal degradation [43–47]. This method uses high-velocity particle deposition at low temperatures, preserving material integrity and preventing oxidation [48–53]. Studies have demonstrated its capability to produce dense and adherent coatings and restore components exposed to wear and corrosion [54–58].

11.9 CONCLUSION

CSAM represents a giant leap forward in AM, with characteristics such as low thermal input, high deposition rates, and material versatility. It operates at relatively low temperatures in contrast to traditional thermal-based additive methods, which allow it to preserve intrinsic material properties without thermal degradation or oxidation. The solid-state process accelerates the rapid buildup of materials, thereby enhancing the efficiency of manufacturing and making it possible to process materials that vary from metals, polymers, and composites. Despite its advantages, CSAM has some challenges. These include strong particle adhesion, optimizing process parameters, and managing high equipment costs. However, research is continuing to address these issues through advancements in process control, novel materials development, and the integration of CSAM with other manufacturing techniques. Constant improvement in CSAM technology is widening its industrial applications from aerospace and automotive to biomedical and electronics [43].

REFERENCES

1. Vary A, Scott J, Mahmood N. A review of cold spray additive manufacturing: Fundamentals, techniques, and applications. *J Mater Sci.* 2020;55(12):5089–5103.
2. Nascimento G, Lima T, Santos M, et al. Cold spray additive manufacturing: A promising alternative for high-performance metal coatings. *Surf Coat Technol.* 2021;419:127332.
3. Moridi A, Chaharmahali R, Marzban A, et al. Cold spray additive manufacturing of aluminum alloys: Process optimization and mechanical properties. *Mater Sci Eng A.* 2020;794:139856.
4. Balamurugan K, Sendhil Kumar R. Review on cold spray additive manufacturing and its applications in aerospace industry. *Mater Today Proc.* 2021;44:204–209.
5. Wang Z, Zhang C, Li Y, et al. Advances in cold spray technology for additive manufacturing: A review. *J Manuf Process.* 2022;73:256–270.
6. Ikeuchi D, Vargas-Uscategui A, Wu X, King PC. Data-efficient neural network for track profile modelling in cold spray additive manufacturing. *Appl Sci.* 2021 Feb 12;11(4):1654.
7. Verma PK, Singh S, Kapoor M, Singh S. A review on the surface topography and corrosion behavior of Mg-alloy coatings for biomedical implants. *Res Surf Interfaces.* 2024 May 9:100227.
8. Benyounis KY, Hahn W, Vasiliev A, et al. Cold spray additive manufacturing: A review of its applications and material systems. *J Mater Process Technol.* 2022;295:117183.
9. Kim J, Lee J, Park Y, et al. Recent advancements in cold spray technology: Enhancements in process parameters and material systems. *Adv Mater Interfaces.* 2021;8(14):2100081.
10. Faghihi S, Sadat D, Gholamzadeh M. Mechanics of particle deposition in cold spray additive manufacturing: A review. *J Manuf Process.* 2020;54:235–249.
11. Gupta A, Chennakeshavulu R, Sahu A, et al. Review on solid-state bonding mechanisms in cold spray additive manufacturing. *Mater Sci Eng A.* 2023;844:143173.
12. Pathak S, Saha GC. Development of sustainable cold spray coatings and 3D additive manufacturing components for repair/manufacturing applications: A critical review. *Coatings.* 2017 Aug 14;7(8):122.
13. Zhanga Z, Dingb J, Qu X, et al. Effects of process parameters on the deposition efficiency of cold spray additive manufacturing. *Surf Coat Technol.* 2022;428:127951.
14. Shokrieh MM, Fadaei H, Mohammadi M, et al. Fabrication of metal matrix composites using cold spray additive manufacturing: A review. *J Alloys Compd.* 2023;934:167531.

15. Kang C, Feng J, Liu Z, et al. Cold spray additive manufacturing of aluminum coatings for aerospace applications: A review. *J Mater Sci.* 2023;58(4):2365–2381.
16. Zhu X, Wei Q, Yang J, et al. Applications of copper coatings in cold spray additive manufacturing. *Surf Coat Technol.* 2024;425:127885.
17. Fathi M, Afshari M, Naderi G, et al. Properties and applications of PEEK and PTFE in additive manufacturing. *Polymers.* 2022;14(3):576.
18. Ghasemi S, Mohammadi S, Rabiee F, et al. The role of polymers in cold spray technology: Opportunities and challenges. *J Mater Sci.* 2023;58(5):2390–2405.
19. Tan Y, Zhang Y, Wang J, et al. Metal matrix composites produced by cold spray: A review of the current state. *J Compos Mater.* 2023;57(2):255–274.
20. Wang C, Chen L, Zhang H, et al. Enhancing the properties of polymer matrix composites through cold spray additive manufacturing. *Compos Part A.* 2023;142:106253.
21. Prakash G, Singh K, Gupta R, et al. Advances in cold spray additive manufacturing: A review. *J Manuf Process.* 2022;79:201–215.
22. Roberts C, Khosrovi A, Patel D, et al. Thermal behavior of materials in cold spray processes. *Materials.* 2023;16(2):362.
23. Lee S, Park J, Kim H, et al. Economic advantages of cold spray additive manufacturing in aerospace applications. *J Manuf Sci Eng.* 2024;146(6):061002.
24. Mohanty S, Gokuldoss Prashanth K. Metallic coatings through additive manufacturing: A review. *Materials.* 2023 Mar 14;16(6):2325.
25. Zhang Z, Li X, Yang J, et al. Advantages of cold spray additive manufacturing in producing complex geometries. *J Manuf Process.* 2023;79:184–196.
26. Kumar P, Banerjee D, Bhattacharya A, et al. Repair techniques in aerospace components using cold spray technology. *Aerosp Sci Technol.* 2023;128:106215.
27. Bellows A, Johnson S, Martin T. Surface preparation techniques for additive manufacturing: Importance and methods. *J Mater Sci.* 2024;59(3):1500–1512.
28. Lee J, Park C, Shin D, et al. Process control in cold spray additive manufacturing: Challenges and solutions. *Addit Manuf.* 2024;39:101837.
29. Li Y, Wei Z, Zhao W, et al. Limitations of cold spray additive manufacturing for brittle materials. *Compos Part A.* 2023;147:106128.
30. Harper N, Nelson R, Kumar A. Economic assessment of cold spray technology: Challenges and opportunities. *J Manuf Sci Eng.* 2024;146(4):041004.
31. Yang H, Wang F, Li M, et al. Applications of CSAM in the aerospace industry: Current status and future prospects. *J Aerosp Eng.* 2023;36(1):04022042.
32. Smith T, Brown K, Johnson L. Cold spray technology in automotive applications: A review. *J Mater Process Technol.* 2024;313:117920.
33. Green M, Anderson J, Lewis S. Cold spray technology in biomedical applications: Advances and challenges. *Biomed Mater.* 2023;18(2):025005.
34. Choi H, Kim Y, Shin J. Cold spray for electronic devices: Coating techniques and applications. *J Electron Mater.* 2023;52(7):1564–1571.
35. Patel R, Kumar V, Nair A, et al. Recent advancements in cold spray additive manufacturing. *J Manuf Process.* 2024;82:48–64.
36. Thompson G, McDonald A, Rogers D. Material developments in cold spray additive manufacturing. *Compos Part B.* 2023;229:109565.
37. Baker R, Lewis T, Chen Y. Hybrid manufacturing: Integrating cold spray with traditional processes. *J Manuf Sci Eng.* 2024;146(5):051006.
38. Zhao J, Li Q, Yang K, et al. Simulation models for cold spray additive manufacturing: A review. *Addit Manuf.* 2024;40:101873.
39. Martin P, White R, Jones E. Future directions for cold spray additive manufacturing: Challenges and opportunities. *Mater Today Adv.* 2023;12:100203.
40. Roberts C, Smith T, Johnson L. Scalability challenges in cold spray technology: A review. *J Manuf Process.* 2024;82:300–311.

41. Kim S, Park J, Lee C. Environmental impacts of cold spray technology and sustainability practices. *J Clean Prod.* 2023;369:132922.
42. Lewis B, Chen A, Walker R. Emerging applications of cold spray technology: A future perspective. *J Mater Sci.* 2024;59(4):1624–1636.
43. Nelson T, Rodriguez M, Kim S. Cold spray additive manufacturing: A new era in materials processing. *J Mater Process Technol.* 2024;314:117887.
44. Prashar G, Vasudev H. A review on the influence of process parameters and heat treatment on the corrosion performance of Ni-based thermal spray coatings. *Surf Rev Lett.* 2022;29(01):2230001.
45. Majji BGR, Vasudev H, Bansal A. A review on the oxidation and wear behavior of the thermally sprayed high-entropy alloys. *Mater Today Proc.* 2022;50:1447–1451.
46. Parkash J, Saggu H, Vasudev H. A short review on the performance of high velocity oxy-fuel coatings in boiler steel applications. *Mater Today Proc.* 2022;50:1442–1446.
47. Singh P, Bansal A, Vasudev H, Singh P. In situ surface modification of stainless steel with hydroxyapatite using microwave heating. *Surf Topogr Metrol Prop.* 2021;9(3):035053.
48. Yedida VS, Vasudev H. A review on the development of thermal barrier coatings by using thermal spray techniques. *Mater Today Proc.* 2022;50:1458–1464.
49. Prashar G, Vasudev H, Thakur L. High-temperature oxidation and erosion resistance of Ni-based thermally-sprayed coatings used in power generation machinery: A review. *Surf Rev Lett.* 2022;29(03):2230003.
50. Vasudev H, Singh P, Thakur L, Bansal A. Mechanical and microstructural characterization of microwave post processed Alloy-718 coating. *Mater Res Express.* 2020;6(12):1265f5.
51. Singh G, Vasudev H, Bansal A, Vardhan S. Influence of heat treatment on the microstructure and corrosion properties of the Inconel-625 clad deposited by microwave heating. *Surf Topogr Metrol Prop.* 2021;9(2), 025019.
52. Prashar G, Vasudev H. Surface topology analysis of plasma sprayed Inconel625-A12O3 composite coating. *Mater Today Proc.* 2022;50:607–611.
53. Sunitha K, Vasudev H. A short note on the various thermal spray coating processes and effect of post-treatment on Ni-based coatings. *Mater Today Proc.* 2022;50:1452–1457.
54. Vasudev H. Wear characteristics of Ni-WC powder deposited by using a microwave route on mild steel: Microwave cladding of Ni-WC. *Int J Surf Eng Interdis Mater Sci (IJSEIMS).* 2020;8(1):44–54.
55. Prashar G, Vasudev H. Hot corrosion behavior of super alloys. *Mater Today Proc.* 2020; 26:1131–1135.
56. Prashar G, Vasudev H. Structure–property correlation of plasma-sprayed Inconel625-Al_2O_3 bimodal composite coatings for high-temperature oxidation protection. *J Therm Spray Technol.* 2022;31(8):2385–2408.
57. Vasudev H, Prashar G, Thakur L, Bansal A. Electrochemical corrosion behavior and microstructural characterization of HVOF sprayed inconel718-Al_2O_3 composite coatings. *Surf Rev Lett.* 2022;29(02):2250017.
58. Prashar G, Vasudev H, Thakur L, Bansal A. Performance of thermally sprayed hydroxyapatite coatings for biomedical implants: A comprehensive review. *Surf Rev Lett.* 2023;30(01):2241001.

12 Bimodal/Nano-Structured Composite Coatings for Protection against Wear and Corrosion

Praveen Kumar Verma and Hitesh Vasudev

12.1 INTRODUCTION

Wear and surface degradation are one of the major problems affecting most industrial applications, indicating considerable economic losses, operational inefficiencies, and even safety risks. Mechatronic components and tools as well as surface coatings are exposed to severe environments that accelerate wear and corrosion and thus adversely affect the length of serviceable time and functionality of the equipment. An evident example here is aerospace components, which suffer extreme mechanical stresses, high temperatures, and corrosive atmospheres. Manufacturing tools suffer repeated impact and abrasive conditions. These stressors require coatings to make the underlying materials resistant to wear and deterioration. Unfortunately, conventional coatings often fail to provide the required amount of protection under such demanding conditions [1].

Considering the intensification of operating conditions and the evolution of industries, surface protection technologies in innovative manners have currently gained much demand. This has accelerated meaningful surface engineering developments, mainly focused on bimodal and nano-structured composite coatings to engender more impressive functionality based on an engineered microstructure facilitating enhanced wear resistance and mechanical properties [2]. Unlike conventional coatings, bimodal and nano-structured composite coatings contain multiple phases. Bimodal and nano-structured composite coatings are engineered materials that combine different phases or structures at multiple length scales, typically including nanosized grains in a matrix of larger grains or incorporating hard nanoparticles in a softer matrix [3]. The hierarchical microstructure of these coatings takes advantage of the unique properties of the individual phase and provides a striking balance between hardness, toughness, and wear resistance. This combination makes bimodal and nano-structured composite coatings highly attractive for industrial applications ranging from aerospace and automotive to manufacturing and biomedical sectors [4]. Bimodal and nano-structured composite coatings are used in a vast range of applications. The wear resistance and thermal stability properties are improved in aerospace parts where it is exposed to rigorous conditions, thereby increasing the

DOI: 10.1201/9781003542957-12

lifespan of critical parts. Likewise, in the automotive field, these coatings can be used for engine parts to avoid wear and tear and enhance their performance to improve fuel efficiency and lower emissions. The manufacturing sector benefits through the longer lifespan of cutting tools and machinery, while the biomedical field can use these advanced coatings for implants and prosthetics where aspects of biocompatibility and wear resistance are of prime importance [5].

Furthermore, bimodal and nano-structured composite coatings may also have important implications for sustainability and resource efficiency. With an extended component life and a decrease in replacement cycles, material consumption goes down, accompanied by less waste production. This fits the growing international interest in green practices and resource saving, so these coatings become economically advantageous as well as environmentally responsible. The hierarchical microstructure of bimodal and nano-structured composite coatings has significance in their functionality. The combination of different phases enables the exploitation of properties unique to each component. The bimodal and nano-structured composite coatings have come to symbolize tremendous progress in surface engineering, which thereby provides a valid alternative to the prevalent limitations of standard coatings. Such coatings represent an exciting combination of increased wear resistance, improved mechanical properties, and applicability in diversified industries due to their special microstructural characteristics [6]. Figure 12.1 shows the influencing parameters of nano-structured coating formation.

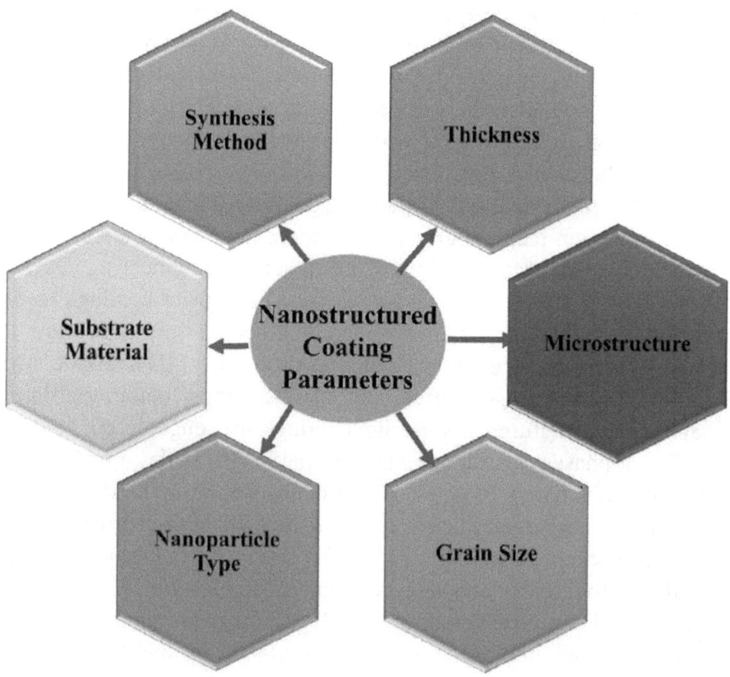

FIGURE 12.1 Various parameters influence nano-structured coating formation [7]. (CC By 4.0 Attribution.)

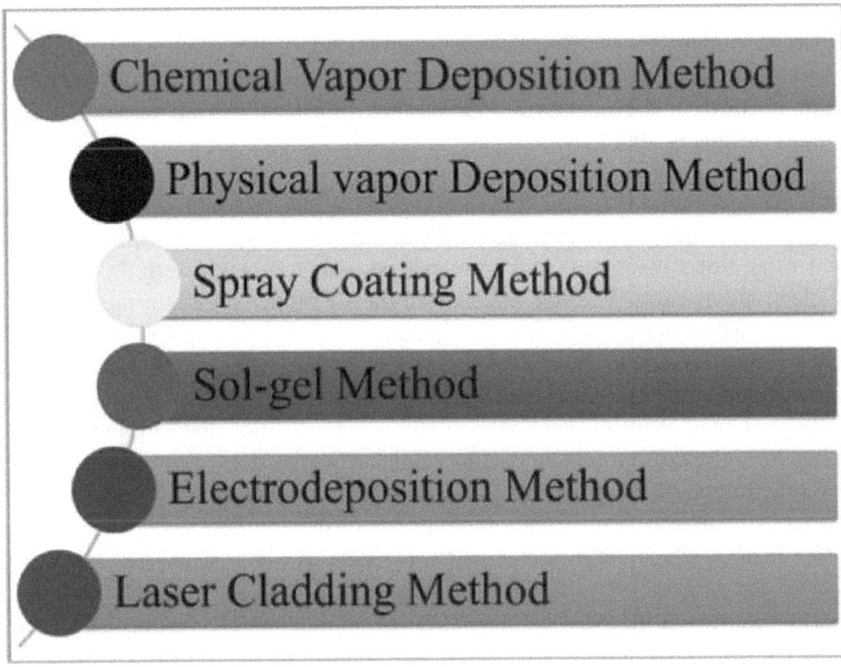

FIGURE 12.2 Different methods to produce nano-structured coatings [7]. (CC By 4.0 Attribution.)

12.2 SYNTHESIS METHODS

Main applications of high-performance materials in industries have promoted research on advanced surface coatings, mainly in bimodal and nano-structured composite architectures. These coatings offer superior wear protection relative to conventional coatings. Their engineered microstructures take advantage of the intrinsic properties of their constituents to optimize high-performance capabilities in challenging conditions. This report discusses the four main synthesis methods used for these coatings: Electrodeposition, Thermal Spray, Chemical Vapor Deposition (CVD), Laser cladding, and Sol-Gel Processing [8]. Figure 12.2 shows the various methods to produce nano-structured coatings.

12.2.1 Electrodeposition

Electrodeposition is the most widely used route toward bimodal and nano-structured composite coatings because of its economy, versatility, and aptitude to deposit coatings on complex geometries. This technique essentially comprises the electrochemical reduction of metal ions from an electrolytic solution onto the conductive substrate, thereby depositing a layer of material [9,10]. The method extensively offers tight control over coating composition and structure. The nano-crystalline matrix doped with micron-sized particles in the bimodal coatings often includes improvements in both hardness and toughness. The main process of formation begins with dispersion in the electrolyte of particles that are nano- to micro-sized, thereby preventing inhomogeneity in distribution, which might lead to non-uniform particle dispersal within the solution. They could be ceramics, carbon-based materials, or metal nanoparticles. In many instances, stabilization of dispersions through the use of stabilizing agents or surfactants is added. After dispersion, the particles must migrate to the electrode surface. Convection, diffusion, and migration in an electric field where the kinetics of the particles are known to significantly affect the rate and uniformity of deposition further down along with this process tend to govern it. Once in the vicinity of the electrode, particles become attached to the electrode surface in the process of forming attachment sites. The quality and adhesion of the coating will depend on the nature of this interaction. At this stage, the electrostatic forces and the van der Waals interactions play a significant role [11].

The current of electrodeposition then causes the collection and entrapment of particles within the emerging metallic film on the substrate. The resulting coatings can have fine microstructures with engineered hardness and toughness, which can significantly enhance wear resistance and other mechanical properties.

Coatings in which ceramic nanoparticles, such as alumina (Al_2O_3) and silicon carbide, and metals like nickel and copper are added to the mixture exhibit considerably improved mechanical properties. In addition to increasing the hardness of the matrix, this addition of nanoparticles enhances the resistance to wear with the strength of the constituent materials being higher. For example, silicon carbide nanoparticle incorporation has resulted in large increases in hardness and wear life with nickel electrodes, proving an advantage in industrial applications requiring high durability under friction [12].

12.2.2 Thermal Spray

Thermal spraying is another highly established technique for the deposition of bimodal and nano-structured coatings that are apt for applications in harsh environments. The technique encompasses a variety of processes involving acceleration of the material to a high velocity after heating followed by its deposition on a substrate as a protective coating. Composite coatings with bimodal or nano-structured architecture can be deposited on substrates using techniques such as plasma spraying and high-velocity oxy-fuel (HVOF) spraying. Such methods are suited to producing thick coatings with good adhesion and uniformity. Coating material usually enters the deposition area in powder or wire form, which is then melted or partially molten by a thermal source such as a flame, electric arc, or plasma. Feedstock material can be highly varied, from traditional metals to advanced ceramic composites. After heating, particles are accelerated to high velocities at the substrate using compressed air or other gas streams. The impacts form a coating through the mechanical deformation of the particles as they "splat" onto the surface. Upon hitting the substrate, the particles rapidly cool down, thereby solidifying and adhering to the surface mechanically by interlocking. It creates strong adhesion, which is vital for wear and corrosion resistance [13].

Thermal sprayed coatings are widely applied in industries that require high wear resistance, such as the automotive and aerospace industries. Coatings developed from

WC-Co systems exhibited hardness and wear resistance sufficient for drilling and machining tools. In addition, the inclusion of nanosized feedstock particles has dramatically enhanced the mechanical properties of the coating to withstand extreme wear conditions [14].

12.2.3 Chemical Vapor Deposition and Physical Vapor Deposition

CVD is an extremely complex method of depositing bimodal and nano-structured coatings based on chemical reactions in the vapor phase. The CVD process allows for exact control over the composition and microstructure of the coating, making it applicable to advanced material applications. Gaseous precursors containing the elements to be deposited are introduced into a reaction chamber where they diffuse to the substrate surface. These precursors could vary from metal-containing gases or organometallic compounds, with regard to the final coating expected to be obtained [14].

At high temperatures, the reactions between the precursor gases are made more energetic by thermal or plasma energy; this causes the depositing of the solid onto the substrate. The chemical reaction creates the solid by-products, which remain attached to the substrate and form the coating. The reaction chamber is evacuated thoroughly to withdraw the unreacted gaseous by-products of the reaction so that conditions are ideal for deposition. CVD is widely used in semiconductor industries to produce thin films using materials such as silicon, silicon nitride, and others. In case of wear-resistant coatings, it is exploited to the fullest to create excellent films of titanium carbide and titanium nitride. Such coatings provide hardness that is simply impressive. It is because of CVD's unique ability that permits the deposition of coatings with a controlled microstructure, which can be useful for enabling bimodal coatings where two different-sized phases coexist and enhance overall performance [15].

The advantages of CVD are uniform and conformal coatings over complex geometries, which makes it highly suited to electronic or industrial applications. The control over chemical composition and the ability to manipulate the microstructural features are critical benefits. Challenges would include high costs associated with equipment, vacuum system requirements, and even safety issues with toxic precursor gases [16].

12.2.4 Sol-Gel Processing

The sol-gel method is an effective way to prepare bimodal and nano-structured composite coatings. Chemical precursors in solution are mixed to yield a gel, which further can be processed to obtain solid coatings. Alkoxide or metal salt precursors hydrolyze and undergo polycondensation to yield the sol. The process transforms from a sol to a gel as particles grow and fuse together. This material can be coated or spun onto substrates and thus used to create thin, uniform films. Upon coating or spinning, the gel dries out to remove any solvents present and then is heat-treated to achieve the densification and phase changes required to enhance the structural integrity and mechanical properties of the materials produced. Nano-structured coatings formed by the sol-gel process have been quite remarkably resistant to wear and corrosion in a wide range of applications, from aerospace to electronics. Therefore, it has been proved that silica-based coatings prepared by the sol-gel method hold better mechanical qualities as well as chemical durability [17].

Although the sol-gel process provides excellent control over the chemical composition and microstructure of coatings, it is often beset by problems associated with drying rates, shrinkage, and cracking during the transition from gel to solid. The processing conditions and material choices can significantly influence the dimensional stability and efficacy of coatings [18].

12.2.5 Laser Cladding

Laser cladding is an advanced surface modification technique widely applied to improve wear and corrosion resistances of materials. It's very important benefits for these bimodal and nano-structured composite coatings are such that the laser-cladded layers microstructure and properties can be controlled accurately in the depositing process. In the present technique, a powder of coating material is fused with the substrate surface by using the high-power laser beam melt of the material. This implies fine-grained microstructures since high rates of cooling characterize the processes in laser cladding [19].

Metallurgical bonding that the laser cladding develops between a coating and its substrate is excellent and usually results in high adhesion strength that precludes the delamination of coatings when exposed to mechanical stress for a very long period of time. Another advantage of the laser cladding process is that it can control the thickness and composition during the deposition of coatings that can form a bimodal structure, composed of both hard phases and tough phases. One of the benefits of a bimodal structure is the uniform load distribution across the coating, usually with reduced localized wear and higher overall wear resistance. Moreover, the excellent control of energy input and cooling rate in the process of laser cladding allows the formation of nano-structured coatings with excellent hardness through grain boundary strengthening. In nanosized grains with a higher grain boundary area, the availability of larger barriers tends to retard the movement of dislocations and enhance the abrasion resistance of the coating. The incorporation of hard reinforcement phases within the coating matrix further improves toughness by obstructing crack propagation, making the coating much more resistant to impact wear [20].

12.3 WEAR PERFORMANCE

Hardness and toughness are two of the properties that influence the wear behavior of bimodal and nano-structured composite coatings. Hardness is a measure of a material's resistance to indentation or scratching, and coatings containing high hardness are particularly effective at preventing wear initiated by abrasive forces. Fine-grained structure in nanocoatings provides a higher concentration of grain boundaries, making the movement of dislocation more difficult and thus showing greater resistance to deformation compared to the bulk analogs. Therefore, their use is highly effective whenever abrasion wear is seen—that is, in most mechanisms as well as components exposed in heavily lubricating, friction-rich environments [21].

This material definition of toughness is the energy absorbability and deformation without cracking of a material, as it is necessary to sustain impact wear. Bimodal coatings, which consist of an integration of different microstructures or hard phases with softer phases, provide a balanced performance between high hardness and high toughness. It allows giving the structure good resistance against wearing. On the other side, when impact forces arise, more favorable softer phases contribute to these impacts and protect the samples from cracking or spalling under high stress. Such a dual phase was especially useful for applications wherein both abrasive and impact wearing are present in heavy-duty industrial tools as well as different components in automobiles [22].

The synergy between hardness and toughness in these coatings makes them function well under extensive variation in conditions of wear. In practice, bimodal coatings are designed such that the hard phases are aligned in a manner that enhances wear resistance without impairing the toughness required to resist shocks. The two-phase hard and tough nature of the coatings allows these coatings to achieve superior performance in wear applications compared with one of these characteristics. Thus, tailoring both the hardness and toughness of bimodal and nano-structured composite coatings becomes crucial in promoting their wear resistance and increasing the lifespan of coated parts [23]. The adhesion strength between the coating and substrate is the other factor that affects the wear performance of bimodal and nano-structured composite coatings, aside from hardness and toughness. The ability to maintain strong adhesion in delamination or failure conditions will be very critical under mechanical stresses due to conditions of wear for the wear performance. It ensures that the coating does not wear off and is continuously maintained intact for its whole service period, even under high stress [24].

12.3.1 Mechanisms of Wear Resistance

Such mechanisms for high wear resistance of bimodal and nano-structured composite coatings include grain boundary strengthening, phase-reinforcing toughness mechanisms, and load-distribution mechanisms that ultimately synergistically enhance the durability and lifetime of the coating [25]. Figure 12.3 shows the (a) tribosystem and (b) magnification of the inter-surface area.

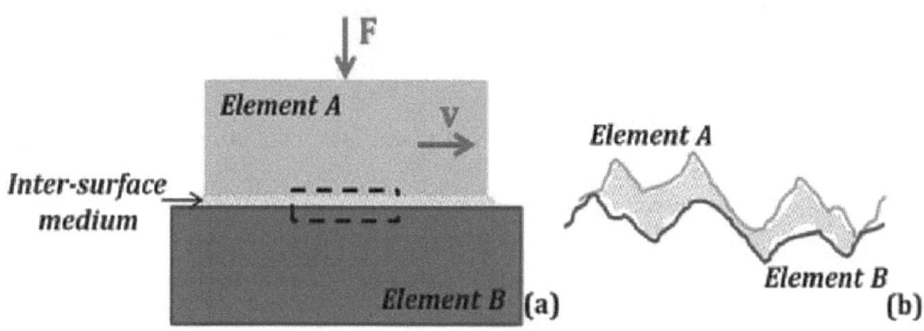

FIGURE 12.3 (a) Schematic representation of a tribosystem and (b) magnification of the inter-surface area marked in (a) [26]. (CC By 4.0 Attribution.)

12.3.2 Grain Boundary Strengthening

Grain boundary strengthening is one of the main reasons behind the high wear resistance observed in nano-structured coatings. In nano-structured coatings, the increase in total grain boundary surface area results from nanosized grains, which play an important role in hindering the movement of dislocations. In general, dislocations are defects within the crystal lattice, which move fairly freely in coarse-grained materials. Under stress, such materials undergo plastic deformation. However, in nano-structured coatings, the grain size is smaller, which increases grain boundaries that act as obstacles to dislocation motion. That means higher resistance to deformation and, therefore, increases the hardness of the material. Thus, nano-structured coatings are more resistant to abrasive wear because harder materials resist scratching or wear. The grain boundary strengthening of the coating produces the overall mechanical strength, whereby it is much less probable to undergo plastic deformation from any heavy-load impact [27].

12.3.3 Reinforcement Phase Toughening

Another important reinforcement mechanism that contributes to the wear resistance of bimodal composite coatings is reinforcement phase toughening. Often, bimodal coatings are designed with a combination of hard reinforcement phases embedded in a softer matrix. These hard phases, often ceramics or metal carbides, act as crack propagation barriers. This crack gets arrested by the reinforcement phases formed in the coating, which hinders the further propagation of the crack and thus does not enter the material. This mechanism helps in the absorption and dissipation of the external stresses, thus enhancing the overall toughness of the coating. The bimodal coatings with the introduction of reinforcement phases balance the hardness and toughness and satisfy the conditions so that the coating does not fail with either abrasion wear or impact loading. The reinforcement phases also impart the required toughness that protects the coatings from spalling or delamination under high-stress conditions, thus giving wear resistance [28].

12.3.4 Load Distribution

The bimodal structure, combining both hard and soft phases, will improve the wear resistance of load distribution across the coating. Such coatings have phases of hard, resisting local deformation and softer phases to be able to absorb and distribute a more constant amount of stress over the entire surface. This uniform distribution of loading reduces stress concentrations in any given area and thus limits the potential for localized wear or cracking, as well as material failure. The bimodal structure spreads out the applied load across the surface, preventing the introduction of wear tracks and hotspots that could allow for premature coating failure. Moreover, the interaction between hard and soft phases ensures that the coating will possess a balance of wear resistance and toughness, which can ensure performance in an array of wear environments—from those involving abrasion to impact-type forces [29].

Together, these mechanisms—grain boundary strengthening, reinforcement phase toughening, and load distribution—provide bimodal and nano-structured composite coatings with significant wear resistance. The ability to combine hardness with toughness while distributing stresses uniformly and inhibiting crack propagation makes such coatings apt for performance under aggressive, wear-prone conditions over long periods [30].

12.4 CORROSION BEHAVIOR

Corrosion behavior is a process related to the degradation of materials, usually metals, by chemical reactions with their environment, which leads to material degradation. Such action is associated with a series of factors, including the nature of the material, corrosive agents present, temperature, humidity, and protective coatings. The main modes of corrosion are uniform attack, pitting, galvanic corrosion, and stress corrosion cracking. These modes have different modes of action, behavior, and environments. The information below helps in the design of appropriate preventive strategies against corrosion that include corrosion-resistant alloys, surface treatments, and inhibitors. In addition, advances in monitoring technology and predictive modeling enhance our ability to assess and monitor corrosion risk and thus extend the service life of structures while lowering maintenance costs [31].

12.4.1 Titanium Oxide Nano-Structured Coatings

Nano-structured coatings of titanium oxide (TiO_2) have drawn increasing attention due to their prominent chemical stability, high resistance to environmental degradation, and ability to create a passive oxide layer. Synthesis of TiO_2 coatings can be realized through several techniques, such as the sol-gel process, CVD, and anodization, which offer different advantages concerning coating thickness, porosity, and structural features. TiO_2 has a wide bandgap of 3.2 eV and photocatalytic properties, making it highly efficient under UV light exposure due to its ability to enhance corrosion resistance by causing the formation of a stable oxide layer. In addition, coatings with TiO_2 exhibit superior mechanical properties in terms of hardness and wear resistance, thus making them suitable for use under harsh conditions. Moreover, the TiO_2 coatings have the self-healing capability through recovery from minor damage, where, consequently, the protective barrier is restored and the onset of corrosion is prevented. The interaction of the coating with the corrosive environment also depends on the surface morphology and crystallinity of TiO_2, which can be controlled to achieve optimal corrosion resistance. Nano-structured TiO_2 coatings offer significant protection against corrosion with improved surface area and electrochemical properties for industries involving aeration and biomedical, marine, and aerospace applications [32].

Figure 12.4 shows scanning electron microscope images of an uncoated AA2024 aluminum alloy substrate and a substrate with TiO_2 nanocoatings deposited at time durations of 40 and 80 s.

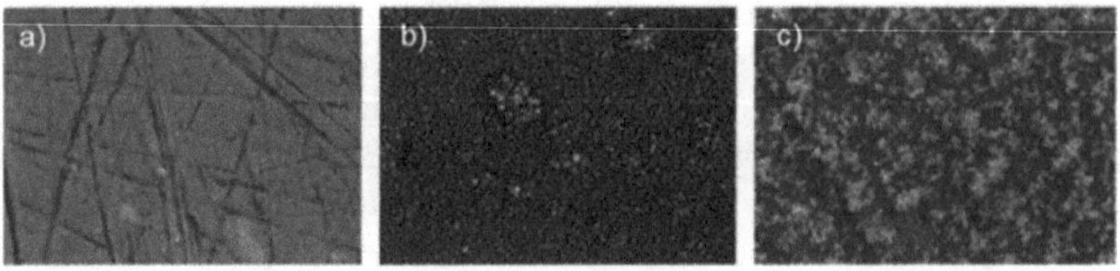

FIGURE 12.4 SEM images of (a) uncoated AA2024 sample, (b) TiO_2-coated AA2024 with a deposition time of 40 s, and (c) TiO_2-coated AA2024 with a deposition time of 80 s [33]. (CC By 4.0 Attribution.)

12.4.2 Alumina Nano-Structured Coatings

Al_2O_3 nano-structured coatings have been highly applied due to excellent corrosion, hardness, and thermal stability. Such coatings are successfully processed using several techniques, including plasma spraying, the sol-gel method, and electrodeposition. It also presents a high melting point and, more importantly, a very strong bond with the metallic substrate, giving very good resistance to wear and chemical attack, making this of particular use in industries related to aerospace, automotive, and energy. A very dense and protective oxide layer prevents corrosive species like chlorides and sulfates from diffusing effectively, thereby protecting the underlying material. The nano-structured Al_2O_3 coatings have improved barrier properties over the bulk material due to higher surface area and a uniform protective coating layer. To increase the mechanical strength and toughness of the coatings without losing their protecting function, nanoparticulate Al_2O_3 is added. Doping elements such as silica or zirconia with Al_2O_3 improve the latter's corrosion resistance due to increased stability and anti-corrosive performance under severe conditions. Incorporating Al_2O_3 coatings on metallic substrates will also reduce maintenance costs with an extension of the service life of equipment exposed to corrosive environments, among which are heat exchangers, pipelines, and turbine blades. Continued development of these coatings includes advancements in nano-structuring methods and hybrid coatings with Al_2O_3 hybridized with other materials to enhance both corrosion protection and mechanical properties [34].

12.4.3 Ceramic Nano-Structured Coatings

These coatings, due to their intrinsic properties such as high hardness and excellent wear resistance, have found more and more applications in corrosion protection. The materials used in such forms are silicon carbide, zirconia, and alumina. These materials, when applied in the form of nanoparticles, exhibit markedly improved performance compared to traditional bulk ceramics. Nano-structured ceramics have better mechanical properties due to a higher surface area and finer grain structure, such as improvement in toughness and corrosion resistance. They also provide shielding against aggressive chemical attack, high temperatures, and environmental degradation. Techniques such as sol-gel, CVD, and plasma spraying will allow for the controlled deposition of fine uniform layers on metallic and non-metallic substrates and will be applied in ceramics. Nano-structured ceramic coatings prevent corrosion due to their dense inert surface, which prevents the penetration of corrosive agents such as water, oxygen, and chloride ions. There is also the added benefit of ceramic coatings having oxidation and wear resistance—an aspect that is highly valuable for applications exposed to abrasive or high-temperature conditions. In addition to their self-healing ability, certain types of ceramic coatings contain an active phase combined with a passivation phase, enhancing their efficiency in extending the lifespan of components exposed to corrosive environments [35].

12.4.4 Tantalum Nitride Nano-Structured Coatings

It has been found that the bimodal and nano-structured composite coatings offer good wear resistivity. For instance, extensive testing of their wear performance on industrial components has been conducted. The grain boundary strengthening mechanisms in the nano-structured coatings involve the presence of nanometer-sized grains. It is because of such grain size that the grain boundary area increases enormously, pinning down the movement of the dislocation and increasing the hardness. This has resulted in an improvement in the abrasion resistance due to the increased hardness. Moreover, bimodal coatings with hard reinforcement phases dispersed within a softer matrix introduce a toughening mechanism, as these toughened phases behave as barriers to crack propagation, improve toughness, and enhance wear resistance due to better blocking of crack growth under mechanical load without affecting the coating excessively. Another benefit these coatings produce is the bimodal structure, which facilitates more uniform stress over the whole coating surface. Such uniformity reduces localized stress concentrations—the primary sites for wear and failure initiation. Bimodal coatings minimize the appearance of wear tracks and hotspots by evenly distributing the applied load, thereby enhancing wear resistance [36,37].

The bare Ti-6Al-4V bears the highest current density in the curve at a given potential, indicating lower corrosion resistance. The pure Ta takes the second best, with current densities lower than those of Ti-6Al-4V, indicating higher corrosion resistance. The Ta_2N coating presents the lowest current densities across the whole range of potentials studied, demonstrating the best corrosion resistance. A remarkable current density decrease for Ta_2N presents evidence that this material has better protection effectiveness, most probably due to a steady oxide-passivation layer insensitive to corrosive interaction. In general, from the graph above, the highest resistance against corrosive penetration occurs in the Ta_2N coatings, followed by pure Ta, while the lowest resistance holds for bare Ti-6Al-4V [38]. Figure 12.5 illustrates the potentiodynamic polarization curves of bare Ti-6Al-4V, pure Ta, and tantalum nitride (Ta_2N) coating.

12.5 CHALLENGES IN SCALABILITY AND UNIFORMITY

Despite all such promising benefits from bimodal and nano-structured composite coatings in effectively enhancing material characteristics such as corrosion resistance, wear resistance, and thermal stability, several challenges persist, particularly regarding their scalability and uniformity. The synthesis complexity is one

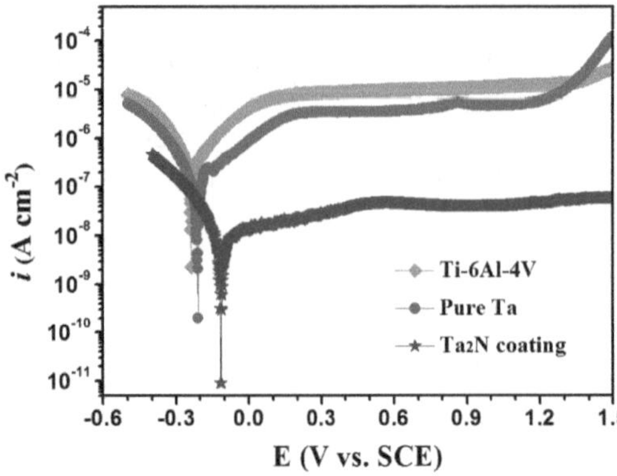

FIGURE 12.5 Potentiodynamic polarization curves of bare Ti-6Al-4V, pure Ta, and Ta_2N coating [38]. (CC By 4.0 Attribution.)

of the significant difficulties in their production. Precise processing conditions are necessary for producing these coatings, especially with complex microstructures. Advanced deposition techniques, such as physical vapor deposition or CVD, must generally be employed to achieve the nano-structured features desired. Such techniques do employ expensive equipment and typically require trained personnel to operate in a reproducible manner. These techniques are therefore obviously of high cost and high processor intensity, making large-scale production extremely difficult at best and not a very economical endeavor for industrial exploitation [39].

Another major difficulty is the problem of scalability. While pilot-scale experiments can result in small batches of fine-quality coatings, the step from pilot-scale preparation to larger industrial production is critical, as achieving uniformity over large areas as well as large lengths is fraught with problems. For example, trying to ensure a consistent coating thickness over large and shaped substrates having complex geometries or varying topographies is a difficult task. Any variation in the coated thickness or microstructure leads to direct variations in the material's properties, and thus, it exhibits erratic protective properties, which is unacceptable in industrial applications. Furthermore, the energy consumption and the time taken in the scaling-up processes frequently lead to costlier production costs, and consequently, these coatings are less commercially viable for mass production [40]. Another critical issue in bimodal and nano-structured composite coatings is their long-term performance under varying environmental conditions. Though such coatings show excellent properties in controlled lab tests, their actual durability and stability to survive real-world conditions in the form of high temperatures, corrosive chemicals, and mechanical stress need more detailed study. All these testing criteria include thermal stability, corrosion performance, fatigue characteristics, and long-term exposure to harsh environments and must therefore be strictly tested to ascertain whether or not such coatings will remain protective for a long time. Without an adequate understanding of their long-term behavior, their large-scale adoption in industries requiring high reliability remains minimal. Overcoming these challenges would be essential to realizing the full potential of these coatings in industrial use [41].

12.6 FUTURE DIRECTIONS

Future research into bimodal and nano-structured composite coatings would help address the current limitations of such systems along with developing new avenues toward a superior performance of composite coatings across an array of applications. Among the types above identified, one promising direction is that of multi-scale modeling, which will significantly advance our understanding of behavior at different scales from atoms to the macroscopic scale. Advanced computational techniques, including molecular dynamics simulations and finite element analysis, shall unveil this information regarding the microstructural evolution, phase transformation, and mechanisms of wear under various environmental and operational conditions of these coatings. Thus, it will be possible to optimize compositions, structures, and processing parameters in order to enhance durability, corrosion resistance, and mechanical properties in coating systems through real-world scenarios that can be simulated. These findings may inform the development of high-performance coatings tailored to specific industrial needs, avoiding trial and error often attached to experimental methods [42].

Another promising area for further improvement of the composite coatings is hybrid coatings. Again, it allows combining bimodal and nano-structured features with others of advanced coating technology types, such as self-lubricating or self-healing coatings. This can allow the creation of multi-functional coatings that offer enhanced wear resistance and corrosion protection, longevity, and other benefits. For instance, the integration of self-healing nanomaterials could lead to coatings that repair microcracks or wear from mechanical degradation, thereby delaying the onset of effectiveness loss in harsh environments. Hybrid coatings, which can provide lubrication, could be applied to reduce friction and thus reduce wear rates and service lifetimes of components coated in applications such as automotive engines and aerospace structures. These multi-functional coatings can revolutionize industries since they offer more efficient, cost-effective, and durable solutions in dealing with wear and corrosion problems [43].

Finally, sustainable synthesis routes are critical for large-scale consumption of bimodal and nano-structured composite coatings. The traditional synthesis methods, such as CVD and electroplating, are costly and environmentally taxing because they involve dangerous chemicals and high energy expenditure. Eco-friendly synthesis routes and cost-effective synthesis routes will be critical to scale up the coating for industrial applications. The environment could be saved further by the research carried out using green chemistry-based synthesis routes or low-temperature processing methods, making the coatings less expensive and reaching a much wider market of various industries. Therefore, future research can be directed to achieve more sustainability when developing coatings that are high-performance yet consistent with the emerging demand for ecologically conscious and economically viable alternatives [44]. Bimodal and nano-structured composite coatings provide effective protection against wear and corrosion due to their enhanced hardness and reduced porosity [45–52]. These coatings are widely used in aerospace, power generation, and marine industries to resist surface degradation. Studies have shown that advanced deposition techniques, such as thermal spraying, help produce coatings with improved bonding strength and thermal stability [53–57]. The performance of these coatings is significantly influenced by process parameters and heat treatment, which ensure their durability under extreme environmental and mechanical conditions [58–60].

12.7 CONCLUSION

Bimodal and nano-structured composite coatings represent a significant advancement in surface engineering, offering enhanced protection against wear through improved mechanical properties and tailored microstructures. While challenges remain in terms of synthesis, scalability, and long-term performance, ongoing research and technological advancements hold promise for their future development and industrial application. By addressing these challenges and exploring new research directions, the potential of bimodal and nano-structured composite coatings in wear protection can be fully realized.

Bimodal and nano-structured composite coatings are a promising solution for dealing with the increasing requirement of advanced protective materials in industries prone to wear, corrosion, and other degradation forms. The coatings combine the superior mechanical, thermal, and chemical properties offered by nanomaterials with the heightened functionalities ensured by bimodal structures that ought to lead to better performance in specific harsh environments. The excellent hardness, wear resistance, corrosion protection, and durability that these coatings demonstrate suggest that they are highly suitable for a wide range of industrial applications, especially in the aerospace, automotive, marine, and energy sectors.

Significant challenges remain in achieving synthesis complexity, scalability, and longer-term performance, despite the great progress made in their development. Making such coatings is often very demanding in terms of control, which often requires rather expensive, specialized equipment and processes, making them unsuitable for widespread industrial-scale applications. Uniformity and consistency across large areas of coatings are also difficult to achieve since minor microstructural differences have an immediate impact on the effectiveness of a coating. There is a need for further research into the long-term stability and performance of these coatings, especially in fluctuating temperatures, aggressive chemical exposure, and mechanical stress conditions, to ascertain their reliability in real-world conditions [45].

REFERENCES

1. Stack MM, Stott FH. Impact of high-temperature erosion–oxidation on the performance of engineering materials. *Tribol Int.* 1998;31(10):653–662.
2. Luo Q, Sui Y, Li N, Ding Y, Cui H, Peng Y. Recent advances in bimodal nanostructured materials and their applications in catalysis. *Catal Sci Technol.* 2015;5(5):3019–3039.
3. Padovano E, Bartuli C, Veca A, Tagliaferri V, Ucciardello N, Ambrogio G, et al. Bimodal structures: a solution for improving the strength-ductility trade-off in metals and alloys. *Mater Sci Eng A.* 2021;822:141636.
4. Chawla N, Deng X. Microstructure and mechanical behavior of nanocrystalline materials. *Mater Sci Eng A.* 2005;409(1–2):168–173.

5. Zhang Z, Sakka Y. Fabrication of functionally graded materials by spark plasma sintering: a review. *Sci Technol Adv Mater.* 2014;15(4):043501.
6. Zhang L, Chen Z, Lin Y, Yu P, Guo J, Zhang Q. Hierarchical nanocomposite coatings for enhanced wear resistance and mechanical properties. *Surf Coat Technol.* 2018;347:216–224.
7. Farooq SA, Raina A, Mohan S, Arvind Singh R, Jayalakshmi S, Irfan Ul Haq M. Nanostructured coatings: review on processing techniques, corrosion behaviour and tribological performance. *Nanomaterials.* 2022 Apr 12;12(8):1323.
8. Sudarshan TS, Mani BN. Advanced surface coatings: a critical review of their manufacturing processes and applications. *Mater Sci Eng A.* 2017;681:63–80.
9. Eliaz N, Gileadi E. Electrodeposition of nanostructured materials. *J Alloys Compd.* 2008;468(1–2):546–555.
10. Sudagar J, Lian J, Sha W. Electroless nickel, alloy, composite and nano coatings – a critical review. *J Alloys Compd.* 2013;571:183–204.
11. Benea L, Bonora PL, Borello A, Martelli S, Wenger F. Effect of particle distribution on the corrosion behavior of Ni and NiCo matrix composite coatings. *Electrochim Acta.* 2003;48(19):2795–2803.
12. Luo J, Xie ZH, Datta MK, Shen Z, Hong RY, Tan HL. Effect of SiC nanoparticles on mechanical and wear properties of electroless Ni–P coating. *Surf Coat Technol.* 2011;205(20):5269–5275.
13. Pawlowski L. *The Science and Engineering of Thermal Spray Coatings.* 2nd ed. John Wiley & Sons; 2008.
14. Guo X, Wu W, Li H, Zhang L. Development of advanced wear and corrosion resistant coatings for harsh environment applications. *Surf Coat Technol.* 2016;307 (Pt A):684–690.
15. Ohring M. *Materials Science of Thin Films.* 2nd ed. Academic Press; 2002.
16. Kessel CM, Jordan E, Evans HE. Chemical vapour deposition: an integrated engineering design for advanced coatings. *Mater Des.* 1997;18(6):323–337.
17. Schuler T, Plank JA. Sol-gel derived nanocomposite coatings: a review. *J Sol-Gel Sci Technol.* 2013;66(2):256–273.
18. Toyserkani E, Khajepour A, Corbin SF. *Laser Cladding.* CRC Press; 2005.
19. Thawari G, Neelam S, Dhawan AK. Wear and corrosion resistance of laser clad coatings: a review. *Mater Sci Forum.* 2015;794–796:158–173.
20. Skandan G, Kear BH, Sadangi RK. Nanostructured coatings and their wear resistance. *JOM.* 2002;54(9):29–32.
21. Lavernia EJ, Srivatsan TS. The rapid solidification processing of materials: Science, principles, technology, advances, and applications. *J Mater Sci.* 2010;45(2):287–325.
22. Wang Y, Li S, Zhang T, Wei Y. Enhanced wear resistance of bimodal structured coatings: microstructure and properties. *J Alloys Compd.* 2017;693:45–53.
23. Pawar VS, Pandey K, Sharma SC. Adhesion strength and wear behavior of nanocomposite coatings. *Mater Today Proc.* 2021;46:1514–1520.
24. Ramakrishnaiah R, Frazer RQ, Weirong L, Flake CJ. Mechanisms of wear resistance in nano-structured composite coatings. *Surf Coat Technol.* 2015;276:307–315.
25. Verma PK, Singh S, Kapoor M, Singh S. A review on the surface topography and corrosion behavior of Mg-alloy coatings for biomedical implants. *Res Surf Interfaces.* 2024 May 9:100227.
26. Psyllaki PP. An introduction to wear degradation mechanisms of surface-protected metallic components. *Metals.* 2019 Sep 28;9(10):1057.
27. Chawla N, Chawla KK. *Metal Matrix Composites.* New York: Springer; 2013. Chapter 2, grain boundary strengthening mechanisms in nanostructured coatings; p. 47–56.
28. Shao Y, Sun Y, Shen B, Hu W. Toughening mechanisms in composite coatings: the role of reinforcement phases. *Wear.* 2017;376–377:1234–1240.

29. Pathak S, Mishra RS. Mechanistic aspects of load distribution in bimodal composite coatings. *Surf Coat Technol.* 2019;370:243–255.
30. Prakash L, Singh B, Katiyar JK. Wear resistance in nanostructured and bimodal coatings: a synergistic effect. *J Mater Sci.* 2018;53(10):7152–7160.
31. Jones DA. *Principles and Prevention of Corrosion.* 3rd ed. Upper Saddle River: Pearson; 2015. Chapter 4, Corrosion behaviour and electrochemical mechanisms; pp. 134–167.
32. Zhang H, Chen C, Li D, Li X. Advances in TiO_2 nanostructured coatings for corrosion protection. *Prog Org Coat.* 2020;139:105446.
33. De Falco G, De Filippis G, Scudieri C, Vitale L, Commodo M, Minutolo P, et al. Nano-TiO2 coating layers with improved anticorrosive properties by aerosol flame synthesis and thermophoretic deposition on aluminium surfaces. *Materials.* 2021 May 28;14(11):2918.
34. Singh R, Singh P, Prakash S. Development and performance of alumina (Al_2O_3) nanostructured coatings for corrosion and wear resistance. *Surf Coat Technol.* 2016;291: 402–412.
35. Guo S, Zhang X, Zhang G, Yang Y. Ceramic nanostructured coatings for enhanced corrosion protection. *J Eur Ceram Soc.* 2018;38(10):3267–3279.
36. Manivasagam G, Rajendran AR, Akella V, Ganesan M. Wear and toughening mechanisms in bimodal and nanostructured coatings. *Tribol Int.* 2020;150:106376.
37. Zhang C, Xiao H, Wang S, Wang L. Corrosion resistance of tantalum nitride (Ta2N) coatings on Ti-6Al-4V alloys. *Corros Sci.* 2022;197:110093.
38. Cheng J, Xu J, Liu LL, Jiang S. Electrochemical corrosion behavior of Ta2n nanoceramic coating in simulated body fluid. *Materials.* 2016 Sep 10;9(9):772.
39. Martin P, Vázquez AJ, Fernández A. Challenges in the scale-up and uniformity of nanostructured coatings: a review of synthesis techniques. *Thin Solid Films.* 2019;687:137382.
40. Rao K, Singh SP, Singh R. Issues and strategies for scalability in nanostructured coating production. *Mater Today Proc.* 2021;45:2672–2679.
41. Liu X, Zhang D, Shen Y. Environmental durability and long-term performance of bimodal and nanostructured coatings. *Corros Sci.* 2022;198:110186.
42. Zhu W, Tang Q, Huang Z, Chen Y. Multi-scale modeling approaches for composite coatings: predicting performance in complex environments. *Mater Des.* 2020;192:108722.
43. Luo J, Wang P, Fan X, Qiao L. Development of hybrid and multifunctional coatings for enhanced wear and corrosion resistance. *Prog Org Coat.* 2021;150:105986.
44. Greenhalgh EJ, Smith KS, Johnson D. Sustainable synthesis techniques for nanostructured coatings: environmental impact and industrial viability. *J Clean Prod.* 2022;330:129962.
45. Hu Z, Feng Y, Zhao Q. Long-term stability and industrial application of nanostructured composite coatings: a comprehensive review. *Surf Coat Technol.* 2023;458:129422.
46. Singh P, Bansal A, Vasudev H, Singh P. In situ surface modification of stainless steel with hydroxyapatite using microwave heating. *Surf Topogr Metrol Prop.* 2021;9(3):035053.
47. Prashar G, Vasudev H, Thakur L. High-temperature oxidation and erosion resistance of Ni-based thermally-sprayed coatings used in power generation machinery: a review. *Surf Rev Lett.* 2022;29(03):2230003.
48. Prashar G, Vasudev H. Surface topology analysis of plasma sprayed Inconel625-Al2O3 composite coating. *Mater Today Proc.* 2022; 50:607–611.
49. Singh G, Vasudev H, Bansal A, Vardhan S. Influence of heat treatment on the microstructure and corrosion properties of the Inconel-625 clad deposited by microwave heating. *Surf Topogr Metrol Prop.* 2021;9(2):025019.
50. Vasudev H, Prashar G, Thakur L, Bansal A. Electrochemical corrosion behavior and microstructural characterization of HVOF sprayed inconel718-Al2O3 composite coatings. *Surf Rev Lett.* 2022;29(02):2250017.

51. Yedida VS, Vasudev H. A review on the development of thermal barrier coatings by using thermal spray techniques. *Mater Today Proc.* 2022;50:1458–1464.
52. Prashar G, Vasudev H. Hot corrosion behavior of super alloys. *Mater Today Proc.* 2020;26:1131–1135.
53. Prashar G, Vasudev H. A review on the influence of process parameters and heat treatment on the corrosion performance of Ni-based thermal spray coatings. *Surf Rev Lett.* 2022;29(01):2230001.
54. Majji BGR, Vasudev H, Bansal A. A review on the oxidation and wear behavior of the thermally sprayed high-entropy alloys. *Mater Today Proc.* 2022;50:1447–1451.
55. Vasudev H, Singh P, Thakur L, Bansal A. Mechanical and microstructural characterization of microwave post processed Alloy-718 coating. *Mater Res Express.* 2020;6(12):1265f5.
56. Parkash J, Saggu HS, Vasudev H. A short review on the performance of high velocity oxy-fuel coatings in boiler steel applications. *Mater Today Proc.* 2022;50:1442–1446.
57. Prashar G, Vasudev H. Structure–property correlation of plasma-sprayed Inconel 625-Al2O3 bimodal composite coatings for high-temperature oxidation protection. *J Therm Spray Technol.* 2022;31(8):2385–2408.
58. Sunitha K, Vasudev H. A short note on the various thermal spray coating processes and effect of post-treatment on Ni-based coatings. *Mater Today Proc.* 2022;50:1452–1457.
59. Prashar G, Vasudev H, Thakur L, Bansal A. Performance of thermally sprayed hydroxyapatite coatings for biomedical implants: a comprehensive review. *Surf Rev Lett.* 2023;30(01):2241001.
60. Vasudev H. Wear characteristics of Ni-WC powder deposited by using a microwave route on mild steel: microwave cladding of Ni-WC. *Int J Surf Eng Interdis Mater Sci (IJSEIMS).* 2020;8(1):44–54.

13 Characterization and Wear Behavior of Additively Manufactured Coatings in High-Temperature Conditions

Praveen Kumar Verma, Hitesh Vasudev, and Bhupinder Singh

13.1 INTRODUCTION

Additive manufacturing (AM), or 3-D printing, is a revolutionary technology around high-performance coatings that creates unprecedented levels of customization and performance enhancement under extreme conditions, including high-temperature conditions. AM-based coatings have emerged as critical innovations in different fields, from aerospace to automobile and energy, where components are subjected to extreme thermal and mechanical stresses during service. In these environments, cracking, delamination, or rapid degradation of conventional coatings can compromise their performance, require frequent maintenance, or even lead to catastrophic component failure. AM solutions can resolve the constraints by enabling the development of coatings designed to resist conditions of interest with control over the thickness, microstructure, and composition. Such developments are sending out frontiers for new possibilities in coating systems that both resist wear and are thermally stable and oxidation-resistant to meet the stringent demands of high-temperature applications [1].

The first important benefit of AM is the possibility of making coatings with higher adhesion and homogeneity via controlled deposition processes that minimize the defects typically seen in traditional coating techniques. Techniques such as thermal spray, laser cladding, and cold spray deposition make it possible to deposit coatings with microstructures fine-tuned for improved wear resistance and high-temperature stability. High-velocity oxygen fuel (HVOF) and plasma spraying are representatives of a class of thermal spray processes that accelerate molten or semi-molten particles to hypersonic velocities to form dense, strong films in applications, such as turbine

DOI: 10.1201/9781003542957-13

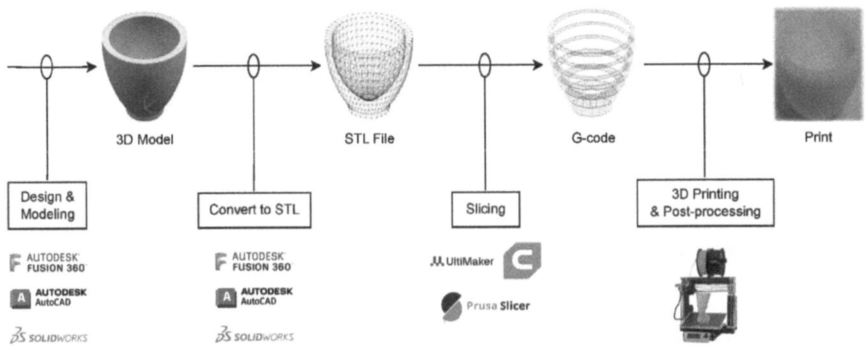

FIGURE 13.1 The overall workflow of the additive manufacturing process [4]. (CC By 4.0 Attribution.)

blades or combustion chamber linings on jet engines, in which toughness and thermal barrier properties are critical [2].

Laser cladding: Another AM process, laser cladding extends the capability of coating applications further by melting a coating material onto the substrate with a laser beam, creating strong metallurgical bonds and minimizing dilution of the coating material. Laser cladding is prized for its high precision and potential compositionally graded coatings, which can be tailored for definite regions to have properties of hardness and thermal resistance. For instance, the application of laser cladding coatings on an automotive exhaust system and cylinder head provides oxidation and wear resistance under cyclic thermal loads [3]. The workflow of the AM process is shown in Figure 13.1.

Cold spray is another low-temperature AM process that employs a supersonic nozzle to accelerate powder particles in the direction of a substrate, where they hit in fully solid-state processing. This process is favored when dealing with materials that are prone to damage from high temperatures, maintaining their inherent properties, and limiting phase changes. They have high density and exhibit strong adhesion as well, thus ideal for components that undergo both corrosive and abrasive environments. Such properties are especially of relevance within the energy sectors where component designs resisting rough operating conditions in power-generation plants are required [5].

The other important characteristic relevant to high-temperature applications is concerning AM coatings, namely, the precise ability to control composition as well as microstructure. Subsequently, grain sizes, phases, and layer thicknesses can be developed in the coating by precisely adjusting the process parameters involved, thus optimizing thermal stability and resistance to oxidation and wear. In coatings containing ceramics, for example, aluminum oxide (Al_2O_3) or zirconium oxide (ZrO_2), stable oxides at high temperatures may be formed, which then create a protective layer ensuring less subsequent oxidation. Conversely, in the case of coatings that contain phases of carbide such as tungsten carbide (WC) or silicon carbide, exceptional hardness is achieved, thus enhancing the wear resistance in environments dominated by abrasive forces. The other potential of AM coatings is also the ability to achieve compositionally graded layering, in which gradual material properties across the thickness of the coating are available. This is very helpful in the case of thermal expansion mismatches between the coating and substrate, which usually happens at high temperatures and can trigger delamination or spalling. The gradient produced by the AM coatings can reduce thermal stresses, thereby increasing its endurance time and reliability under the imposed fluctuating thermal cycles. Such a concept has extensively been researched for application in gas turbines, where the temperature gradients are steep, and the management of thermal expansion is essential for sustained performance [6,7].

More than any of the structural benefits, AM coatings contribute substantially to enhancing the corrosion resistance of components in high-temperature settings. In applications such as fossil fuel power plants where materials are exposed at elevated temperatures to corrosive gases, coatings must resist both high-temperature oxidation and aggressive corrosion. Materials like nickel and cobalt alloys, commonly used in AM coatings, demonstrate excellent resistance to oxidation and carburization, thus enhancing the service life of parts under those demanding conditions. In areas such as energy and automotive fields, where efficiency and longevity are of tremendous importance, corrosion resistivity provided through AM coatings will reduce maintenance costs and increase lifetimes in operation. This high level of control of the deposition and composition of the coatings allows the optimization of other critical properties, including thermal conductivity and electrical insulation. In electronic applications, AM coatings can be finely adjusted for either thermal or electrical insulation or conduction, thus providing a versatile tool for managing heat and electrical currents in high-temperature devices [8].

AM has also brought about a revolution in the design and application of high-performance coatings used at high temperatures. Its ability to precisely tailor the microstructure, composition, and layering of coatings has made AM an extremely valuable tool in industrial sectors facing extreme thermal and mechanical demands on conventional materials. With this new development of AM technologies and materials, the scope of applications for such coatings will open continuously, hence providing engineers with the ability to develop solutions that satisfy and surpass the requirements of new high-temperature operations [9].

Selective laser melting (SLM) is, by far, one of the most advanced AM methods for metallic materials. Compared with traditional manufacturing, this unique layer-wise process boasts a great degree of freedom in customization to fabricate nearly net-shaped parts with complex geometries [10]. Titanium, particularly titanium alloy, has been favored in aerospace, aviation, and biomedical industries because of the advantages of low density, high specific strength, and excellent corrosion resistance. Among all the available alloys, Ti_6Al_4V, one type of dual-phase alloy, demonstrates good mechanical properties. Poor oxidation resistance at high temperatures limits the use of Ti_6Al_4V alloys as essential strength members in advanced heat engines [11]. Figure 13.2 shows the SLM machine.

13.2 OVERVIEW OF ADDITIVE MANUFACTURING TECHNIQUES FOR HIGH-TEMPERATURE COATINGS

High-temperature coatings by AM involved techniques like thermal spray, laser cladding, and cold spray, all of which provided distinct properties affecting wear behavior. The wear behavior of such coatings prepared by thermal spray methods,

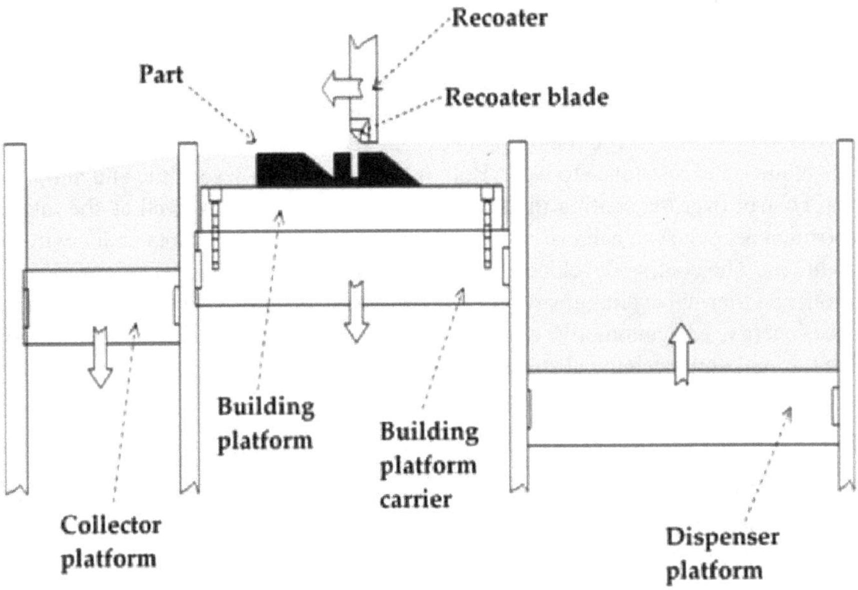

FIGURE 13.2 A schematic of the SLM machine [11]. SLM, selective laser melting. (CC By 4.0 Attribution.)

for example, plasma spraying as well as HVOF, often involves impacts by heated particles traveling at high velocities, thus leading to dense, solid coatings. Such coating methods are very effective in providing coatings with a high hardness and excellent adhesion capabilities, thereby making them resistant to abrasive wear and thermal stresses [12].

Laser cladding, instead, uses a laser beam to melt onto substrates coating materials while achieving metallurgical bonds that are strong and with minimal dilution, thus favorable for high-wear, high-temperature environments. The process is sufficiently precise in the formation of compositionally graded coatings, an improvement from thermal resistance and lower wear. In low-temperature AM, cold spray accelerates solid-state particles onto surfaces, retaining the inherent properties of the material and forming a dense, oxidation-resistant coating. This is particularly beneficial in operating environments that are simultaneously hot and corrosive. Using the proper AM technique, coatings can be formulated to optimize hardness, adhesion, oxidation resistance, and thermal stability to meet unique wear challenges in high-temperature industry applications [13].

13.2.1 Thermal Spray Techniques

HVOF and plasma spraying are among the most widely used high-temperature techniques in thermal spray because of their ability to produce coatings with higher adhesion at higher density with superior hardness. HVOF works on heating the powdered material that is to be coated to extremely high temperatures within the oxygen-fueled flame and then propelling it onto a substrate at such velocities that the coating particles tend to bond in an entirely dense, well-adhered form, with very little porosity. However, in plasma spray, a plasma arc at a high temperature is used to melt the coating materials. These are then sprayed onto the surface. This makes material deposition possible, whose melting point is high, such as ceramics, increasing thermal stability and resistance to wear. Both methods are highly versatile and allow for great control over the coating thickness, velocity of particles, as well as the rate of deposition necessary to achieve optimal conditions for wear behaviors under extreme conditions. The coating developed is extremely durable with excellent abrasion, corrosion, and thermal cycling resistance, making them ideal for operations in the aerospace, energy, and automotive sectors where components experience significantly high thermal and mechanical stresses [14,15].

13.2.2 Laser Cladding

Laser cladding is a high-accuracy additive process that can form custom coatings with specified composition gradients. This technique focuses a laser beam to melt the coating material along with a thin layer of the substrate. Such a procedure, due to metallurgical bonding, results in minimal dilution of original material properties and encourages a strong bond that increases the adhesion and wear resistance of the coated layer, which can be used particularly in high-temperature applications involving large thermal gradients in components [16]. Figure 13.3 shows the schematic diagram of the laser cladding system.

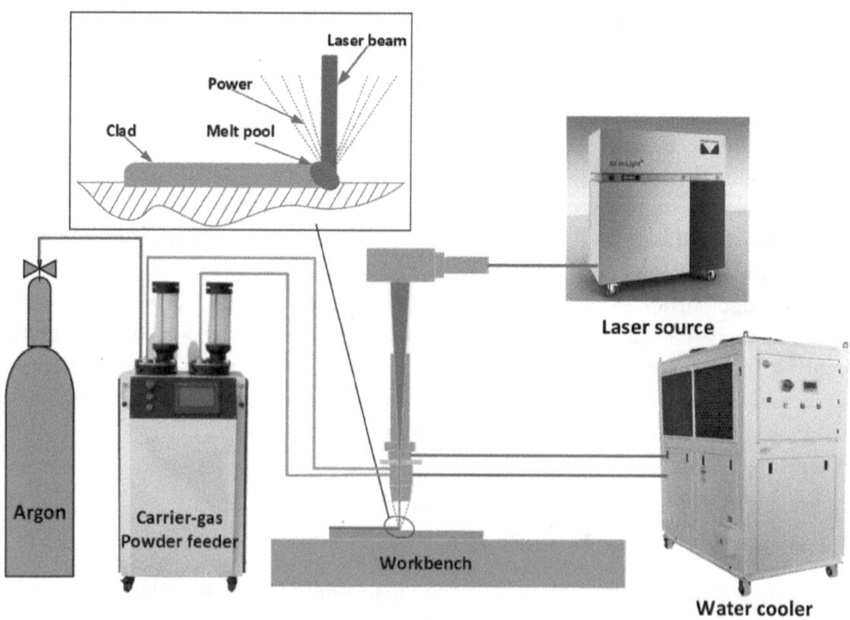

FIGURE 13.3 Schematic diagram of the laser cladding system [17]. (CC By 4.0 Attribution.)

High precision offered by laser cladding provides the possibility of depositing coatings that feature complex, controlled microstructures and composition profiles, optimizing specific properties such as hardness, wear resistance, or thermal stability. Gradients in composition, for example, can be applied to diminish thermal expansion mismatches that otherwise would lead to cracking or spalling under thermal cycling. Another aspect of laser cladding is the flexibility in material selections: the metal-based coatings can be carried out with any metal or alloy base, as well as ceramic-based coatings, tailored to resist oxidation, corrosion, and wear phenomena. Such technological flexibility positions the laser cladding process as a useful solution in those sectors of the industry that require critical performance and extended component life under extreme conditions, such as aerospace, energy, and automotive sectors [18].

13.2.3 Cold Spray

Cold spray is an AM process that is based on solid-state technology, where the particles are accelerated to a very high velocity and then impacted onto a substrate to form a coating (inset Figure 13.4). This low-temperature process serves as a means for preserving the intrinsic properties of the coating materials, thereby providing some specific advantages when handling temperature-sensitive or oxidation-prone materials, such as certain metals, alloys, and composite materials. Since it does not involve high temperatures, as typical of many thermal spraying techniques, it avoids oxidation and other undesirable phase changes within the material. This results in coatings with increased corrosion resistance, ductility, and mechanical integrity. Moreover, cold spray produces coatings that have high density and excellent adhesion, which

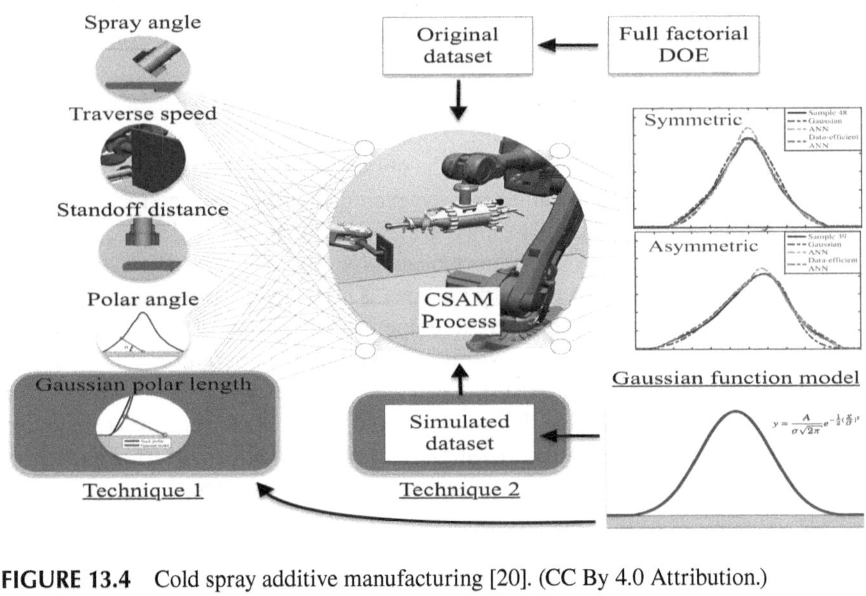

FIGURE 13.4 Cold spray additive manufacturing [20]. (CC By 4.0 Attribution.)

are essential features that ensure bearing wear resistance and structural integrity in harsh environments [19].

Cold spray is now popularly applied on applications that require strong and wear-resistant coatings on materials that can degrade with the presence of high processing temperatures. In this regard, aerospace, automotive, and electronics industries employ the technique in its potential to create layers that are tough and corrosion-resistant yet undetrimental to the coating material's performance. Therefore, it finds its application in components used under mechanical stresses combined with high-temperature exposure [21].

13.3 WEAR MECHANISMS IN HIGH-TEMPERATURE ENVIRONMENTS

The extreme changes of wear mechanisms of AM coatings during their service life and reliability in practical industries are reflected by high-temperature conditions. Elevated temperatures alter the physical and chemical properties of AM coatings, and therefore display different kinds of wear behaviors compared to room temperature conditions. Understanding these changes is essential to optimize the performance of AM coatings under conditions that often characterize those hot environments, such as aerospace, automotive, and cutting tool applications. The most notable wear mechanisms that are affected by high temperatures are abrasive wear, adhesive wear, and oxidation. High-temperature conditions might be conducive to abrasive wear due to material softening, which will decrease the material hardness and ability for wear resistance. Materials become pliable at high temperatures, tending to deform and get removed easily while under frictional contact. There is an impact on adhesive wear behavior; at high temperatures, this is tantamount to the increased diffusion of atoms at the interface between contacting surfaces, leading to greater material transfer and thus deterioration in wear rates. Such interaction can result in wear debris and add to further wear processes [22].

Oxidation is a very important factor in high-temperature wear, and for most materials, especially metallic coatings, oxidation produces an outer layer of oxide. Although that outer layer might prevent the abrasive action from the initial surface interactions, raising the temperature may weaken the oxide layer and increase spalling or detachment, which increases wear rates. Besides, the underlying oxidized layer may also generate new wear mechanisms by secondary particles, which may become abrasive and affect the wear properties. These changes in wear mechanisms affect overall service life due to the high-temperature exposure of AM coatings. AM coatings possessing greater wear resistance at ambient temperatures do not necessarily possess the same characteristics under thermal stress. Therefore, careful consideration must be given both to the chemical composition and to the microstructure and processing conditions of the coatings to ensure optimum performance at elevated temperatures. Advances include selenide and telluride doped with refractory elements or materials that should form stable protective oxide layers, where strong improvements can be expected in their wear resistance and service life at high temperatures [23].

13.4 WEAR IN ADDITIVE MANUFACTURING MATERIALS

Wear in AM materials is an area of study that is crucial because wear determines not only the functionality but also the shelf life of components created using these innovative techniques. AM, which includes layer-by-layer construction methods, permits not only the creation of complex geometries and customized parts but also introduces unique challenges related to wear. The most evident wear mechanisms of AM materials are abrasion, adhesion, and fatigue. Each of them can cause material degradation as applied during operational service. Abrasive wear is caused by interaction between hard particles and the surface; consequently, it results in material removal. Adhesive wear is associated with material transfer between contact surfaces under loads. Fatigue wear arises from the stress cycle with subsequent failure through surface cracks. Therefore, understanding these mechanisms can help one design AM materials to be able to withstand high demands in various applications, from automotive to aerospace [24].

AM materials' wear resistance is also affected by the material composition, microstructure, and specifics of the manufacturing process. Besides, it has been proven beyond doubt that reinforcement of hard phases titanium carbide (TiC) or WC considerably increases the wear resistance through a considerable increase in hardness and hardness-dependent durability. The microstructure is also important; finer microstructures generally lead to a better microstructure since they give high load-bearing capacity and remove the susceptibility of propagation. To control the microstructure and density of the final product, the process parameters in the laser process, that is, laser power, scanning speed, and layer thickness, are tuned. The better the mechanical properties, the better the wear properties will be, for example. Optimized processing conditions may contribute to improved mechanical properties and reduced porosity [25].

13.4.1 The Influence of Material Microstructure on Wear

In AM, it is seen that the microstructure of the material becomes the sole deciding factor in determining the wear behavior of materials. The factors governing the same are grain size and morphology, phase composition, porosity, residual stress, and texture. Microstructure has a major effect on the mechanical properties of the materials, such as strength, hardness, and toughness, which further govern the wear rate and resistance of materials [26].

A promising attribute in the field is to optimize the microstructure of AM materials by changing grain morphology so that mechanical properties may be amplified. Further research is necessary in this field to determine methods for creating ideal microstructures. In lubricated conditions, stainless steel 316L was produced via SLM as well as through traditional routes. The results indicate that the friction and wear properties of SLM samples in contact with brass are slightly better than for conventionally processed samples, which may be attributed to the enhanced wear resistance resulting from the finer grains in SLM samples. However, when contacted by harder materials, the samples underwent large plastic deformations. Considering that the micromorphology of SLM samples contains finish particles, it is obvious that their tribological properties should be superior to those processed using traditional methods [27].

The Ag coating after AM effectively alleviates the downward penetration of shear stress and strain, which proves the better wear resistance of the processed Ag coating [28]. A smaller grain size and uniform phase composition can enhance the strength and hardness of materials to improve their toughness to mitigate wear (Figure 13.5).

Wang et al. [30] have performed oxidation analysis on a thick TiO_2 and Al_2O_3 coatings on the surface of the Ti_6Al_4V alloy. A Ti_6Al_4V-1.0 wt.% Y_2O_3 alloy shows high oxidation resistance, and its thickness is much smaller than that of the Ti_6Al_4V alloy. The oxidation kinetics of the Ti_6Al_4V and Ti_6Al_4V-1.0 wt.% Y_2O_3 alloys at 600°C and 800°C follow a parabolic rule, whereas the oxidation of the Ti_6Al_4V and Ti_6Al_4V-1.0 wt.% Y_2O_3 alloys at 1,000°C follows the linear law. The oxidized surface morphologies of as-printed Ti_6Al_4V and Ti_6Al_4V-1.0 wt.% Y_2O_3 samples at 600°C, 800°C, and 1,000°C for 24 hours are shown in Figure 13.6.

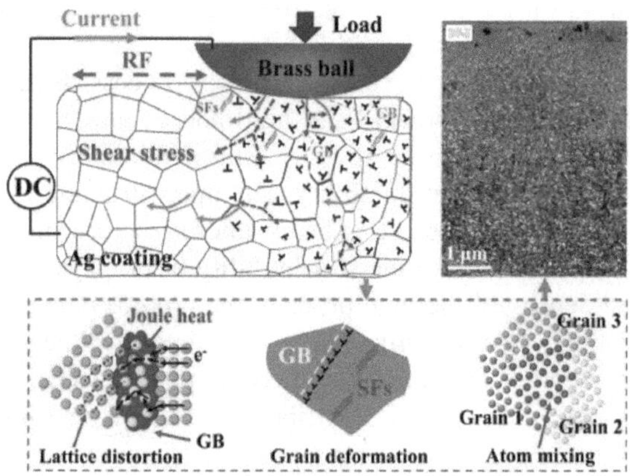

FIGURE 13.5 Schematic diagram of the microstructure evolution mechanism of the silver coating through current-carrying friction [29]. (CC By 4.0 Attribution.)

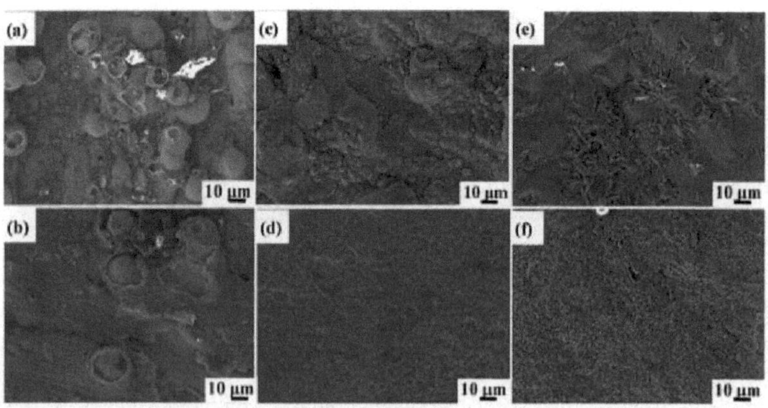

FIGURE 13.6 SEM of surface morphologies of Ti_6Al_4V and Ti_6Al_4V-1.0 wt.% Y_2O_3 samples after oxidation for 24 hours (a, b) 600°C, (c, d) 800°C, (e, f) 1,000°C [30]. SEM, scanning electron microscopy. (CC By 4.0 Attribution.)

13.5 WEAR IN ADDITIVE MANUFACTURING PROCESSES

Wear in AM processes is a critical factor that significantly influences both the service and endurance lives of components produced by these new techniques. AM, or successive layer formation during the generation of an object, enables the production of complex geometries and prototype parts. However, the intrinsic nature of AM processes presents wear issues that need to be grasped to enhance the durability of the component. The key wear mechanisms in AM processes are abrasion, adhesion, and fatigue. Abrasive wear arises from hard particles rubbing against surfaces that cause erosion of material. Adhesive wear takes place in the transfer of material from one contact surface to the other under high loads. Fatigue wear, on the contrary, occurs directly because of the cyclic mechanical stress. This leads to cracks in the surface, which may lead to failure. Understanding these wear mechanisms is crucial to design and use AM materials in demanding applications, such as aerospace, automotive, and tooling industries [31].

The wear resistance of materials produced via AM is owed to several factors. Material composition is certainly one, where the inclusion of hard phases such as TiC or WC can increase wear resistance through overall increased hardness and stability under stress in operational use. Microstructural characteristics of the AM material are essentially prone to manufacturing parameters; therefore, these significantly influence wear performance. Fine-grained structures are usually preferred, as they favor a more uniform load distribution and inhibit crack initiation and propagation. The process parameters, such as layer thickness produced, scanning speed, and energy input, all play important roles in the microstructure and density of the resultant product in an AM process. All these factors can be optimized in such a way that the porosity defects directly related to the improvement in mechanical properties as well as wear resistance of materials can be reduced in manufacture [32].

13.5.1 Influence of Coating Composition on Wear Behavior

The formulation of an AM coating was critical to determining its high-temperature properties, which in turn dictate wear resistance, thermal stability, and overall durability of the coating. The coating can be formulated with a wide range of materials, each possessing unique properties that may influence the behavior of the coating at high temperatures. Understanding these impact trends of different compositions on performance under thermal stress is critical for the development of coatings that would be suitable for extreme conditions of respective industries, such as aerospace, automotive, and power generation [33].

The nature of the base materials on which AM coatings are used has a major impact on their heat conduction and wear properties. Examples include coatings from ceramic materials in zirconia and alumina that have generally higher thermal stability and are less likely to be subject to thermal shock, hence appearing to be more suited for high-temperature applications. These ceramics may maintain their integrity even up to elevated service temperatures. This leads to lower wear rates compared to metal-based coatings, which otherwise soften or deform under similar conditions. In addition, the incorporation of alloying elements that are known to resist high temperatures, such as cobalt, nickel, or chromium, into metal-based coatings improves the oxidation resistance and mechanical strength of these coatings. Some high-temperature alloys are specially designed to form stable oxides, protecting the underlying alloy from further degradation [34].

13.5.2 Effect of Heat Treatment on Wear

Heat treatment for wear resistance is another important aspect of materials science that concerns elements that operate under extreme conditions. The heat treatment processes used to change the microstructure and mechanical properties are hardening, annealing, and tempering, leading to improved wear resistance. This paper discusses the mechanism of heat treatments and their effects on the wear properties of different materials by focusing attention on the critical importance in applications for aerospace, automotive, and tooling industries. Heat treatment is a process applied to materials to create desired mechanical properties through controlled heating and cooling of the material. The fundamental mechanisms that occur during heat treatment include phase transformations, grain refinement, and modification of residual stresses. Thus, for instance, hardening treatments, such as quenching, transform austenite into martensite in a steel alloy and ultimately produce a considerably harder and wear-resistant material. However, this increase in hardness often happens at the expense of toughness [35].

The wear resistance of heat-treated materials is dependent on the associated hardness, microstructure, and presence of phases. For instance, well-tempered martensitic steel can exhibit outstanding wear resistance without experiencing catastrophic failure, as it achieves a good balance between hardness and toughness. Coatings and surface treatments can be applied post-heat treatment to further enhance wear resistance. High-performance materials often bet better in high-stress applications, and properly heat-treated carbide or nitride phases may show superior wear properties. Heat treatment temperature and time would also be of key importance. Overheating may lead to undesirable grain coarsening and softening, which detracts from wear resistance. Thus, the heat treatment parameters must be kept controlled precisely so that the wear performance can be optimized [36].

13.6 CHALLENGES AND FUTURE DIRECTIONS

Even with such wonderful improvements in AM coatings for high-temperature wear applications, several challenges remain, which limit their more widespread use and performance optimization. Ensuring uniformity in the coating across complex geometries is one of the significant challenges. Over-the-layer thickness or composition inconsistencies may lead to performance inconsistency and localized failure at high stress and elevated temperatures. However, the parameters of deposition control, such as laser power, scanning velocity, and powder flow rate, can be controlled precisely; these are still important factors to consider addressing the problems discussed above. These defects include porosity, cracks, and residual stresses that are often introduced by rapid cooling and solidification. Hence, using advanced process monitoring and feedback systems with in situ sensors and high-speed machine learning algorithms can become a promising route to detect and mitigate defects in the fabrication processes. Another major concern is the long-term stability of the AM coatings at high temperatures. Such exposure brings on oxidation, thermal fatigue, and diffusion-induced degradation affecting their microstructure and, thus, their performance over time. Therefore, it is a must to carry out a comprehensive study via accelerated aging tests correlated with actual field data to evaluate and enhance the longevity of such coatings [37].

Optimizing the material composition and layering strategies of AM coatings will remain a large research avenue soon and is likely to lead to significant further improvement in high-temperature performance. Functionally graded materials are also promising, as graded transitions through wear-resistant surfaces to a more ductile substrate would represent attractive opportunities for solutions that can be developed to enhance thermal shock resistance and toughness. New material inclusions, such as ceramic-metal composites or high-entropy alloys, in AM also promise improvements in better-wearing coatings with superior thermal stability. Another promising direction is hybrid AM approaches that include a combination of several AM technologies or traditional techniques. Coupling AM with thermal spraying or laser shock peening improves the adhesion, density, and surface properties of the coatings. Progress in computational modeling and simulation has led to opportunities in the precise design of microstructures and the prediction of coating performance in extreme conditions [38].

Such challenges will be overcome by interdisciplinary research and innovation, thereby realizing the full potential of AM coatings for high-temperature wear environments. This approach shall also lead to more reliable and longer-life components and support sustainable practices by developing minimal waste of materials and extending the life cycles of critical systems [39]. Additively manufactured coatings have demonstrated remarkable performance under high-temperature conditions, offering enhanced wear resistance and oxidation stability [40–47]. These coatings exhibit unique microstructural features, such as refined grains and homogeneous composition, contributing to their durability in extreme environments. Studies emphasize the role of post-processing techniques, including heat treatment, in further improving mechanical and tribological properties [48–52]. Techniques like thermal spraying and microwave-assisted processing have been instrumental in achieving dense, defect-free coatings with superior adhesion [53–55]. Such advancements ensure the extended service life of components in power generation and aerospace applications.

13.7 CONCLUSION

AM has transformed the development of advanced coatings in terms of wear resistance and is specially designed for high-temperature applications. The aerospace, automotive, and energy sectors will also depend on the applications of coating that would have a definite technique from SLM or laser cladding—techniques that can provide coatings with complex geometries, graded compositions, and microstructures with characteristics that could not be easily achieved via conventional means. Nickel-based superalloys, ceramic-metal composites, refractory materials, etc., have been used and shown successful adaptation to building wear-resistant coatings of extreme thermal, mechanical, and corrosive conditions.

Advanced characterization techniques form the basis for supporting the development of high-performance coatings through AM. The advanced characterization techniques provide critical insights into the microstructure, phase distribution, and mechanical properties that may determine their quality and performance. Despite these progresses, huge challenges prevail in the scalability of AM-produced coatings for large-scale components and cost-effectiveness [40].

REFERENCES

1. Gibson I, Rosen DW, Stucker B. *Additive Manufacturing Technologies: 3D Printing, Rapid Prototyping, and Direct Digital Manufacturing.* 2nd ed. New York: Springer; 2015.
2. Pawlowski L. *The Science and Engineering of Thermal Spray Coatings.* 2nd ed. Chichester: Wiley; 2008.
3. Toyserkani E, Khajepour A, Corbin S. Laser Cladding. Boca Raton, FL: CRC Press; 2005.
4. Zhou L, Miller J, Vezza J, Mayster M, Raffay M, Justice Q, Al Tamimi Z, Hansotte G, Sunkara LD, Bernat J. Additive manufacturing: A comprehensive review. *Sensors.* 2024 Apr 23;24(9):2668.
5. Dethlefsen M, Gärtner B, Wenzel M. Cold spray for repair and coating applications. *J Therm Spray Tech.* 2014;23(5):809–816.
6. Koller J, Bogdanoff C, Kim Y, et al. Compositionally graded coatings for high temperature applications: A review. *J Mater Sci.* 2013;48(15):4905–4920.
7. Lee J, Lee K, Kim Y, et al. Thermal expansion management in gas turbine coatings: Gradients in composition and structure. *Mater Sci Eng A.* 2016;657:101–113.
8. Shi Y, Xu J, Zhao Q, et al. Corrosion and wear behavior of additively manufactured coatings in high-temperature conditions. *Corros Sci.* 2017;122:68–77.
9. Zhang W, Wang H, Yu W, et al. The role of additive manufacturing in the development of high-performance coatings. *Int J Adv Manuf Technol.* 2020;107(5–6):2025–2035.
10. Gu D, Meiners W, Wissenbach K, et al. Selective laser melting of metallic components: Materials and applications. *Int J Adv Manuf Technol.* 2012;56(1–4):431–446.
11. Verma PK, Singh S, Kapoor M, Singh S. A review on the surface topography and corrosion behavior of Mg-alloy coatings for biomedical implants. *Res Surf Interfaces.* 2024 May 9:100227.
12. Boulger G, Li X, Ellis T, et al. Wear behavior of coatings prepared by thermal spray methods in high-temperature environments. *J Therm Spray Tech.* 2017;26(2):432–440.
13. Huang J, Song X, Zhang Z, et al. Additive manufacturing for high-temperature wear-resistant coatings: A review of techniques and properties. *J Mater Sci.* 2019;54(1):34–52.
14. Della Corte A, Sorrentino L, Teti R, et al. High-velocity oxygen fuel and plasma spraying for high-performance coatings in extreme environments. *Surf Coat Technol.* 2015;276:349–359.
15. He Z, Xue F, Li Y, et al. Advances in thermal spray coating technologies for high-temperature applications. *Mater Sci Eng A.* 2018;719:290–305.
16. Song J, Zhang Y, Li Q, et al. Laser cladding for high-temperature applications: Mechanisms and improvements. *J Laser Appl.* 2017;29(6):062014.
17. Li X, Xu J, Wang J, Lu Y, Han J, Guo B, Xie T. Prediction of geometric characteristics of laser cladding layer based on least squares support vector regression and crested porcupine optimization. *Micromachines.* 2024 Jul 16;15(7):919.
18. Liu Z, Zhang Z, Zhang Y, et al. Laser cladding for high-performance coatings: Precision, flexibility, and applications. *Surf Coat Technol.* 2018;349:476–485.

19. Nunez P, Zhang X, Li Y, et al. Cold spray coatings for high-temperature wear and corrosion resistance: A review. *J Therm Spray Tech.* 2020;29(1):157–169.
20. Ikeuchi D, Vargas-Uscategui A, Wu X, King PC. Data-efficient neural network for track profile modelling in cold spray additive manufacturing. *Appl Sci.* 2021 Feb 12;11(4):1654.
21. Verma A, Lee H, Singh A. Cold spray technology for wear-resistant coatings in aerospace and automotive applications. *J Therm Spray Tech.* 2019;28(4):820–832.
22. Chou K, Wang X, Li W. Influence of high-temperature conditions on wear behavior of AM coatings: Mechanisms and challenges. *Wear.* 2018;410–411:118–128.
23. Pan X, Chen Y, Xu J. Oxidation and wear resistance of additive manufacturing coatings at elevated temperatures: The role of doped refractory elements. *Mater Des.* 2020;185:108249.
24. Zhang D, Wang H, Xu L. Wear mechanisms in additive manufacturing materials: Challenges and solutions. *Addit Manuf.* 2019;28:292–302.
25. Zhao W, Zhang Q, Zhao J. Influence of process parameters on the wear resistance of additive manufacturing materials: A study on TiC and tungsten carbide reinforcement. *J Mater Process Technol.* 2019;273:116223.
26. Liao Y, Yang Z, Liu H. Effect of microstructure on the wear behavior of additive manufacturing materials: Grain size, phase composition, and porosity. *Tribol Int.* 2018;127:249–259.
27. McGregor H, Walker C, Smith M. Effect of microstructure on the wear behavior of SLM stainless steel 316L. *Mater Sci Eng A.* 2019;759:395–406.
28. Jiang W, Yang H, Wang Y. Wear resistance of Ag coatings in additive manufacturing: Influence of shear stress and strain. *J Mater Sci.* 2020;55(4):1857–1867.
29. Du J, Lu M, Fang J, Li W, Chen D. Current-carrying friction of Ag coatings by additive manufacturing: Uncovering the role of electric current. *Mater Res Lett.* 2024 Jun 2;12(6):459–466.
30. Wang Q, Song P, Niu W, Li N, Hu N. High temperature oxidation behavior of additive manufactured Ti6Al4V alloy with the addition of yttrium oxide nanoparticles. *Materials.* 2024 May 24;17(11):2544.
31. Verma PK, Vasudev H, Singh G, Singh S. A review on the microstructural and biomedical properties of implants manufactured using additive manufacturing. *J Inst Eng (India): Series D.* 2024 May 18:1–5.
32. Liu Z, Zhang W, Wang Y. Effect of process parameters on the wear resistance of AM materials: The role of composition and microstructure. *J Manuf Sci Eng.* 2020;142(10):101701.
33. Wang X, Chen B, Liu H. Influence of coating composition on wear behavior of AM coatings: High-temperature performance and thermal stability. *Wear.* 2018;410–411:165–175.
34. Smith J, Walker A, Patel R. High-temperature alloys and their influence on oxidation resistance in AM coatings. *Mater Sci Eng A.* 2019;765:138104.
35. Zhang H, Xie L, Liu Y. Effect of heat treatment on wear resistance in additive manufacturing materials. *J Mater Sci.* 2017;52(8):4709–4721.
36. Alavi M, Jun Y, Kim Y. Heat treatment effects on the microstructure and wear resistance of AM coatings. *Metall Mater Trans A.* 2019;50(4):1563–1574.
37. Thompson R, Xu Q, Zhao X. Long-term stability of AM coatings under high-temperature wear conditions: A study of oxidation and thermal fatigue. *J Eng Mater Technol.* 2020;142(6):061010.
38. Zhang Z, Zhang D, Li Q. Hybrid additive manufacturing approaches for high-temperature coatings: Combining thermal spraying and AM technologies. *J Mater Process Technol.* 2020;278:116489.
39. Lee S, Choi K, Kim D. Advancing AM coatings for high-temperature applications through interdisciplinary research: Challenges and solutions. *Mater Des.* 2019;160:57–66.

40. Zhao P, Huang S, Wang Z. Challenges and future directions for the scalability and cost-effectiveness of AM coatings. *Int J Adv Manuf Technol.* 2020;108(5):1753–1764.
41. Prashar G, Vasudev H, Thakur L. Surface topology analysis of plasma sprayed Inconel625-Al_2O_3 composite coating. *Mater Today Proc.* 2022;50:607–611.
42. Singh G, Vasudev H, Bansal A, Vardhan S. Influence of heat treatment on the microstructure and corrosion properties of the Inconel-625 clad deposited by microwave heating. *Surf Topogr Metrol Prop.* 2021;9(2):025019.
43. Yedida VS, Vasudev H. A review on the development of thermal barrier coatings by using thermal spray techniques. *Mater Today Proc.* 2022;50:1458–1464.
44. Vasudev H.. Wear characteristics of Ni-WC powder deposited by using a microwave route on mild steel: Microwave cladding of Ni-WC. *Int J Surf Eng Interdis Mater Sci (IJSEIMS).* 2020;8(1):44–54.
45. Vasudev H, Singh P, Thakur L, Bansal A. Mechanical and microstructural characterization of microwave post processed Alloy-718 coating. *Mater Res Express.* 2020;6(12):1265f5.
46. Majji BGR, Vasudev H, Bansal A. A review on the oxidation and wear behavior of the thermally sprayed high-entropy alloys. *Mater Today Proc.* 2022;50:1447–1451.
47. Parkash J, Saggu HS, Vasudev H. A short review on the performance of high velocity oxy-fuel coatings in boiler steel applications. *Mater Today Proc.* 2022;50, 1442–1446.
48. Prashar G, Vasudev H. A review on the influence of process parameters and heat treatment on the corrosion performance of Ni-based thermal spray coatings. *Surf Rev Lett.* 2022;29(01):2230001.
49. Prashar G, Vasudev H, Thakur L. High-temperature oxidation and erosion resistance of Ni-based thermally-sprayed coatings used in power generation machinery: A review. *Surf Rev Lett.* 2022;29(03):2230003.
50. Prashar G. Vasudev H. Hot corrosion behavior of super alloys. *Mater Today Proc.* 2020;26:1131–1135.
51. Prashar G, Vasudev H. Structure–property correlation of plasma-sprayed Inconel625-A12O3 bimodal composite coatings for high-temperature oxidation protection. *J Therm Spray Technol.* 2022;31(8):2385–2408.
52. Prashar G, Vasudev H, Thakur L, Bansal A. Performance of thermally sprayed hydroxyapatite coatings for biomedical implants: A comprehensive review. *Surf Rev Lett.* 2023;30(01):2241001.
53. Singh P, Bansal A, Vasudev H, Singh P. In-situ surface modification of stainless steel with hydroxyapatite using microwave heating. *Surf Topogr Metrol Prop.* 2021;9(3):035053.
54. Sunitha K, Vasudev H. A short note on the various thermal spray coating processes and effect of post-treatment on Ni-based coatings. *Mater Today Proc.* 2022;50:1452–1457.
55. Vasudev H, Prashar G, Thakur L, Bansal A. Electrochemical corrosion behavior and microstructural characterization of HVOF sprayed inconel718-Al_2O_3 composite coatings. *Surf Rev Lett.* 2022;29(02):2250017.

14 Corrosion Resistance in Gas Turbines Using Oxide-Based Coatings

Hitesh Kumar, Gaurav Prashar, and Hitesh Vasudev

14.1 INTRODUCTION

Metals and alloy structures undergo the ubiquitous electrochemical/chemical phenomena of corrosion, which transforms their states into the most stable possible chemical forms. Corrosion is an unavoidable process, although it can be avoided or reduced by employing corrosion-resistant materials, cathode protection, correct material design, and corrosion-resistant coatings. This necessitates effective corrosion mitigation solutions in the chemical, aerospace, power, textile, automotive, biomedical, electronics, steel, petroleum, cement, petrochemical, and construction sectors [1]. Various organic coating systems, such as epoxy, polystyrene, polyaniline, polyamide, polymethylmethacrylate, polypropylene, polyurethane, alkyd, polydimethylsiloxane, polydopamine, and silane, are frequently utilized in metal anticorrosion due to their exceptional property of functioning as a physical barrier. However, the high cross-linking density of the organic coatings causes certain micro-cracks and pores during application, resulting in certain permeability to O_2, H_2O, and Cl^- in corrosive conditions. Naturally, the penetration of these materials reduces the coating's ability to prevent corrosion and quickens the metals' rate of corrosion. For this reason, standard coatings cannot offer lasting protection. Fillers and reinforcements (e.g., oxide ceramic particles) are meant to strengthen the anticorrosion effect by filling the imperfections of the coatings to reduce the occurrence of micro-galvanic corrosion [2]. The application of oxide-based composite coatings has shown promise in improving gas turbine corrosion resistance. These coatings have good oxidation resistance, stability at high temperatures, and the capacity to create protective oxide layers that slow down the deterioration of the material. The reason for their efficacy is that they cause stable oxide scales to form, including chromia (Cr_2O_3) or alumina (Al_2O_3), which function as barriers to stop more corrosion.

This chapter examines the function of oxide-based composite coatings and the significance of corrosion resistance in gas turbines. It addresses the mechanisms by which these coatings preserve turbine materials against corrosion, their potential for future use in increasing gas turbine life and performance, and developments in coating technology.

DOI: 10.1201/9781003542957-14

14.2 OXIDE-BASED COMPOSITE COATINGS

Advanced protective layers, known as oxide-based composite coatings, combine ceramic oxides with other materials to improve surface qualities, especially resistance to oxidation and corrosion in high-temperature conditions. These coatings are intended to shield metallic parts like those found in gas turbines from the damaging elements they encounter, such as high temperatures, corrosive gases, and oxidation [3].

14.2.1 Alumina Coatings

One of the primary causes of thermal barrier coating (TBC) degradation is hot corrosion. Al_2O_3 ceramic coatings are extensively utilized as insulators and wear-resistant coatings. A popular technique for creating alumina coatings is atmosphere plasma spraying or APS. In conventional APS, the particles are injected into the plasma flame. Among the many industrial uses for plasma-sprayed coatings are thermal barriers, wear resistance, and corrosive environments [4].

The hot corrosion analysis of three plasma-sprayed TBCs was assessed: (i) conventional yttria-stabilized zirconia (YSZ), (ii) a particle composite consisting of YSZ and Al_2O_3, and (iii) a layer composite where Al_2O_3 serves as a topcoat on the YSZ layer. The application of an alumina coating as the outer layer in YSZ/Al_2O_3 effectively reduced the penetration of molten salt into YSZ. This modification enhanced the resistance of the TBC against hot corrosion, as evidenced by a significant decrease in the presence of hot corrosion products within the YSZ/Al_2O_3 structure. A particle or layer composite TBC system incorporating an alumina component has the potential to enhance the mechanical properties of the coating and provide resistance to hot corrosion. Al_2O_3 exhibits insolubility in ZrO_2, resulting in the formation of a rigid matrix encasing the ZrO_2 particles. This configuration induces local compressive stress and obstructs the phase transformation of zirconia (ZrO_2) [5]. The study by Arya and Tailor examines the erosion dynamics of APS $Al_2O_3+Sm_2SrAl_2O_7$ composite TBCs across varying temperatures and impingement angles [6].

14.2.2 Zirconia Coatings

Coatings based on ZrO_2 are well known for their superior stability, corrosion resistance, and thermal insulation at high temperatures. Ceramic materials like ZrO_2 are well known for their high melting points, exceptional resistance to chemical corrosion, and superior thermal insulation. Because ZrO_2 coatings act as thermal barriers, shielding metal substrates from extreme heat and averting premature failure from oxidation or thermal stress, this is their main benefit. The power-generating and aerospace sectors make substantial use of these coatings, especially in jet engines and gas turbines where parts are subjected to temperatures exceeding 990°C [7–9].

CMAS penetration may also elevate the coefficient of thermal expansion at the surface, resulting in an increase in strain energy during thermal cycling [10].

The combination of Y_2O_3 and ZrO_2 demonstrates superior resistance to CMAS attack compared to the individual components of ZrO_2 or Y_2O_3. The $La_2Zr_2O_7+YSZ$

duplex coating demonstrates superior thermal shock resistance compared to the YSZ coating. A $La_2Ce_2O_7$ + YSZ duplex coating demonstrates superior cycling life compared to an YSZ coating [10].

14.2.3 Glass-Ceramic-YSZ Coatings

Recently, glass-ceramic coatings have garnered significant interest among researchers due to their straightforward processing techniques and the ability to easily customize their properties. Heat-resistant glass-ceramics demonstrated the potential of glass-ceramics for TBC applications [11].

14.2.4 Spinel ($MgAl_2O_4$)

Spinel, a magnesium-aluminum oxide composite material, has been studied for its elevated melting point, corrosion resistance, and integration with turbine materials.

It is also worth noticing that in the context of the role of different oxides to mitigate corrosion in gas turbine components, the relationship between various oxides should also be taken into account. Figure 14.1 illustrates the schematic diagram depicting the alterations in oxides due to hot corrosion following pre-oxidation. Analysis indicates that following the application of a prefabricated oxidation layer, the surface oxides resulting from hot corrosion exhibit greater completeness compared to those without pre-oxidation. This phenomenon is attributed to the consumption of Na_2SO_4 by the surface oxides to a certain degree, allowing for the incorporation of new oxides within the internal matrix, thereby maintaining the integrity of the oxidation layer. NiO is consumed at the surface, while the more stable Al_2O_3 and Cr_2O_3 are generated internally. The reaction does not exhibit sufficient severity to compromise the protective effect of the oxide layer, as illustrated in Figure 14.1. In summary, the formation of a stable and dense oxide layer can significantly enhance the hot corrosion resistance of alloys, particularly the Al_2O_3 layer [12].

14.3 THERMAL SPRAY COATING TECHNOLOGY USED FOR DEPOSITION OF COMPOSITE COATINGS

Thermal spraying is a widely known process for applying wear and corrosion-resistant coatings in several important industry areas, such as automotive, aerospace, petrochemical, power generation, and offshore [13,14].

The versatility of the materials that may be deposited using thermal spraying techniques is another factor that has led to its widespread adoption [15] (Figure 14.2).

14.3.1 Flame Spray Coating

Flame spray coatings are frequently utilized in industry due to their low cost and convenience of application. In addition to being one of the most cost-effective choices, this method is frequently used since it works with a variety of materials. Although quality can be enhanced by re-melting coatings using a procedure that

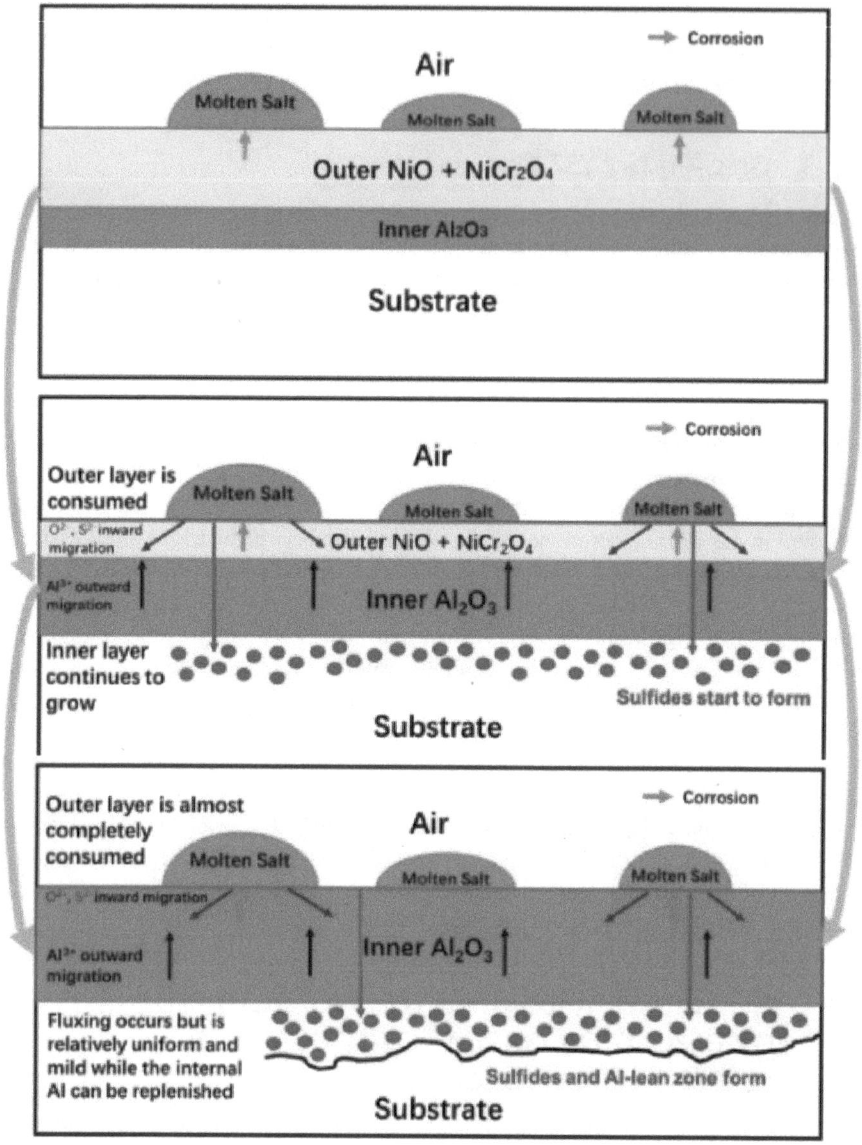

FIGURE 14.1 The schematic diagram illustrates the alterations in oxides during hot corrosion following the pre-oxidation process.

often involves an oxyacetylene flame, significant porosity and poor adhesion to the substrate result in low quality. High porosity (11%–21%) and poor adherence to the substrate (mechanical bonding) are two significant disadvantages of flame spray-deposited coatings. Re-melting the layers significantly lowers porosity, forms a metallurgical bond with the base material, and improves cohesion between the coating's particles [16]. The first thermal spray method created was flame spray. The high

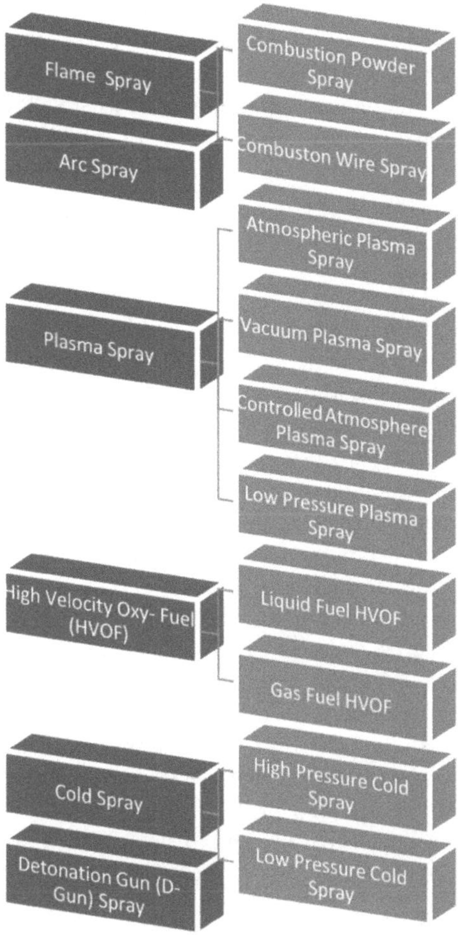

FIGURE 14.2 Classification of thermal spray coating.

particle velocities, which range from 200 to 300 m/s, have altered in modern torches. Acetylene serves as the primary fuel of oxyacetylene torches, which combine it with oxygen to produce high combustion temperatures and particle temperatures of around 2,600°C [17]. The most popular type of torch is an oxygen-acetylene torch, which uses acetylene (C_2H_2) as the primary fuel and oxygen to produce a maximum combustion temperature of about 3,161°C. Ethane, hydrogen (H_2), propylene (C_3H_6), and propane (C_3H_8) are further gases that are used. At the nozzle exit, powders, wires, or rods are axially inserted into the flame through the back of the nozzle. For the majority of metal spraying, reducing (fuel-rich) flames are used to reduce oxidation. The purpose of flame spray torches is to partially scavenge the oxygen required for combustion from cooling air jets or atomizing air. Additionally, this design aids in the production of metal coatings with less oxides. Numerous materials, including ceramics, polymers, and refractory metals, may be deposited using flame spray [18].

This technique offers several advantages, including low costs, ease of operation, simplicity of design, and decreased dust and fume levels. It is commonly used to preserve buildings and machine components against corrosion (for example, offshore platforms, bridges, and LPG bottles). It is employed where a low-cost thermal spray coating is sought and lesser quality is acceptable. Flame spraying is mostly used to preserve buildings and machine components from corrosion, as well as to reclaim worn shafts [19].

14.3.2 Arc Spray Coating

Many businesses use thermal spray (wire arc) for corrosion prevention due to its cost-effective manufacturing and maintenance. Two consumable coating ingredients are fed into a spray gun in wire form using the thermal arc spray technique. The wire can melt by generating an arc by receiving the current, and compressed air at 3–6 bar will speed up the melting process. The high-velocity acceleration will be directed toward the substrate, causing the molten spray to solidify and form a coating on it. The coating material has to be electrically conductive in order for an arc to form and for the substance to melt according to the principle of thermal arc spray function. Customers' needs determine which coating material is best, although often they need zinc alloy, zinc, aluminum, or aluminum alloy [20].

Because wire arc spray coatings produce a coarser coating than plasma or high-velocity oxy-fuel (HVOF) processes, have a larger material output per hour, and have lower operating costs, they are becoming more and more promising. In addition to offering on-site solutions, wire arc coatings are ideally suited for dimensional restoration of damaged and mismachined items. Thus, it provides excellent dependability and adaptability [19]. In order to prevent corrosion in the chemical industry, modern advancements have produced electric arc guns that can function in inert atmospheres by atomizing argon or nitrogen and spraying reactive metals like zirconium and titanium [21].

14.3.3 Plasma Spray Coating

A type of thermal spraying known as "plasma spraying" creates a high-temperature plasma using an electrical arc to melt and spray materials onto a surface. From the micro- to the macro-scale, it is a remarkably versatile and economical way to deposit coatings of different thicknesses [22]. The tungsten cathode and copper anode of the plasma gun are both water-cooled. Between them, a high-frequency arc is ignited. A plasma plume that is several centimeters long forms as a result of the flow of plasma gas (H_2, N_2, He, or mixes) around the cathode and through the anode, which is shaped like a constricting nozzle and ionized. Within the plume, temperatures can rise to 16,000°K. Outside of the gun nozzle, the powdered spray material is fed into the plasma plume, where it melts and is then flung onto the substrate surface by the gas. It could be necessary to construct cooling or spray rate control arrangements in order to keep substrate temperatures between 90°C and 204°C (195°F and 398°F). One benefit of plasma spraying is its ability to spray materials with extremely high melting points, including refractory metals [23].

Plasma spray is used to coat metallic and ceramic materials onto the substrate for corrosion resistance, wear resistance, and thermal barrier application. The greatest option is plasma spray, which works when other methods cannot. Compared to flame spraying, plasma spray often produces coatings that are denser, have less porosity, and have higher adherence. The thermal spray method and spray parameter conditions determine the oxide concentration, bond strength, porosity, wear resistance, and corrosion resistance of the coating. Gas flow rate, standing distance, material flow rate, surface preparation, and coating material morphology are examples of spray parameters [14].

14.3.4 High-Velocity Oxy-Fuel Coating

HVOF thermal spraying has gained popularity in recent years for applying protective overlay coatings to engineering components' surfaces, enabling them to operate in harsh environments. When compared to alternative techniques like plasma spraying, HVOF spraying can provide coatings that are denser and less oxidized [24].

Compared to alternative thermal spraying techniques, the HVOF process's deposits perform better overall. It is possible to apply coatings of materials like cemented carbides using HVOF spraying. When the high-velocity particles strike the substrate's surface, they create a collection of compressed splats that result in coatings with a higher density and uniformity. However, too much heat energy during the HVOF spraying process might change the surface. Therefore, using a discontinuous spraying procedure or certain cooling techniques is crucial for avoiding coating flaws. The HVOF spraying technology is not advised for coating ceramic or refractory materials due to the process temperature restriction. The degree of particle deformation, splat sintering, and coating layer solidification rate all affect the overall microstructure of sprayed coatings. The dynamic state of the gas mixture, which can be modified by spraying factors like the fuel-to-oxygen ratio and the rate of total gas flow, is linked to the behavior of particles while in flight. It is easy to examine the flow and thermal field using simulation techniques of the continuum type [25].

14.3.5 Cold Spray Coating

The cold spray method shows strong coating bond strength and great deposition efficiency. Even after the coating is deposited, the powder's microstructure and characteristics won't change. This method can produce coatings that are both thin (at the micron level) and thick (at the centimeter level), with a porosity level of less than 1% due to the absence of oxide formation. This is the primary justification for using the cold spray procedure when welding is not an option for corrosion and structural repairs of important components. The cold spray procedure can be used to deposit a wide range of materials, including metals, ceramics, composites, and even polymers [26].

Through a de Laval nozzle with a convergent-divergent shape, solid powder particles are accelerated over the sonic velocity in the CS process. On a suitable substrate, particles impact ballistically at velocities between 350 and 1,200 m/s. The ultimate temperature and velocity of sprayed particles are largely determined by the nozzle geometry and the properties of the feeding powders, which are closely linked to the mechanical, physical, and microstructure of the coating. The gas stream's temperature is consistently lower than the melting point of the particle material. Thus, it would be accurate to characterize CS as a solid-state deposition process. Compared to other conventional thermal spray methods, the coating deposition has some very special properties because it is carried out in the solid state [27].

14.3.6 Detonation Gun Spray Coating

Detonation Spray is the best option among the commercially available thermal spray coating processes for achieving the requisite dense, hard, and wear-resistant coatings. The work's goal is to examine how detonation gun (D-gun) spray coating might improve the substrate's surface qualities in order to combat issues including corrosion, fretting fatigue, erosion, residual stress, and thermal behavior. The D-gun spray technique is a thermal spray coating method that produces a coated surface with compressive residual stresses, minimal porosity, and exceptionally high adhesive strength. A tubular barrel with one end closed is used to provide a precisely determined amount of the oxygen and acetylene combustion mixture [28].

A tubular barrel with one end closed is used to feed a combination of acetylene and oxygen. The gas inlets are covered with nitrogen gas to stop the backfire. By traveling down the barrel at high speed, the hot gases produced in the explosion chamber heat the particles to a plasticizing stage (the skin melting of the article) and accelerate them to a speed of 1,200 m/s. After exiting the barrel, these particles strike the object the manipulator is holding to create a coating [29].

14.4 STUDIES ON OXIDE-BASED COMPOSITE COATINGS

Oxide-based thermal spray composite coatings for gas turbines focus on improving the functionality of components subjected to severe thermal and mechanical stresses. Gas turbines function under elevated thermal conditions, leading to potential degradation of components, including blades and combustion chambers, which are susceptible to wear, oxidation, and corrosion. Hence, oxide-based composite coatings have been designed to enhance the reliability, thermal barrier properties, and overall efficiency of these vital components in response to the identified challenges.

By subjecting the created coating to a molten combination of 80 weight percent Na_2SO_4 and 20 weight percent NaCl in air at 1,000°C for 100 hours, the corrosion resistance of the coating was assessed. Parthiban et al. recommended that at 1,000°C, the coating shape altered progressively over time. Within the various TBC layers, no thermally generated oxide layer was visible. The development of a thin reaction layer at the substrate-coating interface improved the coating's adhesion to the substrate. The developed coating shows high promise for usage as a corrosion-resistant coating for real gas turbine engine components made of superalloys [30]. The development of oxide-based coatings has emerged as a critical strategy for enhancing the corrosion resistance of gas turbine components operating in high-temperature environments [31–37]. Research has shown that various coating techniques, including thermally sprayed and microwave-assisted methods, significantly improve the oxidation and erosion resistance of materials like Inconel and Ni-based alloys [38–41]. Additionally, the synergistic effects of composite coatings, as discussed by various authors, further enhance the durability of turbine blades, contributing to increased efficiency and longevity in power generation systems [42–45].

14.5 CONCLUSION & FUTURE SCOPE

Oxide-based composite coatings are a major improvement in improving the corrosion resistance of gas turbine components, especially when applied using thermal spray techniques. These coatings offer a strong defense against thermal cycling, hot corrosion, and high-temperature oxidation—all of which are significant problems in gas turbine settings. It is feasible to get a balance of protective qualities that increase component lifespan and boost operating efficiency by carefully choosing and refining oxide materials and coating techniques. In order to limit the entry of corrosive species, this chapter emphasizes the great effectiveness of thermal spray coatings, such as plasma spray and HVOF coatings, in creating thick, adherent layers with low porosity. Furthermore, adding advanced oxides like YSZ and alumina-based composites improves the coatings' mechanical and thermal durability. In the end, oxide-based composite coatings have a lot of promise for the development of effective, dependable, and long-lasting gas turbines because of their versatility in high-stress situations and room for more innovation. In addition to helping achieve sustainable energy goals by lowering maintenance frequency and boosting fuel economy, continuous research and development in this field are crucial for satisfying the needs of contemporary gas turbines, which operate in more challenging environments.

With prospective avenues to further improve turbine durability and efficiency, oxide-based composite coatings for corrosion resistance in gas turbines have a wide range of potential applications in the future. Coatings with enhanced thermal stability and resistance to high-temperature oxidation may result from research into cutting-edge materials like rare-earth oxides and multi-layered composites. Because they provide greater mechanical and chemical qualities, such as increased toughness, adhesion, and resistance to thermal shock, nanostructured coatings are increasingly gaining popularity. This makes them perfect for harsh situations. Advances in deposition processes, such as cold spray and advanced thermal spray systems, may increase overall protective qualities and enable more precise coating application. By enabling predictive maintenance, lowering unscheduled downtimes, and giving real-time data on coating health, the development of in-situ coating monitoring and diagnostics has the potential to completely transform maintenance. Since future coatings must comply with more stringent standards by reducing emissions and hazardous waste, environmental factors will also be crucial. By combining oxide-based composites with additional protective materials, functionally graded and hybrid coatings provide customized solutions that address numerous types of deterioration, thus expanding their usefulness across different turbine sections. In addition to helping gas turbines, developments in oxide-based coatings may find use in other high-temperature industries such as the petrochemical, power generation, and aerospace sectors.

REFERENCES

1. Jena, G. and Philip, A. 2022. A review on recent advances in graphene oxide-based composite coatings for anticorrosion applications. *Progress in Organic Coatings*. https://doi.org/10.1016/j.porgcoat.2022.107208.
2. Tang, S., Lei, B., Feng, Z., Guo, H., Zhang, P. and Meng, G. 2023. Progress in the graphene oxide-based composite coatings for anticorrosion of metal materials. *Coatings*. https://doi.org/10.3390/coatings13061120.
3. Brunner, G. 2014. Chapter 12: Corrosion in hydrothermal and supercritical water. In *Supercritical Food Science and Technology* (Vol. 5, pp. 591-619). Elsevier. https://doi.org/10.1016/B978-0-444-59413-6.00012-1.
4. Sarikaya, O. 2005. Effect of some parameters on microstructure and hardness of alumina coatings prepared by the air plasma spraying process. *Surface and Coatings Technology*. https://doi.org/10.1016/j.surfcoat.2004.02.007.
5. Afrasiabi, A., Saremi, M. and Kobayashi, A. 2008. A comparative study on hot corrosion resistance of three types of thermal barrier coatings: YSZ, YSZ+ Al_2O_3 and YSZ/Al_2O_3. *Materials Science and Engineering: A*, *478*(1–2), pp. 264–269.
6. Arya, S.B. and Tailor, S. 2022. Erosion behavior of Al_2O_3+ $Sm_2SrAl_2O_7$ composite thermal barrier coatings. *Materials Today: Proceedings*, *66*, pp. 3853–3858.
7. Hung, D.P., Chinh, N.T., Hiep, N.A., Thai, N.X., Lien, L.T.N., Toan, D.H., Giang, H.T.H. and Hoang, T. 2022. Effect of zirconia nanoparticles modified by silane coupling agent on some properties of epoxy coating. *Vietnam Journal of Science and Technology*. https://doi.org/10.15625/2525-2518/16999.
8. Adraider, Y., Pang, Y.X., Nabhani, F., Hodgson, S.N., Sharp, M.C. and Al-Waidh, A. 2013. Fabrication of zirconium oxide coatings on stainless steel by a combined laser/sol–gel technique. *Ceramics International*. https://doi.org/10.1016/j.ceramint.2013.05.089.
9. Moreira, V.B., Puiggalí-Jou, A., Jiménez-Piqué, E., Alemán, C., Meneguzzi, A. and Armelin, E. 2021. Green nanocoatings based on the deposition of zirconium oxide: the role of the substrate. *Materials*. https://doi.org/10.3390/ma14041043.
10. Fergus, J.W. 2014. Zirconia and pyrochlore oxides for thermal barrier coatings in gas turbine engines. *Metallurgical and Materials Transactions E*, *1*, pp. 118–131.
11. Parthiban, K., Bysakh, S. and Ghosh, S. 2024. Novel oxide based anti-corrosion composite coating for gas turbines. *Surface and Coatings Technology*, *484*, p. 130822.
12. Gurrappa, I. 1999. Hot corrosion behavior of CM 247 LC alloy in Na_2SO_4 and NaCl environments. *Oxidation of Metals*, *51*(5), pp. 353–382.
13. Chen, Z., Li, S., Wu, M., Pei, Y., Gong, S. and Zhang, H. 2020. Effects of different surface native pre-oxides on the hot corrosion properties of nickel-based single crystal superalloys. *Materials*, *13*(24), p. 5774.
14. Talib, R.J., Saad, S., Toff, M.R.M. and Hashim, H., 2003. Thermal spray coating technology: A review. *Solid State Science and Technology, 11*(1), pp. 109–117.
15. Tejero-Martin, D., Rezvani Rad, M., Mcdonald, A. and Hussain, T. 2019. Beyond traditional coatings: a review on thermal sprayed functional and smart coating. *Journal of Thermal Spray Technology*. https://doi.org/10.1007/s11666-019-00857-1.
16. González, R., Cadenas, M., Fernández, R., Cortizo, J.L. and Rodríguez, E. 2007. Wear behaviour of flame sprayed NiCrBSi coating remelted by flame or by laser. *Wear*. https://doi.org/10.1016/j.wear.2006.05.009.
17. Habib, K.A., Saura, J.J., Ferrer, C., Damra, M.S., Giménez, E. and Cabedo, L. 2006. Comparison of flame sprayed Al_2O_3/TiO_2 coatings: their microstructure, mechanical properties and tribology behaviour. *Surface and Coatings Technology*. https://doi.org/10.1016/j.surfcoat.2006.02.011.
18. Vouristo, P. 2014. 4.10-thermal spray coating processes. *Comprehensive Materials Processing*. https://doi.org/10.1016/B978-0-08-096532-1.00407-6.

19. Kumar, R. and Kumar, S. 2018. Thermal spray coating: a study. *International Journal of Engineering Sciences & Research Technology*. https://doi.org/10.5281/zenodo.1207005.
20. Abd Malek, M.H., Saad, N.H. and Abas, S.K. 2013. Thermal arc spray overview. *IOP Conference Series: Materials Science and Engineering*. https://doi.org/10.1088/1757-899X/46/1/012028.
21. Herman, H., Sampath, S. and McCune, R. 2000. Thermal spray: current status and future trends. *MRS Bulletin*. https://doi.org/10.1557/mrs2000.119.
22. Wu, S., Liu, X., Yeung, K.W.K., Guo, H., Li, P., Hu, T., Chung, C.Y. and Chu, P.K. 2013. Surface nano-architectures and their effects on the mechanical properties and corrosion behavior of Ti-based orthopedic implants. *Surface and Coatings Technology*. https://doi.org/10.1016/j.surfcoat.2012.10.023.
23. Amin, S. and Panchal, H., 2016. A review on thermal spray coating processes. *International Journal of Current Trends in Engineering & Research (IJCTER)*, *2*(4), pp. 556–563.
24. Sidhu, T.S., Agrawal, R.D. and Prakash, S. 2005. Hot corrosion of some superalloys and role of high-velocity oxy-fuel spray coatings—a review. *Surface and Coatings Technology*. https://doi.org/10.1016/j.surfcoat.2004.10.056.
25. Ghadami, F. and Sabour Rouh Aghdam, A. 2019. Improvement of high velocity oxy-fuel spray coatings by thermal posttreatments: a critical review. *Thin Solid Films*. https://doi.org/10.1016/j.tsf.2019.02.019.
26. Srikantha, A., Mohammed Thalib Basha, G. and Venkateshwarlu, B. 2020. A brief review on cold spray coating process. *Materials Today: Proceedings*. https://doi.org/10.1016/j.matpr.2020.01.482.
27. Moridi, A., Hassani-Gangaraj, S.M., Guagliano, M. and Dao, M. 2014. Cold spray coating: review of material systems and future perspectives. *Surface Engineering*. https://doi.org/10.1179/1743294414Y.0000000270.
28. Singh, L., Chawla, V. and Grewal, J.S. 2012. A review on detonation gun sprayed coatings. *Journal of Minerals & Materials Characterization & Engineering*. https://doi.org/10.4236/jmmce.2012.113019.
29. Tucker, R.C. Jr. 1974. Structure property relationships in deposits produced by plasma spray and detonation gun techniques. *Journal of Vacuum Science and Technology*. https://doi.org/10.1116/1.1312743.
30. Parthiban, K., Bysakh, S. and Ghosh, S. 2024. Novel oxide based anti-corrosion composite coating for gas turbines. *Surface and Coatings Technology*. https://doi.org/10.1016/j.surfcoat.2024.130822.
31. Prashar, G., Vasudev, H. and Thakur, L. 2022. High-temperature oxidation and erosion resistance of Ni-based thermally-sprayed coatings used in power generation machinery: a review. *Surface Review and Letters*, *29*(03), p. 2230003.
32. Singh, G., Vasudev, H., Bansal, A. and Vardhan, S. 2021. Influence of heat treatment on the microstructure and corrosion properties of the Inconel-625 clad deposited by microwave heating. *Surface Topography: Metrology and Properties*, *9*(2), p. 025019.
33. Prashar, G. and Vasudev, H. 2020. Hot corrosion behavior of super alloys. *Materials Today: Proceedings*, *26*, pp. 1131–1135.
34. Vasudev, H., Prashar, G., Thakur, L. and Bansal, A. 2022. Electrochemical corrosion behavior and microstructural characterization of HVOF sprayed inconel718-Al_2O_3 composite coatings. *Surface Review and Letters*, *29*(02), p. 2250017.
35. Singh, P., Bansal, A., Vasudev, H. and Singh, P. 2021. In situ surface modification of stainless steel with hydroxyapatite using microwave heating. *Surface Topography: Metrology and Properties*, *9*(3), p. 035053.
36. Vasudev, H., Singh, P., Thakur, L. and Bansal, A. 2020. Mechanical and microstructural characterization of microwave post processed Alloy-718 coating. *Materials Research Express*, *6*(12), p. 1265f5.

37. Prashar, G. and Vasudev, H. 2022. Structure–property correlation of plasma-sprayed Inconel625-Al2O3 bimodal composite coatings for high-temperature oxidation protection. *Journal of Thermal Spray Technology*, *31*(8), pp. 2385–2408.
38. Prashar, G. and Vasudev, H. 2022. Surface topology analysis of plasma sprayed Inconel625-Al_2O_3 composite coating. *Materials Today: Proceedings*, *50*, pp. 607–611.
39. Prashar, G. and Vasudev, H. 2022. A review on the influence of process parameters and heat treatment on the corrosion performance of Ni-based thermal spray coatings. *Surface Review and Letters*, *29*(01), p. 2230001.
40. Yedida, V.S. and Vasudev, H., 2022. A review on the development of thermal barrier coatings by using thermal spray techniques. *Materials Today: Proceedings*, *50*, pp. 1458–1464.
41. Majji, B.G.R., Vasudev, H. and Bansal, A., 2022. A review on the oxidation and wear behavior of the thermally sprayed high-entropy alloys. *Materials Today: Proceedings*, *50*, pp. 1447–1451.
42. Sunitha, K. and Vasudev, H., 2022. A short note on the various thermal spray coating processes and effect of post-treatment on Ni-based coatings. *Materials Today: Proceedings*, *50*, pp. 1452–1457.
43. Vasudev, H., 2020. Wear characteristics of Ni-WC powder deposited by using a microwave route on mild steel: microwave cladding of Ni-WC. *International Journal of Surface Engineering and Interdisciplinary Materials Science (IJSEIMS)*, *8*(1), pp. 44–54.
44. Prashar, G., Vasudev, H., Thakur, L. and Bansal, A., 2023. Performance of thermally sprayed hydroxyapatite coatings for biomedical implants: a comprehensive review. *Surface Review and Letters*, *30*(01), p. 2241001.
45. Parkash, J., Saggu, H.S. and Vasudev, H., 2022. A short review on the performance of high velocity oxy-fuel coatings in boiler steel applications. *Materials Today: Proceedings*, *50*, pp. 1442–1446.

15 Performance of High-Entropy Alloy Coatings Sprayed Using Plasma Spraying Technique

Praveen Kumar Verma and Hitesh Vasudev

15.1 INTRODUCTION

High-entropy alloy (HEA) coatings represent a new class of materials with a unique composition of multi-principal elements—generally five or more elements, or their equiatomic/near-equiatomic proportions. Such a unique composition brings about extraordinary mechanical properties, thermal stability, and resistance to wear and corrosion, leading it to be a core point of scientific and industrial interest in different industries. The characteristic feature of HEA coatings is the complex microstructure with a high diversity of elements. Unlike most of the conventional alloys, which consist mainly of one or two major constituents, HEAs distribute all their constituents much more evenly. This leads to higher configurational entropy, resulting in the stabilization of various phases and microstructures. Flexibility in composition will allow tailored properties such as enhancements in hardness as well as ductility through variations in elemental ratios. Microstructure analysis of HEA coatings typically exhibits the presence of both face-centered cubic (FCC) and body-centered cubic (BCC) phases, which consequently lead to related properties such as mechanical strength. Techniques for preparation include plasma spraying, laser cladding, and arc melting, each able to yield distinct structures and properties [1,2].

HEA coatings exhibit very good performance characteristics, especially in terms of mechanical toughness and environmental resistance. They exhibit excellent wear resistance due to high hardness values resulting from solid solution strengthening effects and finely distributed precipitates that hinder dislocation movement. For example, it has been well established that HEA coatings can possess microhardness values orders of magnitude greater than those of conventional materials, ensuring extended lifetimes for service in wear-intensive applications. In addition, HEA coatings have been observed to be highly oxidation- and corrosion-resistant and, therefore, hold great promise for high-temperature applications under extreme conditions, such as in aerospace, automotive, or marine applications. This is promoted by the

DOI: 10.1201/9781003542957-15

protective oxide layers formed because of alloying elements stabilizing the structure of the coating [2,3]. The performance of a material in industrial applications is directly related to its surface characteristics. Readily available surface modification technologies enable an economical replacement of poor base material through a coating that shows more desirable surface characteristics and performance. Conditions involving corrosive environments, wear protection, thermal insulation, and even other severe stress applications all hold a significant amount of value through surface modification. Plasma spraying has been discovered to be a potential route to fabricating HEA coatings, which combine HEA's unique properties with the versatility of the plasma spray process. HEAs are multi-element alloys that contain more than three main elements in relatively equal compositions and hence possess significant mechanical, thermal, and chemical properties. HEAs are considered to have great potential for extreme environment coatings. Plasma spraying, particularly atmospheric plasma spraying, makes it possible to deposit HEA coatings on a variety of substrates by melting and accelerating HEA feedstock particles in a high-temperature plasma jet. The process of HEA plasma spraying includes feeding powders of feedstocks into the plasma stream, where powders are partially or completely melted before getting deposited as lamellar splats on the substrate. The resulting HEA coatings generally display refined microstructures and certain specific phase compositions, such as single-phase solid solutions with FCC or BCC phases or mixed-phase compositions, which are associated with the high configurational entropies of these alloys. This phase stability is combined with the rapid solidification characteristic of plasma spraying, which favors superior hardness, wear resistance, and corrosion resistance most often [4–6].

Because of the high-entropy effect, the disordered solid solution can be easily formed. Figure 15.1 shows the HEAs. The atoms in the solid solution are randomly distributed. Each atom may be surrounded by other different atoms. Because of its structural characteristics, each principal element will be different. Such differences will inevitably cause a certain degree of deviation in the lattice atom.

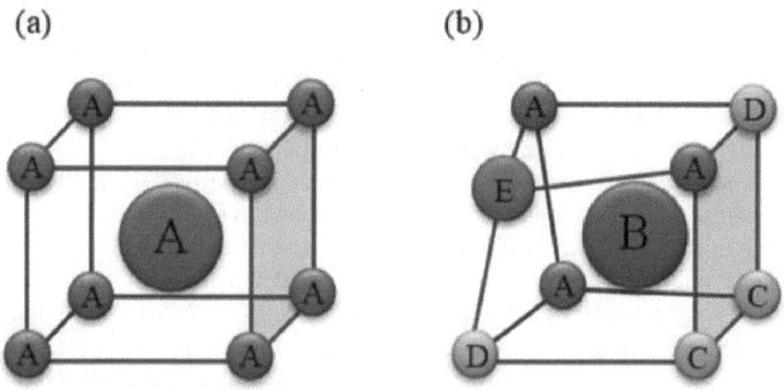

FIGURE 15.1 Schematic diagram of the lattice distortion effect. (a) body-centered cubic (BCC) no lattice distortion, one component alloy; (b) BCC severe lattice distortion, five component alloy [7]. (CC By 4.0 Attribution.)

Then, the lattice distortion is generated. This characteristic is referred to as a lattice distortion effect, which will significantly influence the microstructure and properties of HEAs [7].

Some of the other key process parameters with considerable effects on the HEA coatings' performance are plasma gas composition, power input, stand-off distance, and characteristics of the feedstock. Optimal variations of these parameters can lead to increased density, adhesion, and uniformity of the coating free from porosity or oxidation defects. However, the major drawback encountered in the plasma spraying of HEAs is the in-flight oxidation of metallic components, leading to varying alloy composition, which in turn affects its performance. This can be minimized by using inert gas shrouds or controlled atmospheres during spraying [8]. Plasma-sprayed HEA coatings are extensively developed for protecting advanced wear-resistant surfaces, thermal barrier coatings, and corrosion-resistant layers in severe environments, including marine, chemical, and high-temperature industrial settings. However, the properties of HEAs that are unmatched in hybrid form and the versatility of plasma spraying make the approach appealing for advanced coating solutions; still, processes must be optimized, and problems like oxidation and phase segregation must be overcome by research [9].

15.1.1 Types of Entropy Alloys

HEAs represent a new kind of material, with complex compositions that produce uniquely different properties. The distinctive feature of HEAs is the use of five or more major principal elements. HEAs are unique due to their composition, typically containing five or more principal elements that contribute approximately equal proportions. The HEAs have been defined as alloys that contain at least five elements in near-equal concentrations, generally between 5% and 35% by atomic percent. Examples include the most reported HEA, Cantor alloy (CrMnFeCoNi). The high configurational entropy arising from such an array of elements increases the stability of the solid solution phase, enhances the mechanical properties, increases wear and corrosion resistance, and often leads to improved thermal stability. Microstructures characterized by severe lattice distortion and sluggish diffusion contribute considerably to their strength and hardness [10].

Low-entropy alloys contain primarily one or two major elements, composing the majority part of the alloy. Minor proportions/contents are accounted for by the remaining elements. Low-entropy alloys (LEAs) usually have fewer than five major elements, with one major element dominating and thus lower configurational entropy. For example, steel is a well-known alloy in which iron is the dominant constituent, and alloys containing carbon, along with smaller fractions of elements such as manganese and chromium, exhibit phase behavior and mechanical properties that are very well predicted, as these are well understood in classical alloy theories. The LEAs themselves have good ductility and tensile strength but can lack features of the enhanced properties of their higher-entropy counterparts [11].

Medium entropy alloys (MEAs) have two to four major elements, so the mixing entropy is moderate. MEAs can contain more than one element in significant concentrations than LEAs but less than HEAs. For example, a binary copper-nickel alloy

Low Entropy Alloy	Medium Entropy Alloy	High Entropy Alloy
$\Delta S_{Conf} \leq 1.0R$	$1.0R \leq \Delta S_{Conf} \leq 1.5R$	$\Delta S_{Conf} \geq 1.5R$
One or two dominant elements. Formation of solid solution and/or compounds	Two to four dominant elements	Five or more dominant elements. Single phase solid solution phase formations (FCC/BCC or both)

FIGURE 15.2 Differences between low-, medium-, and high-entropy alloys based on their configurational entropy [13]. (CC By 4.0 Attribution.)

has better mechanical properties than pure metals. With an increase in the number of elements, several mechanical properties that are strength, hardness, corrosion resistance, etc., are greatly improved. MEAs can also exhibit complex phase formations and favorable microstructures [12].

Figure 15.2 shows the grouping of alloys using configurational entropy (ΔS_{conf}), which is proportional to the atomic-level disorder of a material. These alloys can be divided into three types: LEAs, MEAs, and HEAs. The configurational entropy for LEA is $\Delta S_{con} \leq 1.0R$ and contains one or two leading elements. These alloys tend to form either simple solid solutions or ordered intermetallic compounds. MEAs fall under the category of the range $1.0R \leq \Delta S_{con} \leq 1.5R$ with two to four significant elements and show mid-range entropy values leading to a mixture of phases or complex solid solutions. HEAs comprise five or more principal elements in near-equiatomic proportions with $\Delta S_{con} \geq 1.5R$. The high entropy encourages the solid solution to have only one phase, mostly in FCC, BCC, or mixed FCC-BCC structures. Such high entropy stabilizes the solid solution phase over ordered or segregated phases, which results in the excellent mechanical, thermal, and chemical properties of HEAs, such as superior strength, toughness, and resistance to wear, corrosion, and oxidation. This classification framework proves to be very important in formulating alloys for desired properties [13].

It has, however, been stated that the high configurational entropy effect gives the single-phase solid solution structure for HEAs in either disordered or partially ordered form as FCC, BCC, or both. The contribution of enthalpy of mixing and the high configurational entropy is therefore crucial to the forming of the single-phase structure based on Equation (15.1) [14].

$$\Delta G = \Delta H_{mix} - T\Delta H_{conf} \tag{15.1}$$

where ΔG is Gibbs free energy, T is the temperature in Kelvin, and ΔH_{mix} and ΔH_{conf} represent mixing enthalpy and configuration entropy. It has been observed that the free energy lowering makes possible the disordered solid solution microstructure owing to the spontaneous reaction that happened at high temperatures caused by the high configurational entropy. B2- and σ-phase type intermetallic compound formation issues in HEAs are still under discussion. However, the role may be very crucial in desirable properties based on the choice of the composition selection in combining both disordered and ordered structures. Other core effects, including sluggish diffusion, lattice distortion, and cocktail effects, also favor HEAs with desired properties than traditional alloys. HEAs also go by names such as complex concentrated alloys, compositionally complex alloys, multi-component alloys, and baseless alloys [15].

15.2 ELEMENTAL SELECTION STATISTICS OF HIGH-ENTROPY ALLOYS

HEAs exhibit unique features in their multi-principal-element composition. Such compositions include more complexities in interactions among the constituent elements, resulting in unique properties such as strength, high thermal stability, corrosion resistance, and wear resistance. With plasma spraying and any other technique of thermal spray, the selection of elements for HEAs becomes an important issue in the development of coatings with tailored characteristics for specific applications. The feedstock material is, thus, subjected to exposure at extremely high temperatures in the plasma jet and rapid cooling upon deposition during plasma spraying. This thermal cycling can significantly alter the phase stability, microstructure, and properties of the resulting HEA coating. Phase formation tendencies should consider single-phase FCC/BCC versus multi-phase mixtures, melting points, differences in atomic radii, and differences in chemical affinities. Additionally, selected compositions of elements are used to achieve the desired properties. For example, Cr is added for high-temperature oxidation resistance, Ni and Co for excellent thermal stability, and Fe or Al for abrasion resistance. Another major problem in designing HEAs for plasma spraying is finding feedstock that can be effectively exposed to high-temperature plasma without excessive oxidation, decomposition, and phase segregation. Advanced computational tools, like CALPHAD (Calculation of Phase Diagrams) and machine learning models, are being applied to predict phase diagrams and to optimize elemental combinations. Statistical methods further play a crucial role in screening Big Data for optimal composition. Selection and optimization of feedstock elements do, therefore, represent crucial steps in the development of high-performance plasma-sprayed coatings for such extreme conditions of aerospace applications, power generation, and marine industries. This, by its very nature, is a complicated decision-making task that strongly depends on both material science insights combined with process-specific considerations [16,17].

15.3 HIGH-ENTROPY COATING MATERIALS CLASSIFICATION

High-entropy coatings (HECs) are broadly categorized into three groups depending upon their composition and applications, namely, High-Entropy Metallic Coatings, High-Entropy Ceramic Coatings, and High-Entropy Composite Coatings. These coatings take advantage of the special properties of high-entropy materials, like

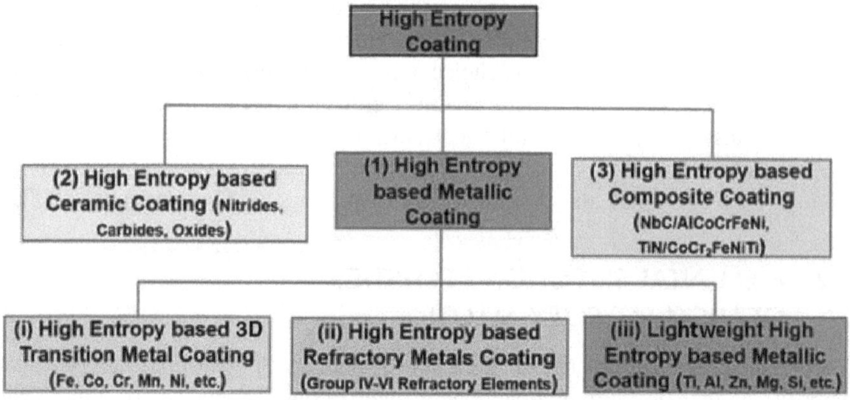

FIGURE 15.3 Classification of high-entropy coatings based on composition [19]. (CC By 4.0 Attribution.)

hardness, wear resistance, corrosion resistance, and thermal stability, to meet the demands of industrial applications [18]. Figure 15.3 shows the classification of HEC.

High-Entropy 3D Transition Metal Coatings: It mainly depends upon the transition metals such as Fe, Co, Cr, Mn, and Ni. These coatings have excellent mechanical properties, wear resistance, and corrosion resistance and are utilized under severe conditions. Coatings in groups IV–VI refractory elements involve Nb, Mo, Ta, W, and V based on high thermal stability and good resistance to oxidation and creep at high temperatures. Coatings of lightweight elements like Ti, Al, Zn, Mg, and Si are applied where low-density coatings are required but strength and corrosion resistance are not compromised [20]. High-entropy ceramic coatings are primarily composed of nitrides, carbides, and oxides. They have high hardness values and are resistant to wear with good thermal stability. Therefore, they are very useful for applications such as cutting tools, thermal barrier coatings, and oxidation-resistant surfaces. High-entropy composite coatings combine metallic as well as ceramic properties, such as NbC/AlCoCrFeNi or TiN/CoCrFeNiTi, creating multi-functional coatings with high-performance attributes. The applications that require a good balance between toughness, wear resistance, and thermal stability are particularly suited to these materials [21].

15.4 MICROSTRUCTURAL CHARACTERISTICS OF PLASMA-SPRAYED HEA COATINGS

Microstructure of thermal spray coatings: The microstructure depends strongly on the method of thermal spraying (TS), the specific TS parameters, and the nature of the feedstock. The TS technique encompasses a wide range of coating parameters, as defined in 'spray tables', including, plasma current and voltage, primary and secondary gas flow rates for plasma spray, carrier gas flow rate, stand-off distance, and the design and dimensions of the nozzle for high-velocity oxy-fuel (HVOF), D-gun, and cold spray processes. The feedstock characteristics also significantly influence the coating microstructure, which is highly susceptible to the feedstock synthesis technique (as discussed in the previous section). Thus, both the feedstock and TS process have to be optimized to obtain a coherent coating also free of undesirable phases and adverse microstructural artifacts. This section elucidates the impact of the TS technique used on the as-sprayed microstructure of HEAs [22].

Many factors affect the microstructure and properties of complex alloys. Considering HEAs have at least five major elements, whereas conventional alloys are based on one or two metallic elements, there are different basic effects between them. From the thermodynamic point of view, the high-entropy effect may influence the development of complex phases. From a kinetical point of view, the slow diffusion effect may slow down the phase transformation. In structure-related aspects, the severe distortion of the lattice may modify the properties to some degree. Concerning properties, the cocktail effect brings a surplus to the predicted mixing amount, which is usually caused by mutual interactions between various atoms and severe lattice distortions. This high-entropy effect drives HEAs and enhances configurational entropy in equimolar or near-equimolar alloys with five or more elements, perhaps favoring solid solution phases rather than intermetallic compounds. The severe lattice deformation effect is generally comparable to single-dominant alloys, where nodes of the lattice are predominantly occupied by the dominant constituent. For multi-component alloys, if the effects of chemical ordering are not considered, then all these elements have an equal probability of occupying the network node. The reason is that the size variations of the atoms concerned lead to considerable lattice defects in these densely packed complex phases. The dislocation at a given point of the atomic lattice would depend on the atom at that specific site and the kinds of atoms that are occupying neighboring sites. Such distortions are more significant in the crystal lattice of an HEA if they happen because of such reasons as opposed to when they occur in a conventional alloy [23,24].

The 'cocktail effect' demonstrates how impressive properties of materials frequently emerge from unsuspected synergies. Indeed, analogously unpredictable, synergistic effects may be observed even in other material behaviors, such as physical properties like near-zero thermal coefficient of expansion, functional properties like thermoelectric response or photovoltaic conversion, or strange combinations of structural properties, such as ultra-high hardness combined with excellent fracture resistance, fatigue resistance, or ductility. In all these cases, the properties are imparted by the composition, microstructure, and electronic structure of the material. Unusual combinations of elements and the microstructures featured by HEA determine associations of properties that present a high grade of novelty [25]. Figure 15.4 shows the effects that influence the formation of HEAs.

In this study, Tian et al. [27] reported that the powder milled for 10 hours was selected as a feedstock for coating deposition due to its proper particle size distribution and good flowability. Figure 15.5 shows the surface morphology of the APS AlCoCrFeNiSi HEA coating. It can be observed from Figure 15.5a that the coating exhibited a rough morphology that is typically observed on the surfaces of atmospheric plasma-sprayed coatings. As can be seen at higher magnification in Figure 15.5b, most of the splats were fully molten and well flattened with obvious outlines of flattened splats. Spherical and droplet-shaped particles less than 10 μm

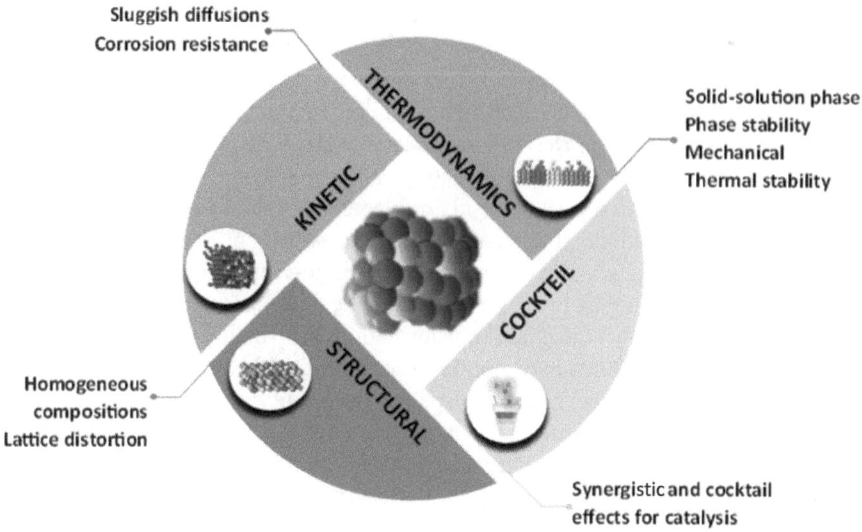

FIGURE 15.4 The four effects that influence the formation of high-entropy alloys [26]. (CC By 4.0 Attribution.)

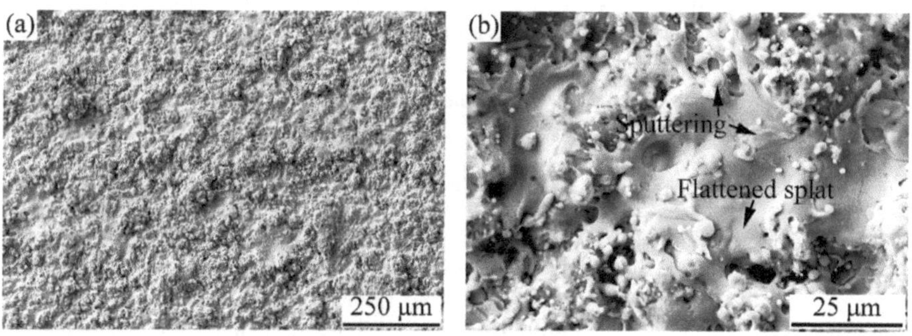

FIGURE 15.5 Surface morphology of the AlCoCrFeNiSi coating at magnifications of 100 × (a) and 1000 × (b) [27]. (CC By 4.0 Attribution.)

in size were also found on the surface of the coating produced, which were directly brought from the sputtering of the flying particles onto the already formed coating surface due to impact.

15.5 PERFORMANCE CHARACTERISTICS OF PLASMA-SPRAYED HEA COATINGS

Among the most important properties affecting the suitability of plasma-sprayed HEA coatings for demanding applications such as wear-resistant surfaces and protective coatings, their mechanical properties stand out. Coatings obtained by plasma spraying are typically characterized by superior hardness, toughness, and adhesion strength due to rapid quenching inherent to the process and the inherent multi-principal-element composition of HEAs. One of the defining mechanical characteristics is the high hardness of HEA coatings sprayed by plasma, derived from the refined microstructure as well as from solid solution strengthening enabled by the high configurational entropy of the alloys. The fast-cooling rates inherent in plasma spraying generally lead to ultra-fine grains or amorphous microstructures, improving hardness and wear resistance. Moreover, multi-phase HEAs can exhibit notable fracture toughness because of the combined effects of ductile FCC phases and hard BCC or intermetallic phases. Besides that, adhesion strength is one major factor since it is associated with durability through mechanical and thermal stresses. Optimization of plasma spraying parameters, including power input, stand-off distance, and particle feedstock size, is required to produce coatings with strong interfacial bonding between the HEA coating and substrate. Despite their superior properties, HEA coatings that were plasma-sprayed sometimes experience problems like porosity or residual stresses, which might have a tiny nuance effect on their mechanical performance [28,29].

Post-deposition treatments, such as annealing or hot isostatic pressing, are generally used to improve bonding strength, reduce porosity, and further enhance mechanical strengthening. The excellent hardness, toughness, and adhesion strength features of the plasma-sprayed HEA coatings can be recommended for applications in harsh environments, such as aggressive surroundings, where wear, impact, and mechanical degradation are the issues. These features indicate an increased interest in HEA coatings for use in the aerospace, automotive, and industrial sectors [30].

15.6 WEAR RESISTANCE

HEA coatings, which have undergone plasma spraying, provided excellent wear resistance properties and easily handled abrasive, adhesive, and erosive wear conditions. The multi-principal element nature of HEA composition and the related solid solution strengthening enhance hardness and contribute to better wear performance. Besides this, the rapid cooling during plasma spraying enables a finer microstructure to form into densely coated material with lamellar splats and few grain boundaries, which further reduces wear rates. HEA coatings' wear resistance is related to phase structure. Typically, coatings can be of a single FCC, BCC, or mixed-phase composition. For instance, the FCC phases ensure ductility, whereas BCC and intermetallic phases are responsible for the hardness and abrasion resistance. Optimized plasma spraying parameters and post-deposition treatments further reduce the porosity and defects, enhancing wear resistance. Thus, they are used in industrial applications such as cutting tools, turbine blades, and highly wear-intensive parts [31].

Noble et al. [32] performed a worn morphology study on annealed AlCoCrFeNi-coated samples that went through the atmospheric and high-temperature wear tests. The samples exhibited enhanced wear resistance, confirmed by the outcomes analyzed from the specific wear rate (SWR) and coefficient of friction (COF.) The worn morphology of samples subjected to the 400°C high-temperature wear test was investigated, as they were seen to be performing better under this condition.

Figures 15.6a and b reveal the worn surfaces of annealed APS-coated samples of AlCoCrFeNi at 15 N and 45 N load conditions, respectively. As illustrated in Figure 15.6a, under a load of 15 N, the wear surface is relatively smoother with fewer cracks and pores due to annealing that improves the topography and promotes the formation of Al and Cr-rich precipitates. The coating showed minimal flaking, low plastic deformation, and enhanced wear resistance. This was attributed to the decrease in contact pressure, especially with the formation of a protective oxide layer. The main wear mechanism is related to abrasion with debris, resulting in minor surface deformation. In contrast, at 45 N (Figure 15.6b), the wear mechanisms changed toward plastic deformation, cracking, and delamination with deeper grooves and pits as well as more extensive surface degradation. The increased stress at higher loads propagates cracks and accelerates delamination, with abrasive wear dominating and causing severe plowing. Figures 15.6c and d show the worn morphology of the coated samples after wear tests at 400°C under similar loads. At 15 N (Figure 15.6c), minor pits and grooves with fewer cracks are observed due to reduced stress concentration and thermal softening, thus showing better grain refinement and higher wear resistance. Predominant grooves, cracks, and plastic deformation, along with the loss of the protective oxide layer, transition to abrasive wear at 45° impact angle (Figure 15.6).

15.7 CORROSION RESISTANCE

Plasma-sprayed HEA coatings are characterized by corrosion resistance, based on the inherent chemical stability of HEAs. These HEA elements, typically incorporating Cr, Ni, Al, and Co, will help form a protective, stable oxide layer on the outer coating surface that guards the material from further destruction. Local corrosion mechanisms, such as pitting and crevice corrosion, are suppressed by the dense microstructure from plasma spraying and the high configurational entropy. HEA coatings showed improved corrosion resistance to the most aggressive environments, which range from harsh marine to chemical and industrial high-temperature processes. However, some technical problems in the form of in-flight oxidation during plasma spraying might alter the surface chemistry before its final characteristics can be attained toward corrosion resistance. All these could be offset by using inert gas shrouds or post-spraying surface treatments [33].

15.8 THERMAL STABILITY AND OXIDATION RESISTANCE

Plasma-sprayed HEA coatings have outstanding thermal stability and oxidation resistance, hence promising for high-temperature applications such as thermal barriers and turbine components. Due to the solid solution phases being stabilized by their highly configurational entropy, HEAs bear the potential to stabilize even at an elevated temperature without experiencing phase segregation and grain growth. Elements Al, Cr, and Si within HEAs combine to form protective oxide scales like Al_2O_3 and Cr_2O_3 oxides, providing advanced oxidation resistance under extreme thermal cycling [34].

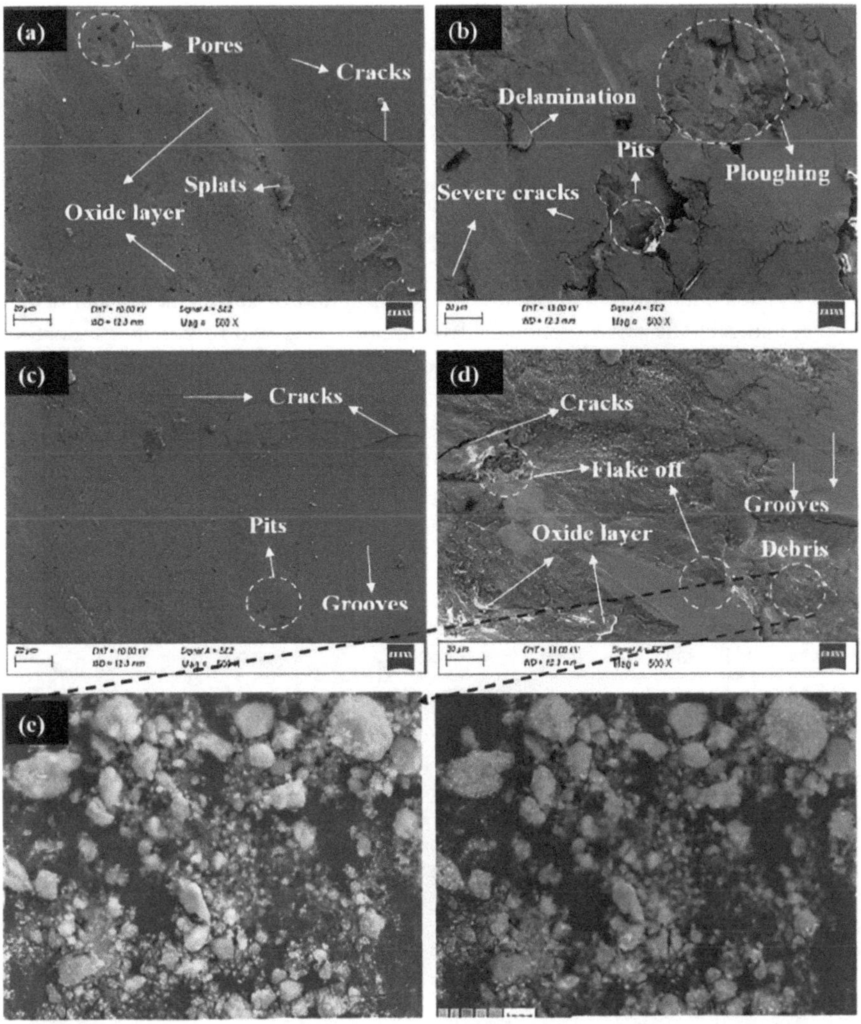

FIGURE 15.6 Worn morphology at atmospheric temperature: (a) annealed AlCoCrFeNi-coated sample at 15N, (b) annealed AlCoCrFeNi-coated sample at 45N 4m/s. Wear test at 400°C, and the (c) annealed AlCoCrFeNi-coated sample at 15N, (d) annealed AlCoCrFeNi-coated sample at 45N, (e) debris analysis [32]. (CC By 4.0 Attribution.)

The plasma spraying process itself leads to the development of a fine-grained microstructure that can withstand thermal stresses and ensure the integrity of the coating over time. However, residual stresses and porosity issues are detrimental to thermal stability by a few percent, which may be improved by either process optimization or post-deposition heat treatments. Overall, the holistic combination of thermal stability, oxidation resistance, and mechanical robustness makes plasma-sprayed HEA coatings an ideal selection for high-temperature and oxidative environments [35].

Vasudev et al. [36] have researched plasma-sprayed Al_xCrCoFeNi HEAs for high-temperature oxidation resistance. Figure 15.7 shows the oxidation behaviors of stainless steel (SS-304) and three alloy coatings, namely, AlCrCoFeNi, $Al_{0.5}$CoFeNi, and $Al_{0.1}$CoFeNi. The bar graph in Figure 15.7 (a) exhibits the weight gain (mg/cm²) after oxidation, showing that SS-304 has achieved maximum weight gain, which is symptomatic of more severe oxidation; in contrast, the weight gains of the alloy coatings are considerably lower, indicating excellent oxidation resistance. Surface morphology scanning electron microscope (SEM) micrographs after oxidation are presented for SS-304 (b), AlCrCoFeNi (c), $Al_{0.5}$CoFeNi (d), and $Al_{0.1}$CoFeNi (e) followed by elemental mapping to check the distribution of key elements such as Cr, Co, Fe, and Ni. The cross-sectional SEM images (f, g, and h) give more detailed information about the substrate, coating region, and oxidized zone of each coating. The AlCrCoFeNi coating offers a prominent coating region with good adhesion and oxidation resistance. In the same way, the $Al_{0.5}$CoFeNi (g) and $Al_{0.1}$CoFeNi (h) coatings serve to separate the substrate from the oxidized zone and thus provide good protection. Overall, it is seen that the alloy coatings, particularly with a higher percentage of aluminum, give better oxidation resistance than bare SS-304.

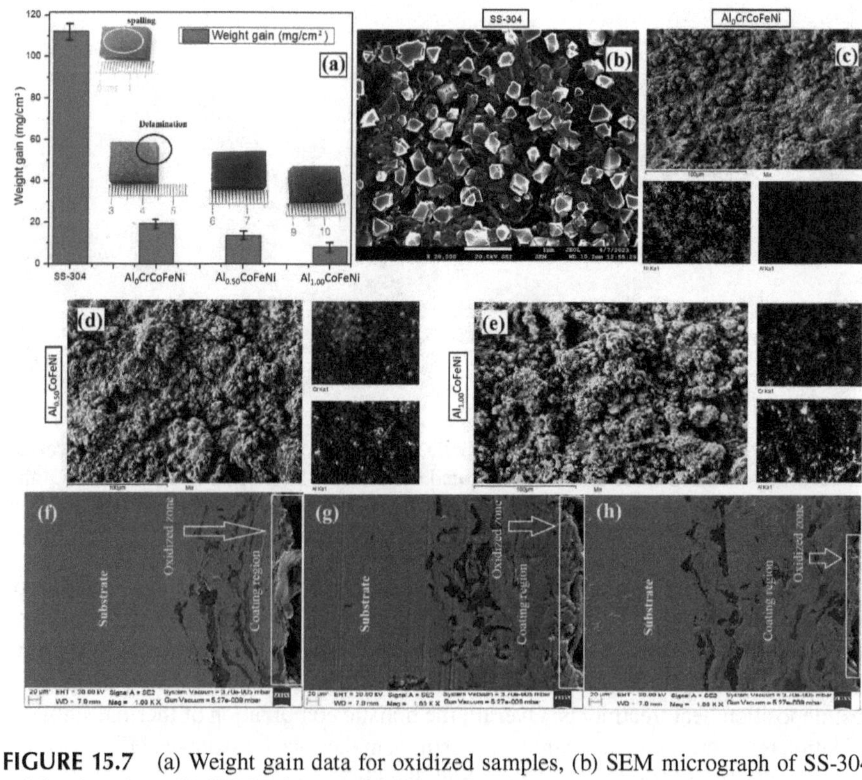

FIGURE 15.7 (a) Weight gain data for oxidized samples, (b) SEM micrograph of SS-304 oxidized surface, (c) SEM micrograph of A1, (d) A2, (e) A3 coating, (f) cross-section of oxidized A1, (g) A2, and (h) A3 coatings [36]. SEM, scanning electron microscope.

15.9 APPLICATIONS OF PLASMA-SPRAYED HEA COATINGS

Plasma-sprayed HEA coatings have found versatile applications across various industries due to their unique combination of properties, such as excellent wear resistance, high thermal stability, and corrosion resistance. In the aerospace and defense sectors, these coatings are used to enhance the durability of components exposed to extreme temperatures and abrasive environments, such as turbine blades, engine parts, and heat shields. Their ability to withstand high thermal stresses makes them ideal for power generation applications, particularly in gas turbines and heat exchangers, where efficient performance under high heat is crucial. In the chemical and petrochemical industries, HEA coatings are applied to reactors, pipelines, and heat exchangers, protecting them from aggressive chemical environments and high-temperature corrosion, which increases the longevity of these critical components. The biomedical field also benefits from plasma-sprayed HEA coatings, where they are used in implants, prosthetics, and surgical tools. Their biocompatibility, along with their resistance to wear and corrosion, ensures that they maintain functionality in the human body for extended periods. Overall, plasma-sprayed HEA coatings offer an effective solution for industries that require materials capable of enduring harsh conditions, thus improving the performance, reliability, and lifespan of critical components. These coatings continue to evolve, pushing the boundaries of material science and expanding their applications in advanced engineering fields [37].

15.10 CHALLENGES AND LIMITATIONS

Several challenges and limitations still need to be addressed for HEAs with such promising attributes in various applications to be optimized for performance and viable for practical large-scale industrial utilization. One primary issue is process-induced defects in the HEAs, which critically affect the material's integrity and performance. Oxidation, for instance, might cause the growth of brittle oxides that are averse to the mechanical properties of HEAs. In addition, many manufacturing procedures inevitably introduce porosity into the material, which can act as initiation sites for cracks or any other degradation mechanisms subjected to stress. Residual stresses are another typical flaw produced by thermal gradients during processing and lead to warping or unexpected failure modes. Scalability is another significant challenge for the large-scale implementation of HEAs in industry. Although excellent properties have been demonstrated by laboratory-scale production methods, the challenge is to scale up the processes into industrially viable quantities uniformly and reproducibly. The compositional complexities of HEAs require very tight control over the processing conditions, which is challenging to maintain at large scales. Thus, significant technical challenges prevail in scaling small-scale experimental setups toward industrial-scale production to satisfy high-quality and consistency requirements [38].

Cost is another limiting factor to the wider application of HEAs. The feedstock used to manufacture HEAs includes several raw materials, which are quite expensive compared to conventional alloys. This is further posed by a substantial requirement for extremely pure elements and by much more complicated processing requirements.

This cost factor is again compounded by additional requirements for special, expensive fabrication and finishing equipment and processes, which increase the overall expense. However, though superior properties of HEAs may justify higher costs in some high-performance applications, the economic viability for broader use in cost-sensitive industries remains questionable. It is by continued research and development that these challenges are going to be addressed, unlocking the full industrial application potential of HEAs [39,40]. HEAs have gained attention for their superior mechanical, thermal, and corrosion-resistant properties, particularly when applied as coatings using plasma-spraying techniques [41–46]. These coatings, including those based on Inconel and other HEAs, offer enhanced performance under extreme conditions, such as high-temperature oxidation and erosion [47–53]. The addition of ceramic phases like Al_2O_3 further improves wear resistance, while heat treatments can significantly influence corrosion properties, as demonstrated in recent works [54,55].

15.11 CONCLUSION

In summary, plasma-sprayed HEA coatings possess properties that are extraordinary for extreme applications. They demonstrate excellent wear resistance, high hardness, and superior oxidation and corrosion resistance due to the special composition of HEAs, involving principal elements in multiple combinations. The coatings, thus, can be deposited well by plasma spraying to ensure good bonding with the substrate as well as a uniform microstructure, which is critical to ensuring the integrity and performance of the coatings under hostile conditions. The studies and experimental data demonstrate the fact that HEA coatings substantially increase the life and reliability of components exposed to extreme environments: either elevated temperatures or corrosive atmospheres. Thus, they can find especially valuable applications within the aerospace, power generation, and chemical processing industries, where conventional materials simply break. Additionally, due to the plasma spraying process, its flexibility gives free rein in determining coating compositions and thicknesses so that optimized solutions can be specified for an application. However, many challenges have to be overcome before such HEA coatings can be fully exploited. Process-induced defects such as oxidation, porosity, and residual stresses must be minimized using optimized spraying parameters and post-processing treatments. The scalability of plasma spraying is also a barrier for large-scale industrial applications. HEA feedstock materials have high-cost implications. A sustained level of research and development challenges will remain crucial for the promotion of HEA coatings application, paving the way for further widespread industry application that might significantly increase performance and durability in critical components.

REFERENCES

1. Cantor B, Chang ITH, Knight P, Vincent AJ. Microstructural development in equiatomic multicomponent alloys. *Mater Sci Eng A*. 2004;375–377:213–218.
2. Yeh JW, Chen SK, Lin SJ, et al. Nanostructured high-entropy alloys with multiple principal elements: novel alloy design concepts and outcomes. *Adv Eng Mater.* 2004;6(5):299–303.

3. Abdullah MR, Peng Z. Review and perspective on additive manufacturing of refractory high entropy alloys. *Mater Today Adv.* 2024 Jun 1;22:100497.
4. Manzoni AM, Daoud H, Mondal S, et al. Investigation of phases in $Al_{23}Co_{15}Cr_{23}Cu_8Fe_{15}Ni_{16}$ and $Al_8Co_{17}Cr_{17}Cu_8Fe_{17}Ni_{33}$ high-entropy alloys and comparison with equilibrium phases predicted by Thermo-Calc. *J Alloys Compd.* 2013;552:430–436.
5. Miracle DB, Senkov ON. A critical review of high entropy alloys and related concepts. *Acta Mater.* 2017;122:448–511.
6. Senkov ON, Wilks GB, Miracle DB, Chuang CP, Liaw PK. Refractory high-entropy alloys. *Intermetallics.* 2010;18(9):1758–1765.
7. Wang Z, Zhang S. Research and application progress of high-entropy alloys. *Coatings.* 2023 Nov 9;13(11):1916.
8. Verma PK, Singh S, Kapoor M, Singh S. A review on the surface topography and corrosion behavior of Mg-alloy coatings for biomedical implants. *Results Surf Interf.* 2024 May 9:100227. https://doi.org/10.1016/j.rsurfi.2024.100227
9. Zhang Y, Zuo TT, Tang Z, et al. Microstructures and properties of high-entropy alloys. *Prog Mater Sci.* 2014;61:1–93.
10. Yeh JW, Chen SK, Lin SJ, et al. Nanostructured high-entropy alloys with multiple principal elements: novel alloy design concepts and outcomes. *Adv Eng Mater.* 2004;6(5):299–303.
11. Miracle DB, Senkov ON. A critical review of high entropy alloys and related concepts. *Acta Mater.* 2017;122:448–511.
12. George, EP, Raabe, D, Ritchie, RO. High-entropy alloys. *Nat Rev Mater* 2019;4:515–534. https://doi.org/10.1038/s41578-019-0121-4
13. Bhaskaran Nair R, Supekar R, Morteza Javid S, Wang W, Zou Y, McDonald A, Mostaghimi J, Stoyanov P. High-entropy alloy coatings deposited by thermal spraying: a review of strengthening mechanisms, performance assessments and perspectives on future applications. *Metals.* 2023 Mar 13;13(3):579.
14. Yeh J-W, Chen S-K, Lin SJ, et al. Nanostructured high-entropy alloys with multiple principal elements: novel alloy design concepts and outcomes. *Adv Eng Mater.* 2004;6(5):299–303. https://doi.org/10.1002/adem.200300567.
15. Gludovatz B, Hohenwarter A, Furuhara T, et al. Microstructure and properties of high-entropy alloys. *JOM.* 2014;66(3):472–479. https://doi.org/10.1016/j.pmatsci.2013.10.001.
16. Zhang Y, Zuo T, Yang X, et al. High-entropy alloys: design, processing and properties. *Mater Sci Eng R.* 2020;131:1–18. https://doi.org/10.1016/j.mser.2018.09.001.
17. Balaji, V, Xavior, A. Development of high entropy alloys (HEAs): current trends. *Heliyon.* 2024;10:e26464 https://doi.org/10.1016/j.heliyon.2024.e26464.
18. Raghavan R, Palani R, Chattopadhyay K, et al. High entropy coatings: a review of their development and applications. *Surf Coat Technol.* 2021;405:126580. https://doi.org/10.1016/j.surfcoat.2020.126580.
19. Arshad M, Amer M, Hayat Q, Janik V, Zhang X, Moradi M, Bai M. High-entropy coatings (HEC) for high-temperature applications: materials, processing, and properties. *Coatings.* 2022 May 18;12(5):691.
20. Pickering EJ, Jones NG. High-entropy alloys: a critical assessment of their founding principles and future prospects. *Int Mater Rev.* 2016 Apr;61(3):183–202.
21. Otto F, Yang Y, Bei H, George EP. Relative effects of enthalpy and entropy on the phase stability of equiatomic high-entropy alloys. *Acta Materialia.* 2013 Apr 1;61(7):2628–2638.
22. Ren MX, Li BS, Fu HZ. Formation condition of solid solution type high-entropy alloy. *Trans Nonferrous Met Soc China.* 2013 Apr 1;23(4):991–995.

23. MacDonald BE, Fu Z, Zheng B, Chen W, Lin Y, Chen F, Zhang L, Ivanisenko J, Zhou Y, Hahn H, Lavernia EJ. Recent progress in high entropy alloy research. *JOM.* 2017 Oct;69:2024–2031.
24. Alaneme KK, Bodunrin MO, Oke SR. Processing, alloy composition and phase transition effect on the mechanical and corrosion properties of high entropy alloys: a review. *J Mater Res Technol.* 2016 Oct 1;5(4):384–393.
25. Salishchev GA, Tikhonovsky MA, Shaysultanov DG, Stepanov ND, Kuznetsov AV, Kolodiy IV, Tortika AS, Senkov ON. Effect of Mn and V on structure and mechanical properties of high-entropy alloys based on CoCrFeNi system. *J Alloys Compd.* 2014 Apr 5;591:11–21.
26. Caramarin S, Badea IC, Mosinoiu LF, Mitrica D, Serban BA, Vitan N, Cursaru LM, Pogrebnjak A. Structural particularities, prediction, and synthesis methods in high-entropy alloys. *Appl Sci.* 2024 Aug 27;14(17):7576.
27. Tian L, Fu M, Xiong W. Microstructural evolution of AlCoCrFeNiSi high-entropy alloy powder during mechanical alloying and its coating performance. *Materials.* 2018 Feb 23;11(2):320.
28. Zirari T, Trabadelo V. A review on wear, corrosion, and wear-corrosion synergy of high entropy alloys. *Heliyon.* 2024 Feb 11.
29. Patel P, Roy A, Sharifi N, Stoyanov P, Chromik RR, Moreau C. Tribological performance of high-entropy coatings (HECs): a review. *Materials.* 2022 May 21;15(10):3699.
30. Liu D, Zhao J, Li Y, Zhu W, Lin L. Effects of boron content on microstructure and wear properties of FeCoCrNiBx high-entropy alloy coating by laser cladding. *Appl Sci.* 2019 Dec 19;10(1):49.
31. Lv S, Zu Y, Chen G, Fu X, Zhou W. An ultra-high strength CrMoNbWTi-C high entropy alloy co-strengthened by dispersed refractory IM and UHTC phases. *J Alloys Compd.* 2019 Jun 5;788:1256–1264.
32. Noble N, Radhika N, Natarajan J. Microstructure and wear behaviour of AlCoCrFeNi-coated SS316L by atmospheric plasma spray process. *Sci Technol Adv Mater.* 2024 Dec 31;25(1):2341611.
33. Meghwal A, Anupam A, Luzin V, Schulz C, Hall C, Murty BS, Kottada RS, Berndt CC, Ang AS. Multiscale mechanical performance and corrosion behaviour of plasma sprayed AlCoCrFeNi high-entropy alloy coatings. *J Alloys Compd.* 2021 Feb 15;854:157140.
34. Kumar S, Narayanan TS, Raman SG, Seshadri SK. Thermal oxidation of CP-Ti: evaluation of characteristics and corrosion resistance as a function of treatment time. *Mater Sci Eng C.* 2009 Aug 1;29(6):1942–1949.
35. Cofer CG, Economy J. Oxidative and hydrolytic stability of boron nitride—a new approach to improving the oxidation resistance of carbonaceous structures. *Carbon.* 1995 Jan 1;33(4):389–395.
36. Majji BG, Vasudev H, Bansal A. Application of plasma-sprayed AlxCrCoFeNi High-Entropy alloys for high-temperature oxidation resistance. *Mater Lett.* 2024 Apr 15;361:136069.
37. Ang AS, Berndt CC, Sesso ML, Anupam A, Kottada RS, Murty BS. Plasma-sprayed high entropy alloys: microstructure and properties of AlCoCrFeNi and MnCoCrFeNi. *Metall Mater Trans A.* 2015 Feb;46:791–800.
38. Rotich SK, Kipkirui NG, Lin TT, Chen SH. Effect of varying plasma powers on high-temperature applications of plasma-sprayed Al0. 5CoCrFeNi2Ti0. 5 coatings. *Materials.* 2022 Oct 15;15(20):7198.
39. Kiplangat RS, Lin TT, Kipkirui NG, Chen SH. Microstructure and mechanical properties of the plasma-sprayed and cold-sprayed $Al_{0.5}CoCrFeNi_2Ti_{0.5}$ high-entropy alloy coatings. *J Therm Spray Technol.* 2022 Apr;31(4):1207–1221.

40. Steinhäuser S, Wielage B, Hofmann U, Schnick T, Ilyuschenko A, Azarova T. Plasma-sprayed wear-resistant coatings with respect to ecological aspects. *Surf Coat Technol.* 2000 Sep 1;131(1–3):365–371.
41. Vasudev H. Wear characteristics of Ni-WC powder deposited by using a microwave route on mild steel: microwave cladding of Ni-WC. *Int J Surf Eng Interdiscip Mater Sci (IJSEIMS).* 2020;8(1):44–54.
42. Prashar G, Vasudev H, Thakur L. High-temperature oxidation and erosion resistance of Ni-based thermally-sprayed coatings used in power generation machinery: a review. *Surf Rev Lett.* 2022;29(03):2230003.
43. Yedida VS, Vasudev H. A review on the development of thermal barrier coatings by using thermal spray techniques. *Mater Today: Proc.* 2022;50:1458–1464.
44. Prashar G, Vasudev H. Structure–property correlation of plasma-sprayed Inconel625-A12O3 bimodal composite coatings for high-temperature oxidation protection. *J Therm Spray Technol.* 2022;31(8):2385–2408.
45. Singh P, Bansal A, Vasudev H, Singh P. In situ surface modification of stainless steel with hydroxyapatite using microwave heating. *Surf Topogr Metrol Prop.* 2021;9(3):035053.
46. Prashar G, Vasudev H. A review on the influence of process parameters and heat treatment on the corrosion performance of Ni-based thermal spray coatings. *Surf Rev Lett.* 2022;29(01):2230001.
47. Singh G, Vasudev H, Bansal A, Vardhan S. Influence of heat treatment on the microstructure and corrosion properties of the Inconel-625 clad deposited by microwave heating. *Surf Topogr Metrol Prop.* 2021;9(2):025019.
48. Parkash J, Saggu HS, Vasudev H. A short review on the performance of high velocity oxy-fuel coatings in boiler steel applications. *Mater Today Proc.* 2022;50:1442–1446.
49. Majji BGR, Vasudev H, Bansal A. A review on the oxidation and wear behavior of the thermally sprayed high-entropy alloys. *Mater Today Proc.* 2022;50:1447–1451.
50. Prashar G, Vasudev H. Surface topology analysis of plasma sprayed Inconel625-A12O3 composite coating. *Mater Today Proc.* 2022;50:607–611.
51. Prashar G, Vasudev H. Hot corrosion behavior of super alloys. *Mater Today: Proc.* 2020;26, 1131–1135.
52. Vasudev H, Singh P, Thakur L, Bansal A. Mechanical and microstructural characterization of microwave post processed Alloy-718 coating. *Mater Res Express.* 2020;6(12):1265f5.
53. Vasudev H, Prashar G, Thakur L, Bansal A. Electrochemical corrosion behavior and microstructural characterization of HVOF sprayed Inconel718-Al_2O_3 composite coatings. *Surf Rev Lett.* 2022;29(02):2250017.
54. Prashar G, Vasudev H, Thakur L, Bansal A. Performance of thermally sprayed hydroxyapatite coatings for biomedical implants: a comprehensive review. *Surf Rev Lett.* 2023;30(01):2241001.
55. Sunitha K, Vasudev H. A short note on the various thermal spray coating processes and effect of post-treatment on Ni-based coatings. *Mater Today Proc.* 2022;50:1452–1457.

16 Case Study on Boiler Steels' Applications for High-Temperature Conditions

Gaurav Prashar, Hitesh Vasudev, and Lalit Thakur

16.1 INTRODUCTION

For the generation and transfer of steam from the boiler to turbines in thermal power plants, the use of materials with high heat resistance is essential. Choosing the right boiler material involves factors such as service requirements, financial considerations, and the availability of the size that is needed [1,2]. Boiler tube failure occurs periodically. Extended overheating and erosion, also referred to as dry erosion, are phenomena linked to the breakdown of boiler tubes. The primary factors contributing to the degradation of boiler tubes utilized in energy production are outlined [3]. The expenses (primarily maintenance) associated with damaged tubes may increase to 54% of the entire production cost [4,5]. The expenses related to erosion can reach approximately US$150 million annually, as significant solid particle erosion substantially diminishes the service life and efficiency of these components. Identifying and rectifying the primary cause of tube failures is essential for minimizing the likelihood of future issues [6–10]. In this study, a failure analysis of the boiler bed tube is presented.

16.2 INVESTIGATION

A comprehensive analysis was conducted to identify the root cause leading to the breakdown of boiler tubes. The burst tube is manufactured from SA210 GRA1 steel. After 12 months of use, the boiler tube exhibited a window-type split in the parent metal, accompanied by secondary signs of erosion. The initial section of the tube has a width of 50 mm (see inset Figure 16.1a) and a length of 101 mm (see inset Figure 16.1b). This study aims to identify the breakdown mechanism and recommend suitable preventive measures through a comprehensive analysis of the failed tube. Table 16.1 presents the details of fuel composition. The specifications for the boiler tubes and the associated operating conditions are outlined below.

DOI: 10.1201/9781003542957-16

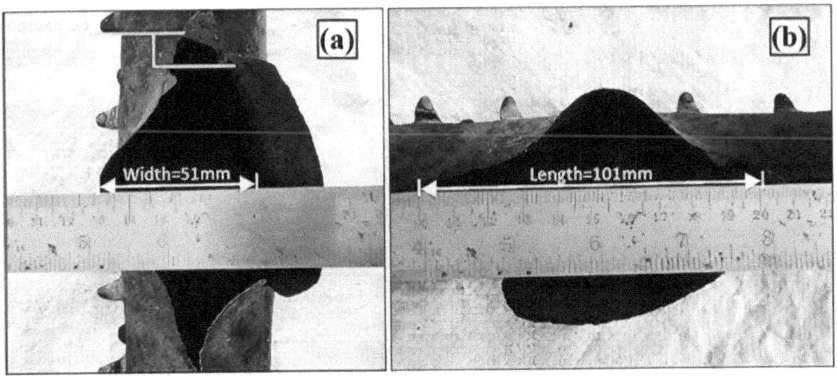

FIGURE 16.1 Measured dimensions of the tube: (a) width and (b) length.

16.2.1 Boiler Type and Service Conditions

1. Boiler type: Water tube
2. Manufacturer/Model: Thermax Ltd./Babcock and Wilcox
3. Boiler temperature: 700°C–850°C
4. Fuels used for steam production: Rice husk/Indian coal
5. Steam temperature & pressure: 480°C ± 5°C/65 kg/cm^2
6. Fly ash velocity: 19–25 m/s

16.3 EXPERIMENTAL METHODOLOGY

The failing tube underwent a failure study, paying particular attention to the bursting portion of the tube. Samples from various parts of the failing tube were prepared for the examination of the inner wall surface topography. The sample microstructures were analyzed using scanning electron microscopy, EDAX, and optical microscopy.

16.4 RESULTS AND DISCUSSIONS

16.4.1 Visual Inspection

The failed tube is shown in Figure 16.1 as reported during a visit to the site. The hole was encircled by the bent metal. The conclusion drawn indicates that the breakdown of bed coils is attributed to the thinning of the coils over time (refer to Figure 16.2). This thinning is a result of the abrasive characteristics of the bed material ash, which impacts the boiler bed coils at high velocities. Consequently, the thinning of bed coils takes place starting from the outer diameter of the tube. Based on this finding, measurements were conducted to assess the wall thickness of the damaged tube. The fire-facing side exhibits a minimum thickness of approximately 3 mm, which is considerably less than the wall thickness observed on the rear side, measured at 9 mm.

The as-received failed tube was cut into three samples and subjected to analysis both at the location of the rupture and a short distance away from it in order to determine the cause of the tube's failure. Two-minute samples, designated Sample (a) and

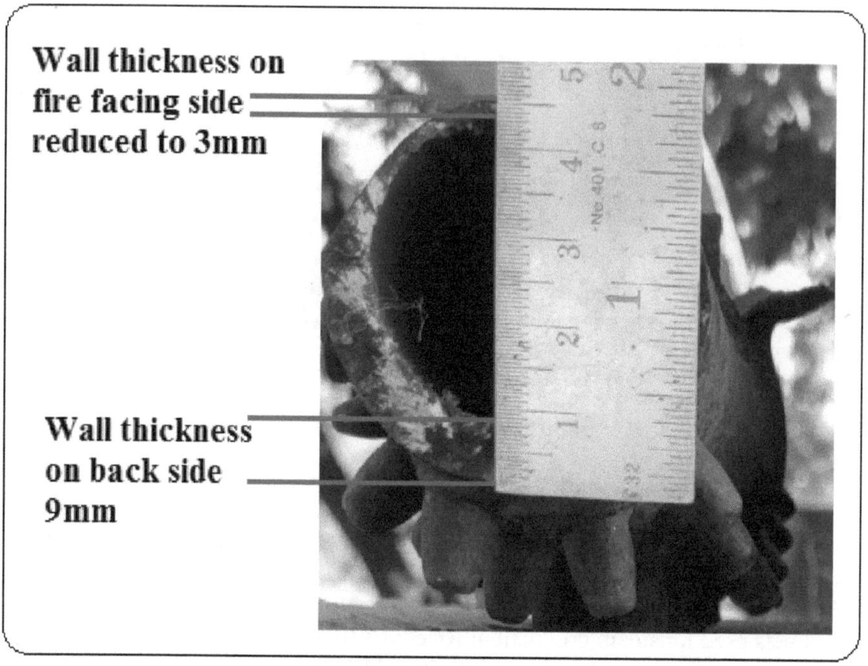

FIGURE 16.2 Failed boiler bed tube inspection.

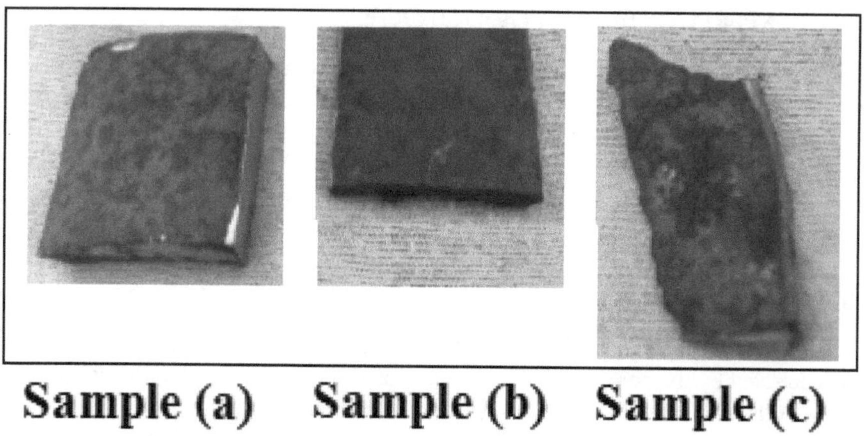

FIGURE 16.3 The surface of the boiler tube on the (a) fire-facing side, (b) the back side, and (c) the cocked-up metal slice around the failure region (c).

Sample (b), were cut from the fire-facing side and the back of the tube, respectively. These samples are depicted in Figures 16.3a and b.

The crooked slice surrounding the burst hole was likewise cut down and given the Sample (c) designation (Figure 16.3c). As can be observed, pits are easily visible on the surface of the fire-facing side (Sample a) and the cocked-up slice of the burst hole (Sample c), but are barely discernible on the wall of the back side (Sample b). It demonstrates that there is more severe corrosion on the failed tube's fire-facing side than on the other side.

16.4.2 Chemical Analysis

Table 16.1 summarizes the details of the chemical Composition of SA210 GrA1. The results indicate that the chemical composition is within acceptable limits.

16.4.3 Hardness Test

The Vickers indentation, manufactured by Qualitech Systems in Punjab, Ludhiana, with a model number QS-1AT-8, was used to determine the microhardness of all samples. The hardness of the undamaged portion (samples A & B) seems to be lower compared to the fractured area (Sample C). It is possible that the fractures are a result of the stresses in the affected area and its surroundings (Table 16.2).

16.4.4 Microstructure Examination

Figure 16.4 depicts the microstructure of samples A and B. Pearlite and ferrite are present in the microstructure.

TABLE 16.1
Measured Chemical Composition of SA210 GrA1 Tube Samples

Elements	Sample A	Sample B	Sample C
C	0.162	0.170	0.155
Si	0.245	0.246	0.250
Mn	0.490	0.500	0.503
S	0.017	0.015	0.015
P	0.013	0.012	0.012
Fe	Balance	Balance	Balance

TABLE 16.2
Measured Hardness Test Values of Tube Samples

Sample	Hardness (Hv)
A	138
B	140
C	150

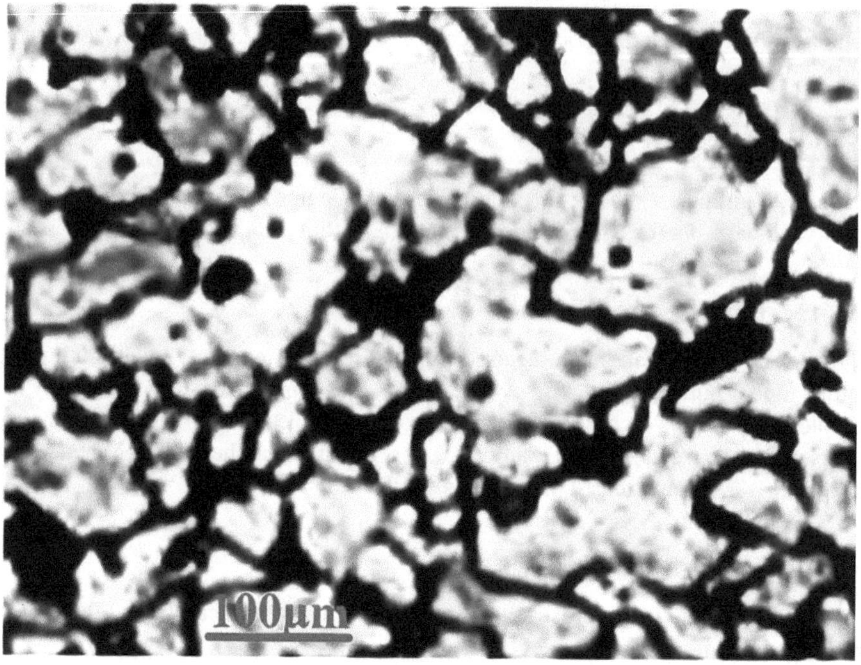

FIGURE 16.4 Microstructure of SA210 GRA1 steel tube samples A & B depicts Ferrite and Pearlite.

The microstructure of the surface near the fractured area of Sample C exhibits a ferritic matrix with a somewhat uneven distribution of pearlite, as depicted in Figure 16.5. The exposed surface of the boiler tube underwent grain growth as a result of excessive heating (as depicted in Figure 16.5). It was noted that the unevenness in grain size and pearlite distribution could potentially contribute to the eventual failure of the boiler tube.

16.4.5 Discussion

Based on the tests conducted, it seems that the boiler tube that has not been affected or damaged does not have any notable material defects. During the service life of the boiler tube, it was noticed that the grain size of the failure region (Sample C) was coarser compared to samples A and B. One possible cause of this issue is the development of scale or deposits of iron oxide on the inner surface of the boiler tube. The deposits or internal scale have a thermal conductivity that is lower than that of the steel tube. There is a noticeable rise in tube metal temperatures, which seems to have caused grain growth and ultimately resulted in early creep failure. The hardness of the fractured region was found to be higher compared to the undamaged region. It seems that this is caused by stresses resulting from hydrogen embrittlement. The formation of oxide scales on the tube surfaces contributes to the occurrence of hydrogen embrittlement.

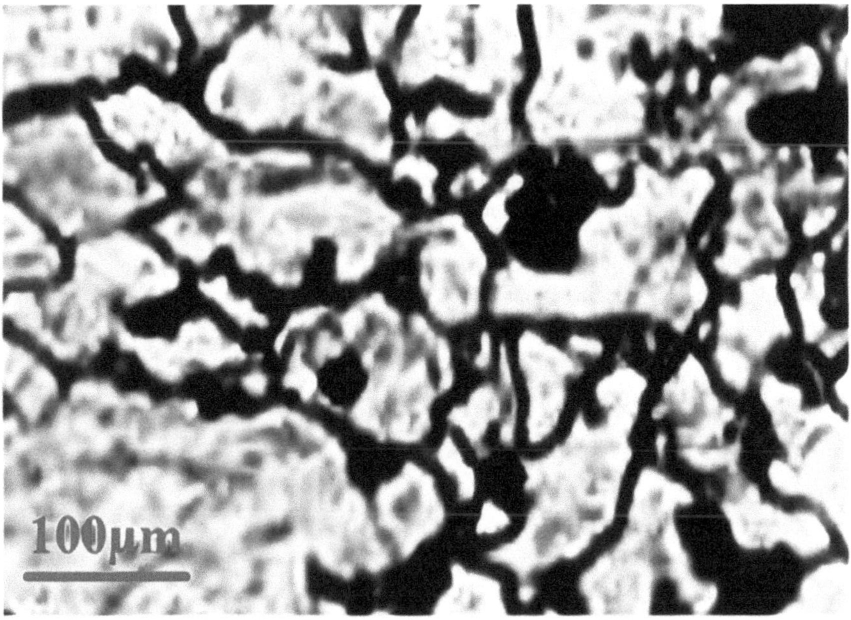

FIGURE 16.5 Microstructure of SA210 GRA1 steel tube Sample C depicting grain growth and non-uniformity of pearlite.

16.5 CONCLUSION

Based on the results and discussion provided, it was determined that the tubes experienced brittle fracture as a result of excessive heating, which led to the formation of a thick iron oxide scale on the inner surface of the boiler tube. Proper maintenance or use of protective coatings can help prevent this issue. Researchers have studied several thermally sprayed coatings, including Ni20Cr, Ni-50Cr, NiCrAlY, WC-Co, Inconel 625-Al_2O_3, Inconel 718-Al_2O_3, NiCrTi, NiCrAlY-B4C, Stellite-21, Stellite-6, and Cr_3C_2-NiCr, to withstand high-temperature failure in boiler tube materials.

REFERENCES

1. Shahzad Baig, K. and Yousaf M., 2017. Coal fired power plants: emission problems and controlling techniques. *Journal of Earth Science & Climatic Change*, *8*(404), p. 2.
2. Soeder, D.J. and Borglum, S.J., 2019. *The Fossil Fuel Revolution: Shale Gas and Tight Oil*. Elsevier.
3. Prashar, G. and Vasudev, H., 2020. Hot corrosion behavior of super alloys. *Materials Today: Proceedings*, *26*, pp. 1131–1135.
4. Mishra, P. and Mishra, S.B., 2023. A review of high-temperature oxidation behaviour of thermally sprayed boiler tube materials in advanced coal-fired thermal power plants. *NanoWorld Journal*, *9*(S1), pp. S374–S379.
5. Prashar, G., Vasudev, H. and Thakur, L., 2023. A comprehensive review on the hot corrosion and erosion performance of thermal barrier coatings. *Protection of Metals and Physical Chemistry of Surfaces*, *59*(3), pp. 461–492.

6. Prashar, G., Vasudev, H. and Thakur, L., 2022. High-temperature oxidation and erosion resistance of Ni-based thermally-sprayed coatings used in power generation machinery: a review. *Surface Review and Letters*, *29*(03), p. 2230003.
7. Prashar, G. and Vasudev, H., 2022. Structure–property correlation of plasma-sprayed Inconel625-Al_2O_3 bimodal composite coatings for high-temperature oxidation protection. *Journal of Thermal Spray Technology*, *31*(8), pp. 2385–2408.
8. Prashar, G. and Vasudev H., 2023. High-temperature oxidation behavior of direct-aged bimodal Al_2O_3-reinforced Inconel625 plasma sprayed composite coatings. *Surface and Coatings Technology*, *475*, p. 130156.
9. Vasudev, H., Thakur, L., Bansal, A., Singh, H. and Zafar, S., 2019. High temperature oxidation and erosion behaviour of HVOF sprayed bi-layer Alloy-718/NiCrAlY coating. *Surface and Coatings Technology*, *362*, pp. 366–380.
10. Prashar, G. and Vasudev, H., 2021. High temperature erosion behavior of plasma sprayed Al_2O_3 coating on AISI-304 stainless steel. *World Journal of Engineering*, *18*(5), 760–766.

Index

MIX
Papier | Fördert gute Waldnutzung
FSC® C083411

Kolibri 360 GmbH
Ferdinand-Jühlke-Straße 7
99095 Erfurt, Deutschland
produktsicherheit@kolibri360.de